SIGNIFICANT CHANGES TO THE

INTERNATIONAL RESIDENTIAL CODE®

2015 EDITION

CENGAGE Learning®

Australia • Brazil • Japan • Korea • Mexico •

CENGAGE Learning®

Significant Changes to the International Residential Code® 2015 Edition
International Code Council

Stephen A. Van Note and Sandra Hyde, P.E.

Cengage Learning Staff :

Executive Director of Professional Technology and Trades Training Solutions: **Taryn Zlatin McKenzie**

Product Manager: **Vanessa Myers**

Associate Content Developer: **Jenn Wheaton**

Director of Marketing: **Beth A. Lutz**

Senior Marketing Manager: **Marissa Lavigna**

Marketing Communications Manager: **Nicole McKasty-Stagg**

Senior Production Director: **Wendy Troeger**

Production Director: **Patty Stephan**

Senior Content Project Manager: **Stacey Lamodi**

Senior Art Director: **Benjamin Gleeksman**

ICC Staff :

Executive Vice President and Director of Business Development: **Mark A. Johnson**

Senior Vice President, Product Development: **Hamid Naderi**

Vice President, Education and Certification: **Doug Thornburg**

Director, Products and Special Sales: **Suzane Nunes**

Senior Marketing Specialist: **Dianna Hallmark**

Cover images courtesy of (left to right):
© Andrew Lebedev/Shutterstock.com;
© Christian De Araujo/Shutterstock.com;
© EPSTOCK/Shutterstock.com

Printed in United States of America
Print Number: 01 Print Year: 2014

Library of Congress Control Number: 2014937026

ISBN: 978-1-305-25473-2

ICC World Headquarters
500 New Jersey Avenue, NW
6th Floor
Washington, D.C. 20001-2070
Telephone: 1-888-ICC-SAFE (422-7233)
Website: **http://www.iccsafe.org**

Cengage Learning
20 Channel Center Street
Boston, MA 02210
USA

Cengage Learning is a leading provider of customized learning solutions with office locations around the globe, including Singapore, the United Kingdom, Australia, Mexico, Brazil, and Japan. Locate your local office at: **international.cengage.com/region**

Cengage Learning products are represented in Canada by Nelson Education, Ltd.

Visit us at **www.ConstructionEdge.cengage.com**
For more learning solutions, please visit our corporate **website at www.cengage.com**

Notice to the Reader
Publisher does not warrant or guarantee any of the products described herein or perform any independent analysis in connection with any of the product information contained herein. Publisher does not assume, and expressly disclaims, any obligation to obtain and include information other than that provided to it by the manufacturer. The reader is expressly warned to consider and adopt all safety precautions that might be indicated by the activities described herein and to avoid all potential hazards. By following the instructions contained herein, the reader willingly assumes all risks in connection with such instructions. The publisher makes no representations or warranties of any kind, including but not limited to, the warranties of fitness for particular purpose or merchantability, nor are any such representations implied with respect to the material set forth herein, and the publisher takes no responsibility with respect to such material. The publisher shall not be liable for any special, consequential, or exemplary damages resulting, in whole or part, from the readers' use of, or reliance upon, this material.

Contents

Preface

The purpose of *Significant Changes to the International Residential Code®, 2015 Edition*, is to familiarize building officials, fire officials, plans examiners, inspectors, design professionals, contractors, and others in the building construction industry with many of the important changes in the 2015 *International Residential Code®* (IRC®). This publication is designed to assist code users in identifying the specific code changes that have occurred and understanding the reasons behind the changes. It is also a valuable resource for jurisdictions in their code-adoption process.

Only a portion of the code changes to the IRC are discussed in this book. The changes selected were identified for a number of reasons, including their frequency of application, special significance, or change in application. However, the importance of the changes not included is not to be diminished. Further information on all code changes can be found in the *Code Changes Resource Collection,* available from the International Code Council® (ICC®). This resource collection provides the published documentation for each successful code change contained in the 2015 IRC since the 2012 edition.

Significant Changes to the International Residential Code, 2015 Edition, is organized into nine parts, each representing a distinct grouping of code topics. It is arranged to follow the general layout of the IRC, including code sections and section number format. The table of contents, in addition to providing guidance in the use of this publication, allows for a quick identification of those significant code changes that occur in the 2015 IRC.

Throughout the book, each change is accompanied by a photograph or an illustration to assist in and enhance the reader's understanding of the specific change. A summary and a discussion of the significance of the change are also provided. Each code change is identified by type, be it an addition, modification, clarification, or deletion.

The code change itself is presented in a legislative format similar to the style utilized for code-change proposals. Deleted code language is shown with a strikethrough, whereas new code text is indicated by underlining.

As a result, the actual 2015 code language is provided, as well as a comparison with the 2012 language, so the user can easily determine changes to the specific code text.

As with any code-change text, *Significant Changes to the International Residential Code, 2015 Edition*, is best used as a companion to the 2015 IRC. Because only a limited discussion of each change is provided, the code itself should always be referenced in order to gain a more comprehensive understanding of the code change and its application.

The commentary and opinions set forth in this text are those of the authors and do not necessarily represent the official position of ICC. In addition, they may not represent the views of any enforcing agency, as such agencies have the sole authority to render interpretations of the IRC. In many cases, the explanatory material is derived from the reasoning expressed by code-change proponents.

Comments concerning this publication are encouraged and may be directed to ICC at significantchanges@iccsafe.org.

About the *International Residential Code®*

Building officials, design professionals, contractors and others involved in the field of residential building construction recognize the need for a modern, up-to-date residential code addressing the design and installation of building systems through both prescriptive and performance requirements. The *International Residential Code®* (IRC), *2015 Edition*, is intended to meet these needs through model code regulations that safeguard the public health and safety in all communities, large and small. The IRC is kept up to date through ICC's open code-development process. The provisions of the 2012 edition, along with those code changes approved through 2013, make up the 2015 edition.

The IRC is one in a family of International Codes® published by ICC. This comprehensive residential code establishes minimum regulations for residential building systems by means of prescriptive and performance-related provisions. It is founded on broad-based principles that make possible the use of new materials and new building designs. The IRC is a comprehensive code containing provisions for building, energy conservation, mechanical, fuel gas, plumbing and electrical systems. The IRC is available for adoption and use by jurisdictions internationally. Its use within a governmental jurisdiction is intended to be accomplished through adoption by reference, in accordance with proceedings establishing the jurisdiction's laws.

Acknowledgments

Grateful appreciation is due to many ICC staff members for their generous assistance in the preparation of this publication. Fred Grable, P.E., ICC Senior Staff Engineer, shared his expertise and provided commentary on the plumbing provisions. Gregg Gress, ICC Senior Technical Staff, provided welcome assistance on the mechanical and fuel gas provisions. Larry Franks, P.E., ICC Senior Staff Engineer, provided insight into updated structural provisions, particularly the foundation chapter. Grateful appreciation also is due to Peter Kulczyk for use of his photos of residential construction in this publication. All contributed to the accuracy and quality of the finished product.

About the Authors

Stephen A. Van Note, CBO
International Code Council
Managing Director, Product Development

Stephen A. Van Note is the Managing Director of Product Development for the International Code Council (ICC), where he is responsible for developing technical resource materials in support of the International Codes. His role also includes the management, review, and technical editing of publications developed by ICC staff members and other expert authors. In addition, Steve develops and presents *International Residential Code* seminars nationally. He has over 40 years of experience in the construction and building code arena. Prior to joining ICC in 2006, Steve was a building official for Linn County, Iowa. Prior to his 15 years at Linn County, he was a carpenter and construction project manager for residential, commercial, and industrial buildings. A certified building official and plans examiner, Steve also holds certifications in several inspection categories.

Sandra Hyde, P.E.
International Code Council
Senior Staff Engineer, Product Development

Sandra Hyde is a Senior Staff Engineer with the International Code Council (ICC), where, as part of the Product Development team, she develops technical resource materials in support of the structural provisions of the International Codes. Her role also includes review and technical editing of publications authored by ICC and engineering associations, and the presentation of technical seminars on the IRC and IBC structural provisions. Prior to joining ICC in 2010, Sandra worked in manufacturing and research of engineered wood products. She is a Registered Civil Engineer in Idaho and California.

About the International Code Council®

The International Code Council is a member-focused association. It is dedicated to developing model codes and standards used in the design, build, and compliance process to construct safe, sustainable, affordable, and resilient structures. Most U.S. communities and many global markets choose the International Codes® (I-Codes®). ICC Evaluation Service (ICC-ES) is the industry leader in performing technical evaluations for code compliance, fostering safe and sustainable design and construction.

ICC Headquarters:
500 New Jersey Avenue, NW, 6th Floor
Washington, DC 20001

Regional Offices:
Birmingham, AL; Chicago, IL; Los Angeles, CA

1-888-422-7233 (ICC-SAFE)
www.iccsafe.org

PART 1

Administration

Chapters 1 and 2

- **Chapter 1** Scope and Administration
- **Chapter 2** Definitions

The administration part of the *International Residential Code* (IRC) covers the general scope, purpose, applicability, and other administrative issues related to the regulation of residential buildings by building safety departments. The administrative provisions establish the responsibilities and duties of the various parties involved in residential construction and the applicability of the technical provisions within a legal, regulatory, and code-enforcement arena.

Section R101.2 establishes the criteria for buildings that are regulated by the IRC. Buildings beyond the scope of Section R101.2 are regulated by the *International Building Code* (IBC). The remaining topics in the administration provisions of Chapter 1 include the establishment of the building safety department, duties of the building official, permits, construction documents, and inspections.

The definitions contained within the IRC are intended to reflect the special meaning of such terms within the scope of the code. As terms can often have multiple meanings within their ordinary day-to-day use or within the various disciplines of the construction industry, it is important that their meanings within the context of the IRC be understood. Most definitions used throughout the IRC are found in Chapter 2, but additional definitions specific to the applicable topics are found in the energy provisions of Chapter 11, the fuel gas provisions of Chapter 24, and the electrical provisions of Chapter 35. ■

R101.2, R202

Scope—Accessory Structures

R104.11

Alternative Materials, Design, and Methods of Construction and Equipment

R105.3.1.1

Existing Buildings in Flood Hazard Areas

R106.1.4

Information for Construction in Flood Hazard Areas

R101.2, R202

Scope—Accessory Structures

CHANGE TYPE: Modification

CHANGE SUMMARY: The maximum height for accessory structures has been increased from two to three stories above grade plane. Technical requirements have been removed from the definition, and accessory structures are now permitted to be unlimited in area.

2015 CODE: R101.2 Scope. The provisions of the *International Residential Code for One- and Two-family Dwellings* shall apply to the construction, alteration, movement, enlargement, replacement, repair, equipment, use and occupancy, location, removal and demolition of detached one- and two-family dwellings and townhouses not more than three stories above grade plane in height with a separate means of egress and their accessory structures not more than three stories above grade plane in height.

SECTION R202
DEFINITIONS

ACCESSORY STRUCTURE. A structure ~~not greater than 3,000 square feet (279 m^2) in floor area, and not more than two stories in height, the use of which~~ that is ~~customarily~~ accessory to and incidental to that of dwelling(s) and ~~which~~ that is located on the same lot.

CHANGE SIGNIFICANCE: In previous editions of the IRC, the definition in Section R202 placed limitations of 3,000 square feet in area and two stories on accessory structures. The 3,000-square-foot limitation was introduced in the 2006 IRC based on a concern of the potential fire load

International Code Council®

Accessory building

in residential accessory buildings. The area limitation has been removed from the 2015 IRC based on the residential setting of these buildings, the need for larger accessory buildings in rural areas, and the fact that dwellings and townhomes constructed under the IRC are unlimited in area. The change also recognizes that zoning regulations typically set limits for area and height of accessory buildings based on the density of housing and other factors unique to the individual jurisdiction. It was judged more appropriate to allow jurisdictions to decide what limits are placed on accessory buildings. For example, in rural areas with large lots and acreages, very large accessory buildings are routinely constructed for vehicle and farm equipment storage and to house hobby shops and workshops. In addition, definitions are not intended to contain technical requirements such as area and height limitations, which should be addressed in the applicable sections in the body of the code. The definition maintains the key elements for permitting accessory buildings to be constructed under the IRC—that they must be accessory to and incidental to that of the dwelling and located on the same lot as the dwelling.

The height limitation for accessory buildings has also been removed from the definition and placed in the scoping provisions of the IRC. The maximum height has increased to three stories above grade plane for consistency with the height limitations for dwellings and townhomes.

R104.11

Alternative Materials, Design, and Methods of Construction and Equipment

International Code Council®

Alternative product

CHANGE TYPE: Addition

CHANGE SUMMARY: When proposed alternatives are not approved, the reason for the disapproval must be stated in writing by the building official.

2015 CODE: **R104.11 Alternative Materials, Design, and Methods of Construction and Equipment.** The provisions of this code are not intended to prevent the installation of any material or to prohibit any design or method of construction not specifically prescribed by this code, provided that any such alternative has been approved. An alternative material, design or method of construction shall be approved where the building official finds that the proposed design is satisfactory and complies with the intent of the provisions of this code, and that the material, method or work offered is, for the purpose intended, ~~at least~~ not less than the equivalent of that prescribed in this code. Compliance with the specific performance-based provisions of the International Codes shall be an alternative to the ~~in lieu of~~ specific requirements of this code ~~shall also be permitted as an alternate~~. Where the alternative material, design or method of construction is not approved, the building official shall respond in writing, stating the reasons the alternative was not approved.

CHANGE SIGNIFICANCE: When a building official denies a proposal for using an alternative material, design or method of construction, the reason for denial must be provided in writing to the applicant. This new requirement mirrors the permit application provisions in Section R105.3.1, which require the building official to state in writing the reasons for rejection of a permit application. This change assumes reasons for responding to the applicant in writing are in order to ensure effective communication and due process of law. The applicant, using a written denial, may determine whether to modify the product or design, substitute a new product or method of construction, or correct errors in application of the alternate. The new language is added to all of the International Codes for consistency of application.

R105.3.1.1

Existing Buildings in Flood Hazard Areas

CHANGE TYPE: Modification

CHANGE SUMMARY: Determination of substantial improvement for existing buildings in flood hazard areas is the responsibility of the building official. The related provisions are now consolidated in Section R105.3.1.1.

2015 CODE: R105.3.1.1 Determination of Substantially Improved or Substantially Damaged Existing Buildings in Flood Hazard Areas. For applications for reconstruction, rehabilitation, addition, alteration, repair or other improvement of existing buildings or structures located in a flood hazard area as established by Table R301.2(1), the building official shall examine or cause to be examined the construction documents and shall make a determination ~~prepare a finding~~ with regard to the value of the proposed work. For buildings that have sustained damage of any origin, the value of the proposed work shall include the cost to repair the building or structure to its predamage condition. If the building official finds that the value of proposed work equals or exceeds 50 percent of the market value of the building or structure before the damage has occurred or the improvement is started, ~~the finding shall be provided to the board of appeals for a determination of substantial improvement or substantial damage. Applications determined by the board of appeals to constitute substantial improvement or substantial damage~~ the proposed work is a substantial improvement or restoration of substantial damage and the building official shall require all existing portions of the entire building or structure to meet the requirements of R322.

~~**R112.2.1 Determination of Substantial Improvement in Flood Hazard Areas.**~~ ~~When the building official provides a finding required in Section R105.3.1.1, the board of appeals shall determine whether the value of the proposed work constitutes a substantial improvement.~~ For the purpose of this determination, a substantial improvement shall mean

R105.3.1.1 continues

International Code Council®

Flood-prone construction

R105.3.1.1 continued

any repair, reconstruction, rehabilitation, addition, or improvement of a building or structure, the cost of which equals or exceeds 50 percent of the market value of the building or structure before the improvement or repair is started. ~~If~~ Where the building or structure has sustained substantial damage, ~~all~~ repairs necessary to restore the building or structure to its pre-damaged condition shall be ~~are~~ considered substantial improvements regardless of the actual repair work performed. The term shall ~~does~~ not include either of the following:

1. Improvements of a building or structure required to correct existing health, sanitary, or safety code violations identified by the building official and which are the minimum necessary to assure safe living conditions; or
2. Any alteration of a historic building or structure, provided that the alteration will not preclude the continued designation as a historic building or structure. For the purposes of this exclusion, a historic building shall be any of the following ~~is~~:
   2.1 Listed or preliminarily determined to be eligible for listing in the National Register of Historic Places.
   2.2 Determined by the Secretary of the U.S. Department of Interior as contributing to the historical significance of a registered historic district or a district preliminarily determined to qualify as an historic district.
   2.3 Designated as historic under a state or local historic preservation program that is approved by the Department of Interior.

CHANGE SIGNIFICANCE: The criteria used to determine substantial improvement or substantial damage for existing buildings in flood hazard areas has been moved from the Building Board of Appeals provisions in Section R112.2.1 to Section R105.3.1.1 related to the building official's action on a permit application. The language requiring the Building Board of Appeals to make a determination of substantial improvement in flood hazard areas has been removed from Section R112.2. In effect, this determination is now a one-step process rather than a two-step process. It relies on the building official to determine whether work on existing buildings in flood hazard areas meets the definitions for "substantial improvement" and "substantial damage," rather than having the building official make a finding and then having the Board of Appeals make a determination based on that finding.

R106.1.4

Information for Construction in Flood Hazard Areas

CHANGE TYPE: Modification

CHANGE SUMMARY: Construction documents for dwellings in Coastal A Zones shall include the elevation of the bottom of the lowest horizontal structural member.

2015 CODE: R106.1.~~3~~4 Information for Construction in Flood Hazard Areas. For buildings and structures located in whole or in part in flood hazard areas as established by Table R301.2(1), construction documents shall include:

1. Delineation of flood hazard areas, floodway boundaries and flood zones and the design flood elevation, as appropriate.
2. The elevation of the proposed lowest floor, including basement; in areas of shallow flooding (AO zones), the height of the proposed lowest floor, including basement, above the highest adjacent finished grade.
3. The elevation of the bottom of the lowest horizontal structural member in coastal high hazard areas (V Zone) and in Coastal A Zones where such zones are delineated on flood hazard maps identified in Table R301.2(1) or otherwise designated by the jurisdiction.
4. If design flood elevations are not included on the community's Flood Insurance Rate Map (FIRM), the building official and the applicant shall obtain and reasonably utilize any design flood elevation and floodway data available from other sources.

R106.1.4 continues

Photo Courtesy of Chesterfield/Danbro

Wave scour of a dwelling foundation

R106.1.4 continued

CHANGE SIGNIFICANCE: Dwellings in areas designated as "Coastal A Zones" are required to meet the requirements of Section R322.3 for dwellings in coastal high-hazard areas (Zone V). A new exception in Section R322.3 allows backfilled stemwall foundations rather than open foundations (pilings or columns) if the foundation is designed to account for wave action, debris impact, and local scour in the Coastal A Zones. These dwellings must have marked on the construction documents the elevation of the bottom of the lowest horizontal member in the structure.

PART 2

Building Planning

Chapter 3

■ **Chapter 3** Building Planning

Chapter 3 includes the bulk of the nonstructural provisions, including the location on the lot, fire-resistant construction, light and ventilation, emergency escape and rescue, fire protection, safety glazing, fall protection, and many other provisions aimed at protecting the health, safety, and welfare of the public. In addition to such health and life-safety issues, Chapter 3 provides the overall structural design criteria for residential buildings regulated by the IRC. Section R301 addresses live loads, dead loads, and environmental loads such as wind, seismic, and snow. ■

TABLE R301.2(1)

Climatic and Geographic Design Criteria

R301.2

Wind Design Criteria

R301.2

Wind Speed Maps

TABLE R301.2(2)

Component and Cladding Loads

R301.2.1.1.1

Sunrooms

R301.2.1.2

Protection of Openings in Wind Borne Debris Regions

R301.2.1.4

Wind Exposure Category

TABLE R301.2.1.5.1

Modifications for Topographic Wind Effects

R301.2.4

Floodplain Construction

R301.3

Story Height

R302.1

Exterior Walls

R302.2

Townhouse Separation

R302.13

Fire Protection of Floors

R303.7, R303.8

Stairway Illumination

R304.1

Minimum Habitable Room Area

R305

Ceiling Height

R308.4.2

Glazing Adjacent to Doors

R308.4.5

Glazing and Wet Surfaces

R308.4.7

Glazing Adjacent to the Bottom Stair Landing

R310

Emergency Escape and Rescue Openings

R310.5, R310.6

Emergency Escape and Rescue Openings for Additions, Alterations and Repairs

R311.1

Means of Egress

R311.7.3, R311.7.5.1

Stair Risers

R311.7.10.1

Spiral Stairways

R311.7.11, R311.7.12

Alternating Tread Devices and Ship Ladders

R311.8

Ramps

R312.1.2

Guard Height

R312.2.1

Window Fall Protection

R314

Smoke Alarms

R315

Carbon Monoxide Alarms

R322.1, R322.2

Flood Hazards

R322.3

Coastal High-Hazard Areas

R325

Mezzanines

Table R301.2(1)

Climatic and Geographic Design Criteria

CHANGE TYPE: Modification

CHANGE SUMMARY: Table R301.2(1) Climatic and Geographic Design Criteria now contains a section to include whether the jurisdiction contains special wind regions or wind borne debris zones.

2015 CODE:

TABLE R301.2(1) Climatic and Geographic Design Criteria

Ground Snow Load	Wind Design				Seismic Design Category[f]
	Speed[d] (Mph)	Topographic effects[k]	Special wind region[l]	Wind borne debris zone[m]	

(Portions of table and footnotes not shown remain unchanged)

l. In accordance with Figure R301.2(4)A, where there is local historical data documenting unusual wind conditions, the jurisdiction shall fill in this part of the table with "YES" and identify any specific requirements. Otherwise, the jurisdiction shall indicate "NO" in this part of the table.

m. In accordance with Section R301.2.1.2.1, the jurisdiction shall indicate the wind-borne debris wind zone(s). Otherwise, the jurisdiction shall indicate "NO" in this part of the table.

CHANGE SIGNIFICANCE: The special wind regions and wind design required regions are shown on a single map for the continental United States in Figure R301.2(4)B. For wind borne debris zones, attempting to interpret wind speed from Figure R301.2(4)B near locations where the contour lines occur can be difficult and may lead to misapplication. The contour lines do not follow county lines or borders. Identification of zones where wind borne debris requirements are applied should be provided by the local jurisdiction to ensure that provisions are applied correctly.

Although the special wind region and wind borne debris requirements do not apply to most of the United States, when applicable they can have a major impact on the design and construction of residential structures. It is important that the designer determine when a project is in one of these regions by contacting the building department.

International Code Council®

Special wind region—Columbia River gorge

R301.2

Wind Design Criteria

Hurricane-force winds

Photo Courtesy of NASA/Jeff Schmaltz, MODIS Land Rapid Response Team

CHANGE TYPE: Modification

CHANGE SUMMARY: Ultimate design wind speed values replace basic wind speed values for 3-sec gust wind speeds in Section R301.2.1. A wind speed conversion table has been added for conversion from ultimate design to nominal design wind speeds.

2015 CODE:

SECTION R202
DEFINITIONS

HURRICANE-PRONE REGIONS. Areas vulnerable to hurricanes, defined as the U.S. Atlantic Ocean and Gulf of Mexico coasts where the ultimate design wind speed, V_{ult}, ~~basic wind speed~~ is greater than 115 ~~90~~ miles per hour (51~~40~~ m/s), and Hawaii, Puerto Rico, Guam, Virgin Islands, and America Samoa.

WIND BORNE DEBRIS REGION. Areas within hurricane-prone regions located in accordance with one of the following: ~~as designated in accordance with Figure R302.1(4)C.~~:

1. Within 1 mile (1.61 km) of the coastal mean high water line where the ultimate design wind speed, V_{ult} is 130 mph (58 m/s) or greater.
2. In areas where the ultimate design wind speed, V_{ult}, is 140 mph (63.6 m/s) or greater; or Hawaii.

R301.2.1 Wind Design Criteria. Buildings and portions thereof shall be constructed in accordance with the wind provisions of this code using the ultimate design ~~basic~~ wind speed in Table R301.2(1) as determined from Figure R301.2(4)A. The structural provisions of this code for wind loads are not permitted where wind design is required as specified in Section R301.2.1.1. Where different construction methods and structural materials are used for various portions of a building, the applicable requirements of this section for each portion shall apply. Where not otherwise specified, the wind loads listed in Table R301.2(2) adjusted for height and exposure using Table R301.2(3) shall be used to determine design load performance requirements for wall coverings, curtain walls, roof coverings, exterior windows, skylights, garage doors, and exterior doors. Asphalt shingles shall be designed for wind speeds in accordance with Section R905.2.4. A continuous load path shall be provided to transmit the applicable uplift forces in Section R802.11.1 from the roof assembly to the foundation.

R301.2.1.1 Wind Limitations and Wind Design Required. The wind provisions of this code shall not apply to the design of buildings where wind design is required in accordance with Figure R301.2(4)B ~~or where the basic wind speed from Figure R301.2(4)A equals or exceeds 110 miles per hour (49 m/s)~~.

Exceptions:

1. For concrete construction, the wind provisions of this code shall apply in accordance with the limitations of Sections R404 and R608.

2. For structural insulated panels, the wind provisions of this code shall apply in accordance with the limitations of Section R610.
3. For cold-formed steel light-frame construction, the wind provisions of this code shall apply in accordance with the limitations of Sections R505, R603, and R804.

In regions where wind design is required in accordance with Figure R301.2(4)B ~~or where the basic wind speed shown on Figure R301.2(4)A equals or exceeds 110 miles per hour (49 m/s)~~, the design of buildings for wind loads shall be in accordance with one or more of the following methods:

1. AF&PA *Wood Frame Construction Manual* (WFCM); or
2. ICC *Standard for Residential Construction in High-Wind Regions* (ICC 600); or
3. ASCE *Minimum Design Loads for Buildings and Other Structures* (ASCE 7); or
4. AISI *Standard for Cold-Formed Steel Framing—Prescriptive Method For One- and Two-Family Dwellings* (AISI S230); or
5. *International Building Code.*

The elements of design not addressed by the methods in Items 1 through 5 shall be in accordance with the provisions of this code.

Where ASCE 7 or the *International Building Code* is used for the design of the building, the wind speed map and exposure category requirements as specified in ASCE 7 and the *International Building Code* shall be used.

R301.2.1.3 Wind Speed Conversion. Where referenced documents are based on nominal design ~~fastest mile~~ wind speeds and do not provide the means for conversion between the ultimate design wind speeds and the nominal design wind speeds, the ultimate design ~~three-second gust basic~~ wind speeds, V_{ult} ~~V_{3s}~~, of Figure R301.2(4)A shall be converted to nominal design ~~fastest mile~~ wind speeds, V_{asd} ~~V_{fm}~~, using Table R301.2.1.3.

TABLE R301.2.1.3 Wind Speed Conversions[a]

V_{ult}	110	115	120	130	140	150	160	170	180	190	200
V_{asd}	85	89	93	101	108	116	124	132	139	147	155

For SI: 1 mile per hour = 0.447 m/s.

a. Linear interpolation is permitted

CHANGE SIGNIFICANCE: This code change brings the wind provisions of the *International Residential Code* (IRC) in line with the 2015 *International Building Code* (IBC) and ASCE 7-10 standard, *Minimum Design Loads for Buildings and Other Structures*. For the 2012 IRC, maps based on the ASCE 7-10 ultimate design wind speed data for 3-second gusts were converted to allowable stress design (ASD) values. Meanwhile, wind speed maps in the 2012 IBC and ASCE 7-10 were printed using strength design or "ultimate design" values. This led to confusion among users working with both codes.

R301.2 continues

R301.2 continued

For the 2015 IRC, winds speeds are ultimate design wind speeds. These values are provided in wind maps. The boundaries between wind speeds moved slightly as wind speeds are rounded to the nearest 5 or 10 mph. This means values on wind speed maps change, but component and cladding loads converted to allowable stress design values remain the same as in the 2012 IRC. This change affects engineered design, but doesn't affect prescriptive requirements. High wind regions, areas that previously required wind design, will continue to require alternate design, whether by use of an alternate design standard or engineered design. Changes in the 2015 IRC update wind design criteria, definitions, and maps, and provide a conversion table from ultimate design wind speeds (V_{ult}) to nominal design wind speeds (V_{asd}). Table R301.2.1.3 gives ultimate design wind speeds and their equivalent nominal design wind speeds for use with standards that have not updated their provisions.

Note, the former IRC term "basic wind speed" is now "nominal design wind speed" and refers to wind values based on allowable stress design, an engineering method to determine loads on a building. The new term "ultimate design wind speed" refers to values based on wind speeds of the 2015 IRC Figures R301.2.4(A) and R301.2.4(B). Although these values are higher, when the adjustment factor of $\sqrt{0.6}$ is applied to them, the value will be approximately the same as the former basic wind speed. The terms "nominal design wind speed" and "ultimate design wind speed" have been added to tables, figures, and code text throughout the IRC to clarify which wind speed is described. Most code provisions in the IRC now use ultimate design wind speed limits.

A Discussion of Engineering and Wind Speed. The following section describes why wind speeds have changed. Understanding why the values changed is not necessary for use of the IRC, but may be of interest.

The most visible aspect of the wind speed modifications is the change in wind speed maps in the 2015 IRC. The maps were updated to match those adopted in ASCE 7-10. Over the past 10 years, new research indicated that the hurricane wind speeds provided in ASCE 7-05 were too conservative and should be adjusted downward. As more hurricane data became available, it was recognized that substantial improvements could be made to the hurricane model used to develop the wind speed maps. The new data resulted in an improved representation of the hurricane wind field.

Changes to the model include:

- Refined modeling of sea–land transition and hurricane boundary layer height
- New models for hurricane weakening after landfall
- Improved statistical modeling for the characteristic controlling wind pressure relationships

Although the new hurricane hazard model yields hurricane wind speeds lower than those given in earlier code editions, the overall rate of intense storms produced by the new model increased compared to those produced by the hurricane model used to develop previous wind speed maps. This means lower wind speeds over land but more frequent storms. As the wind speed model is developed in part by looking at the statistical chance that a hurricane might occur in a given location, equivalent wind speed values on the map may be slightly higher, the same, or lower than previous values.

In developing new maps, it was decided to use strength-design-based maps in conjunction with a wind load factor of 1.0. For allowable stress design (ASD), the wind load factor would then be reduced to 0.6, and thus footnotes referring to adjustment to a nominal design wind speed include a factor of 0.6.

Factors related to more accurate wind load determination were considered, leading to the decision to move to strength-based ultimate-event wind loads in ASCE 7, the IBC, and the IRC. The most pertinent factor for the IRC is that an ultimate event or strength design wind speed map makes the overall approach for wind consistent with the strength-based seismic design procedures. Both wind and seismic load effects are mapped as ultimate events and use a load factor of 1.0 for the strength design load combinations in engineered design.

As a result of the new strength-based wind speed, new terminology is introduced into the 2015 IRC. The former terms "basic wind speed" and "wind speed" are replaced with "ultimate design wind speed" and labeled V_{ult}. The wind speed that is equivalent to the former "basic wind speed" is now called "nominal design wind speed," V_{asd}. The conversion between the two is,

$$V_{asd} = V_{ult} \times \sqrt{0.6}$$

The conversion from V_{asd} to V_{ult} is a result of the wind load being proportional to the square of the velocity pressure and the ASD wind load being 0.6 times the strength level ultimate wind load. Thus,

$$W = V^2$$
$$V_{asd}^2 = 0.6W_{ult}$$
$$V_{asd}^2 = 0.6V_{ult}^2$$
$$V_{asd} = \sqrt{0.6} \times V_{ult}$$

Where,

W = wind load for IBC and ASCE 7 wind load equations
V_{asd} = nominal design wind load
V_{ult} = ultimate design wind load

Note that the term "basic wind speed" in ASCE 7-10 corresponds to references to wind speed or basic wind speed in the IRC now referred to as nominal design wind speed.

Because many different code provisions in the code are based upon wind speed, it was necessary to modify the wind speed conversion section so that the many provisions triggered by wind speed are not changed. The terms "ultimate design wind speed" and "nominal design wind speed" are added to help the code user distinguish between them. Table R301.2.1.3 converts the new ultimate design wind speeds to an equivalent nominal design wind speed. Converted wind speeds vary less than 2 miles per hour from former basic wind speeds.

For example, in a case where the 2012 IRC imposed requirements where the basic wind speed exceeds 85 mph, the 2015 IRC imposes

R301.2 continues

R301.2 continued

requirements where V_{ult} exceeds 110 mph. Use of Table R301.2.1.3 or the conversion factor of the square root of 0.6 converts the 110 mph wind speed to approximately an 85 mph nominal design wind speed.

As a second example, in a metal building standard a nominal design wind speed, V_{asd}, of 100 mph corresponds to an ultimate design wind speed, V_{ult}, of 130 mph.

CHANGE TYPE: Modification

R301.2
Wind Speed Maps

CHANGE SUMMARY: New to the 2015 IRC, wind maps use ultimate design wind speeds. Figure R301.2(4)A shows the new wind speeds within the continental United States, Alaska, and Puerto Rico. A list of wind speeds for special wind regions in Hawaii, Guam, American Samoa, and the Virgin Islands is included. Figure R301.2(4)B shows the regions of the United States that require engineered design of residential structures for high wind loads.

R301.2 continues

2015 CODE:

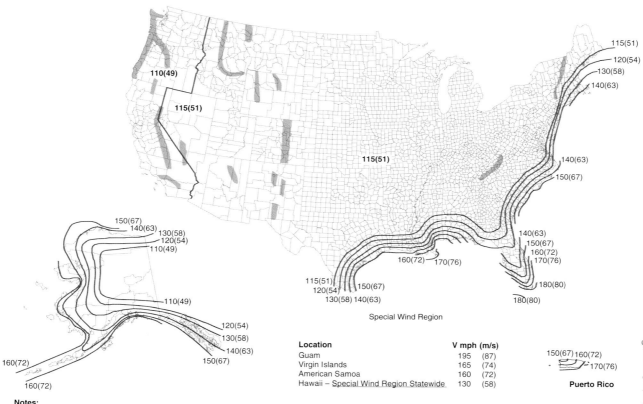

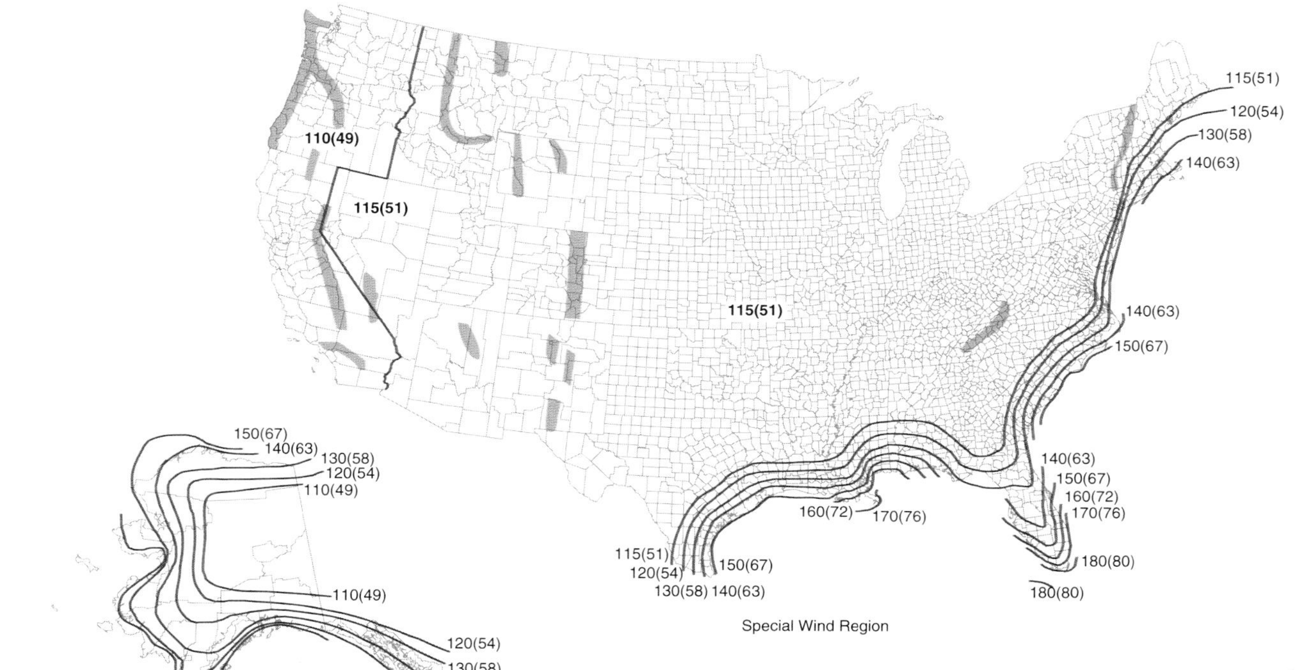

Location	V mph	(m/s)
Guam	195	(87)
Virgin Islands	165	(74)
American Samoa	160	(72)
Hawaii – Special Wind Region Statewide	130	(58)

Notes:

1. Values are ultimate design 3-second gust wind speeds in miles per hour (m/s) at 33 ft (10 m) above ground for Exposure category C.
2. Linear interpolation between contours is permitted.
3. Islands and coastal areas outside the last contour shall use the last wind speed contour of the coastal area.
4. Mountainous terrain, gorges, ocean promontories, and special wind regions shall be examined for unusual wind conditions.
5. Wind speeds correspond to approximately a 7% probability of exceedance in 50 years (Annual Exceedance Probability = 0.00143, MRI = 700 years).

Figure R301.2(4)A Ultimate design wind speeds

R301.2 continued

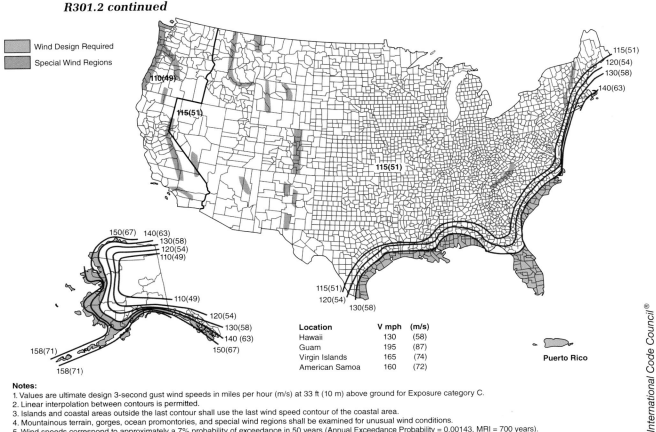

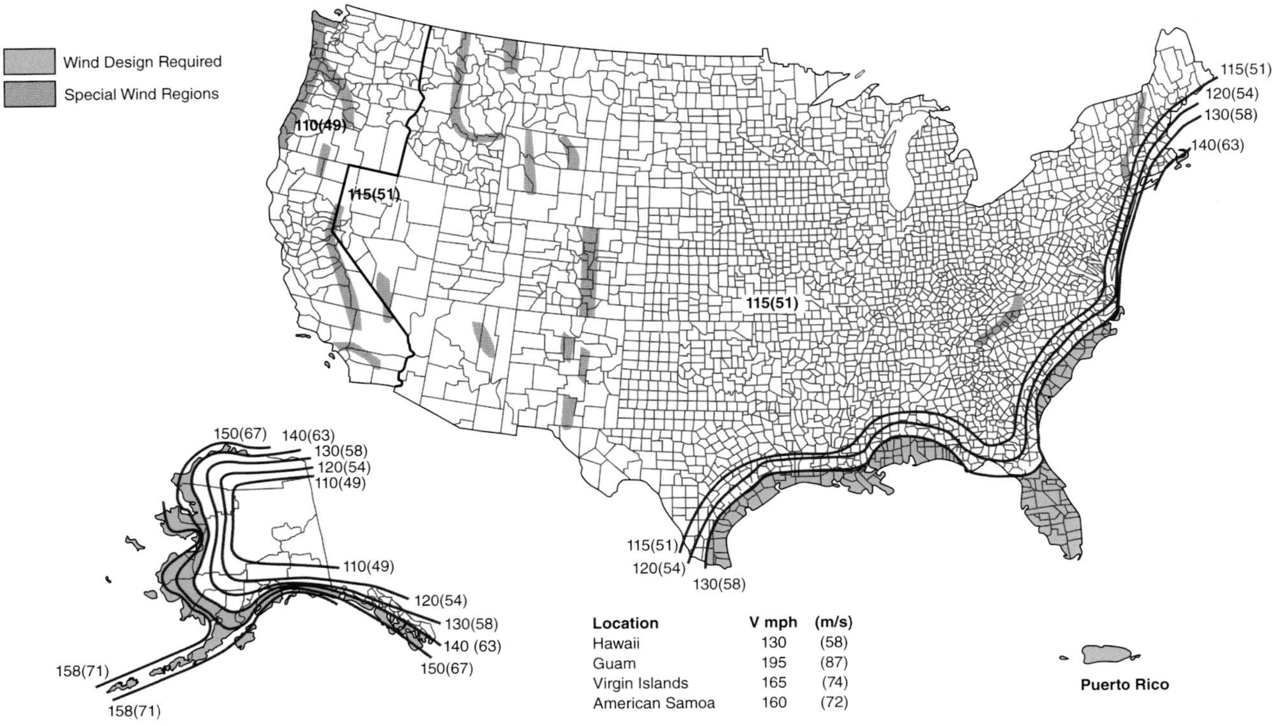

Notes:
1. Values are ultimate design 3-second gust wind speeds in miles per hour (m/s) at 33 ft (10 m) above ground for Exposure category C.
2. Linear interpolation between contours is permitted.
3. Islands and coastal areas outside the last contour shall use the last wind speed contour of the coastal area.
4. Mountainous terrain, gorges, ocean promontories, and special wind regions shall be examined for unusual wind conditions.
5. Wind speeds correspond to approximately a 7% probability of exceedance in 50 years (Annual Exceedance Probability = 0.00143, MRI = 700 years).

International Code Council®

Figure R301.2(4)B Regions where wind design is required

CHANGE SIGNIFICANCE: Maps based on ASCE 7-10 ultimate wind speed data converted to nominal allowable stress design (ASD) values were provided in the 2012 IRC. Changes in 2015 IRC update the wind design criteria, definitions, and wind speed maps. Wind speed maps now show wind using strength design values. Some contour lines in Figure R301.2(4)A are in new locations compared to contour lines in maps of the 2012 IRC. As ultimate design wind speeds lines are drawn for 110, 115, 120, and 130 mph, these lines do not occur in exactly the same location as lines drawn in previous editions of the IRC. The values are roughly equivalent, and for regions that are not near a contour line, no change in required connections will occur. Regions crossing over former or new contour lines may see a change in required minimum fastener values.

The areas in Figure R301.2(4)B where use of alternate prescriptive high-wind standards or engineered design is required are defined using the 130-mph contour along the Gulf Coast and along southern portions of the Atlantic coast from Florida to North Carolina. The 140-mph contour is used for northern portions of the Atlantic coast from Virginia to Maine and Alaska. A 130-mph trigger is also used for Caribbean and Pacific islands considered part of the "hurricane-prone" region. This creates a region requiring engineering that is approximately equal to the region defined by the 110-mph contour in wind maps used in the 2000 through 2009 IRC and the 100-mph hurricane-prone region contour in

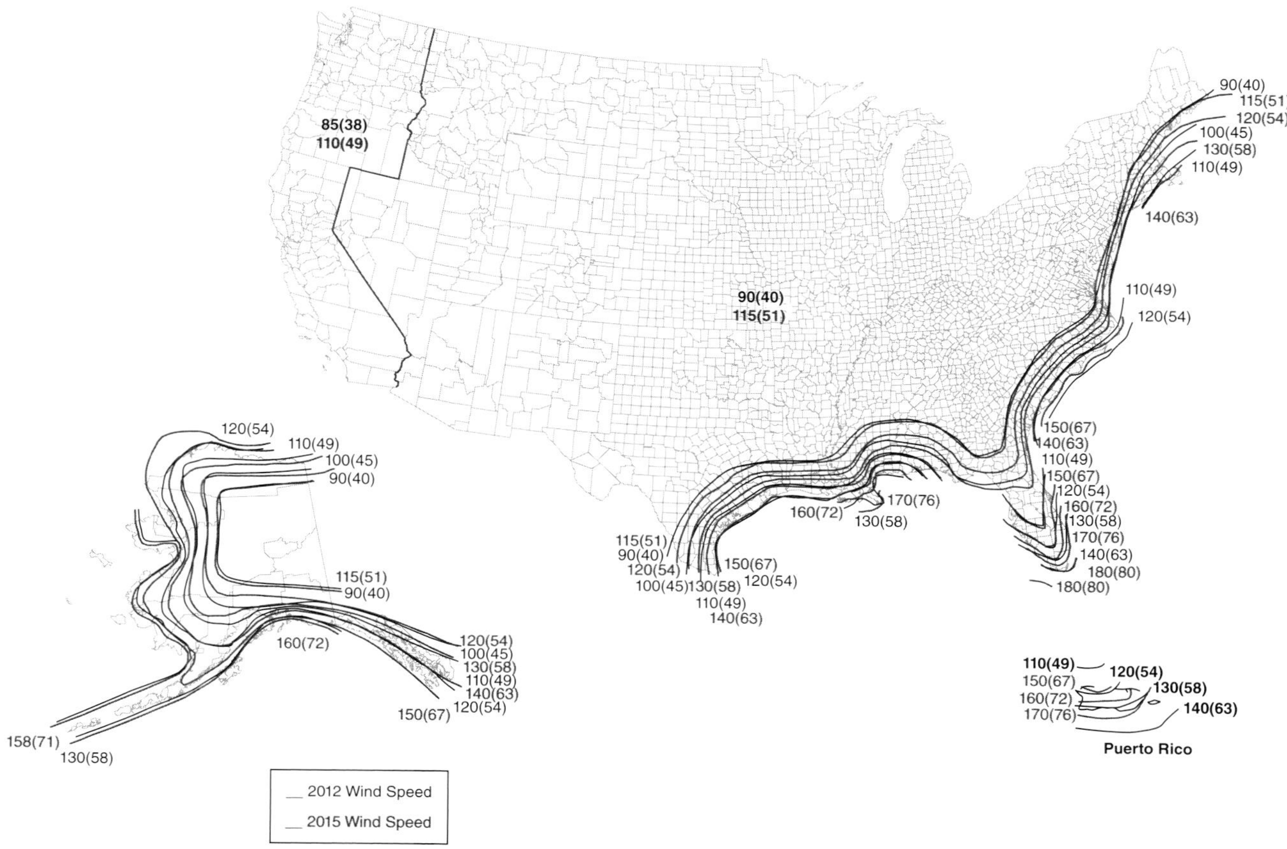

Comparison of wind speed

International Code Council®

the 2012 IRC wind map. The shaded regions maintain areas of Florida and the Gulf Coast traditionally outside prescriptive limits of the IRC, and maintain areas of New England traditionally included within prescriptive limits of the IRC.

Users of the IRC desiring a more accurate determination of wind speed in areas near or along a particular contour line can use the Applied Technology Council's (ATC) Wind Speed by Location website www.atcouncil.org/windspeed to obtain site-specific wind speeds using latitude/longitude or the site address. This website was developed by ATC using the same data used to develop wind maps for ASCE 7, the IBC, and the IRC. The wind speed for Risk Category II is appropriate for use with the *International Residential Code*. Wind speeds in special wind regions will need to be obtained from the local jurisdiction.

Table R301.2(2)

Component and Cladding Loads

CHANGE TYPE: Modification

CHANGE SUMMARY: The component and cladding table uses ultimate design wind speeds in place of former basic wind speeds as limits for loads. Roof slopes are divided into new categories for determining component and cladding loads.

2015 CODE:

TABLE R301.2(2) Component and Cladding Loads for a Building with a Mean Roof Height of 30 Feet Located in Exposure B (ASD)(psf)[a, b, c, d, e]

	Zone	Effective Wind Area (feet^2)	Ultimate Design Wind Speed, V_{ULT} (mph)								
			110	115	120	130	140	150	160	170	180
Roof 0 to 7 degrees											
Roof >7 to 27 degrees											
Roof >27 to 45 degrees											
Wall											

(Table values not shown for brevity and clarity.)

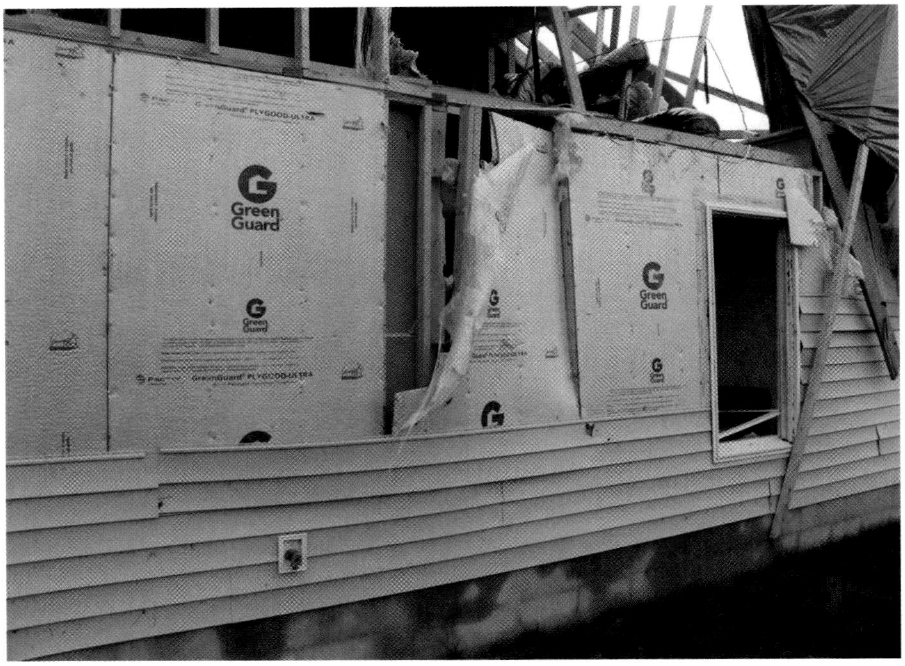

Photo Courtesy of APA

Cladding damage

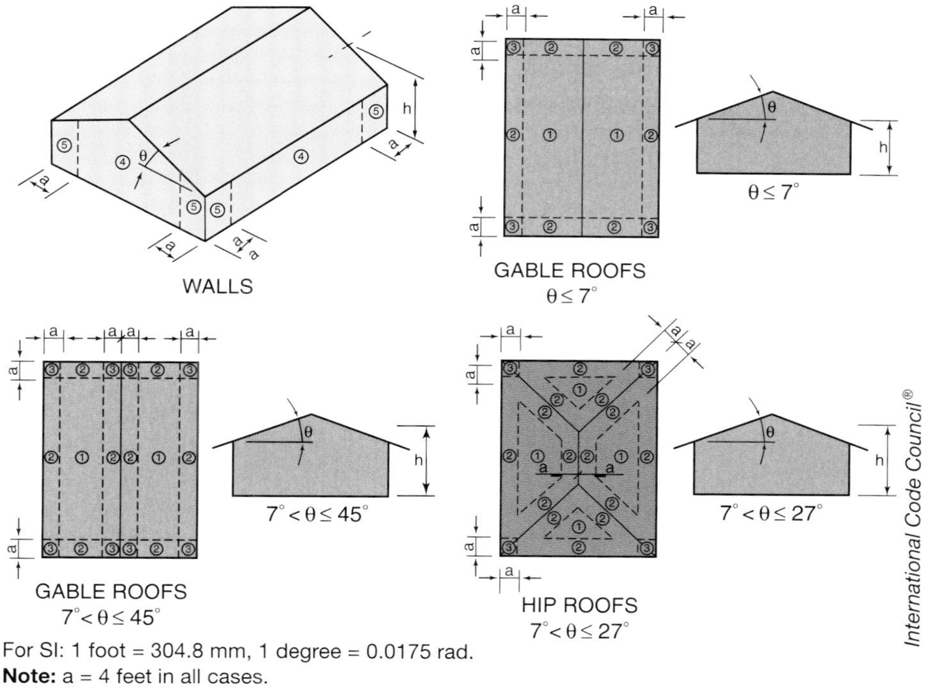

Component and cladding pressure zones

CHANGE SIGNIFICANCE: This code change brings the wind provisions of the *International Residential Code* in line with the 2015 *International Building Code* (IBC) and *Minimum Design Loads for Buildings and Other Structures* (ASCE/SEI 7-10). For more information about the change in wind speed values, see significant change R301.2, Wind Speed Design Criteria.

Values in wind speed maps have changed, but component and cladding loads converted to allowable stress design values remain the same.

The component and cladding pressure table is set up using ultimate design wind speed, but values printed in the table are listed as allowable stress design values. In other words, the listed pressures incorporate a 0.6 multiplier on the wind loads for allowable stress design load combinations shown in Section 1605.3 of the *International Building Code*. This is done to allow simple adaptation of existing designs, construction documents, and guidelines to the 2015 IRC, as loads and pressures are comparable to previous editions of the IRC. Residential product manufacturers typically still use allowable stress design values on their packaging and installation instructions.

This code change also divides roof slopes into new categories for determining component and cladding loads. Roof slope is divided into three categories. The categories are based on component and cladding roof slope divisions in ASCE 7 for hip roofs. Table 3.1 below compares the new categories for roof slope to those in previous editions.

TABLE 3.1 Comparison of Roof Angle Categories

2015 IRC	2000–2012 IRC
• 0 to 7 degrees	• 0 to 10 degrees
• Greater than 7 to 27 degrees	• Greater than 10 to 30 degrees
• Greater than 27 to 45 degrees	• Greater than 30 to 45 degrees

R301.2.1.1.1

Sunrooms

Sunroom

CHANGE TYPE: Addition

CHANGE SUMMARY: The 2015 *International Residential Code* requires sunrooms to comply with AAMA/NPEA/NSA 2100-12. The standard contains requirements for habitable and nonhabitable sunrooms.

2015 CODE: **R301.2.1.1.1 Sunrooms.** Sunrooms shall comply with AAMA/NPEA/NSA 2100.

For the purpose of applying the criteria of AAMA/NPEA/NSA-2100 based on the intended use, sunrooms shall be identified as one of the following categories by the permit applicant, design professional, or the property owner in the construction documents. Component and cladding pressures shall be used for the design of elements that do not qualify as main wind force resisting systems. Main wind force resisting systems pressures shall be used for the design of elements assigned to provide support and stability for the overall sunroom.

Category I: A thermally isolated sunroom with walls that are open or enclosed with insect screening or 0.5 mm (20 mil) maximum thickness plastic film. The space is nonhabitable and unconditioned.

Category II: A thermally isolated sunroom with enclosed walls. The openings are enclosed with translucent or transparent plastic or glass. The space is nonhabitable and unconditioned.

Category III: A thermally isolated sunroom with enclosed walls. The openings are enclosed with translucent or transparent plastic or glass. The sunroom fenestration complies with additional requirements for air infiltration resistance and water-penetration resistance. The space is nonhabitable and unconditioned.

Category IV: A thermally isolated sunroom with enclosed walls. The sunroom is designed to be heated or cooled by a separate temperature control or system and is thermally isolated from the primary structure. The sunroom fenestration complies with additional requirements for water penetration resistance, air infiltration resistance, and thermal performance. The space is nonhabitable and conditioned.

Category V: A sunroom with enclosed walls. The sunroom is designed to be heated or cooled and is open to the main structure. The sunroom fenestration complies with additional requirements for water-penetration resistance, air infiltration resistance, and thermal performance. The space is habitable and conditioned.

CHANGE SIGNIFICANCE: The 2012 *International Residential Code* defined a sunroom as "A one-story structure attached to a dwelling with a glazing area in excess of 40 percent of the gross area of the structure's exterior walls and roof." These structures were typically constructed in one of two manners:

1. Using typical wood framing techniques.
2. Using a stick system that consists of prefabricated framing of aluminum, fiberglass, wood, or other materials, with glass or opaque wall or roof panels, and steel or aluminum connections.

Using the 2012 IRC, the first technique was done in accordance with the provisions of the IRC for wood framed construction. There were no provisions for the second method of constructing a sunroom other than by engineering analysis or demonstrating equivalence to the current provisions of the *International Residential Code*.

By adding reference to the provisions of AAMA/NPEA/NSA 2100-12 *Specifications for Sunrooms* to the IRC, prescriptive construction is easier. Sunrooms designed and constructed in accordance with AAMA/NPEA/NSA 2100 are required within the standard to meet the structural provisions of the IRC or IBC. In addition, the standard establishes specific requirements for these structures based upon their designated category.

AAMA/NPEA/NSA 2100 was first published in 2002 by the American Architectural Manufacturers Association (AAMA), the National Sunroom Association (NSA), and the National Patio Enclosure Association (NPEA). It is the first U.S. standard for the design and specification of sunrooms. The standard creates five categories of sunrooms based upon intended use of the space. Specific design and performance criteria based on end use were added to the standard.

The standard was later revised to meet the requirements of AAMA/WDMA/CSA 101/I.S.2/A440 for the design, testing, and labeling of windows, glass doors, and skylights, and the foundation requirements of the *International Residential Code*.

AAMA/NPEA/NSA 2100, and the five categories of sunrooms it establishes, clarifies criteria for sunrooms with regard to egress, natural ventilation, and resistance of the exterior envelop to air leakage and water penetration.

R301.2.1.2

Protection of Openings in Wind Borne Debris Regions

© Dennis Tokarzewski/Shutterstock.com.

Wood structural panels protecting windows from wind-borne debris

© Dennis Tokarzewski/Shutterstock.com.

Poly panels protecting windows from wind-borne debris

CHANGE TYPE: Modification

CHANGE SUMMARY: Requirements for glazed openings to be protected from wind borne debris have been clarified by the addition of a new section detailing changes to the ASTM E 1996 standard.

2015 CODE: R301.2.1.2 Protection of Openings. Exterior glazing in buildings located in wind borne debris regions shall be protected from wind borne debris. Glazed opening protection for wind borne debris shall meet the requirements of the Large Missile Test of ASTM E 1996 and ASTM E 1886 as modified in Section R301.2.1.2.1 ~~referenced therein~~. ~~The applicable wind zones for establishing missile types in ASTM E 1996 are shown on Figure R301.2(4)C.~~ Garage door glazed opening protection for wind-borne debris shall meet the requirements of an approved impact-resisting standard or ANSI/DASMA 115.

> **Exception:** Wood structural panels with a ~~minimum~~ thickness of not less than 7/16 inch (11 mm) and a ~~maximum~~ span of not more than 8 feet (2438 mm) shall be permitted for opening protection~~ one- and two-story buildings~~. Panels shall be precut and attached to the framing surrounding the opening containing the product with the glazed opening. Panels shall be predrilled as required for the anchorage method and shall be secured with the attachment hardware provided. Attachments shall be designed to resist the component and cladding loads determined in accordance with either Table R301.2(2) or ASCE 7, with the permanent corrosion-resistant attachment hardware provided and anchors permanently installed on the building. Attachment in accordance with Table R301.2.1.2 is permitted for buildings with a mean roof height of 45 ~~33~~ feet (~~10 058~~13 720 mm) or less where the ultimate design wind speed, V_{ult}, is 180 mph (80 m/s) or less. ~~located in Wind Zones 1 and 2 in accordance with Figure R301.2(4)C.~~

R301.2.1.2.1. Application of ASTM E 1996. The text of Section 2.2 of ASTM E 1996 shall be substituted as follows:

> **2.2 ASCE Standard:**
> ASCE 7-10 American Society of Civil Engineers Minimum Design Loads for Buildings and Other Structures
>
> The text of Section 6.2.2 of ASTM E 1996 shall be substituted as follows:
>
> 6.2.2 Unless otherwise specified, select the wind zone based on the ultimate design wind speed, V_{ult}, as follows:
>
> 6.2.2.1 Wind Zone 1—130 mph (58 m/s) ≤ ultimate design wind speed, V_{ult} < 140 mph (63 m/s).
>
> 6.2.2.2 Wind Zone 2—140 mph (63 m/s) ≤ ultimate design wind speed, V_{ult} < 150 mph (67 m/s) at greater than one mile (1.6 km) from the coastline. The coastline shall be measured from the mean high water mark.

6.2.2.3 Wind Zone 3—150 mph (67 m/s) ≤ ultimate design wind speed, V_{ult} ≤ 170 mph (76 m/s), or 140 mph (63 m/s) ≤ ultimate design wind speed, V_{ult} ≤ 170 mph (76 m/s) and within one mile of the coastline. The coastline shall be measured from the mean high water mark.

6.2.2.4 Wind Zone 4—ultimate design wind speed, V_{ult} > 170 mph (76 m/s).

CHANGE SIGNIFICANCE: In the early development of the legacy high wind standard, SBCCI Deemed to Comply, limits were developed for the geometry of structures covered by the standard. These limits included a mean roof height of 33 feet. The 33-foot height was based on zoning regulations of the time, the referenced wind speed height in the contemporary ASCE wind standard, and the height of most anemometers (wind measuring devices). The legacy standard limited wood buildings to two stories in height. As the standard evolved the height limit was changed from a 33-foot mean roof height to simply two stories.

From a wind perspective, the geometry of the structure matters. Its internal structure of floors and walls affect the resistance of the structure to the wind. The "two-story-only" requirement puts artificial limitations on the use of shutter provisions. The requirement has limited the use of shutter provisions in three-story residential structures built on sloped surfaces or with a first story partially embedded in the ground. SBCCI *Deemed to Comply* was the precursor to SBCCI *Standard for Hurricane Resistant Residential Construction* (SSTD-10) and ultimately the ICC *Standard for Residential Construction in High Wind Regions* (ICC 600). The new height limit of 45 feet allows use of shutter provisions with all three-story buildings.

The final change within this section modifies the ASTM standard, ASTM E 1996, which has wind speeds based on basic wind speed values. The modified values in Section R301.2.1.2.1 replace those values with ultimate design wind speed values for wind zones.

R301.2.1.4

Wind Exposure Category

CHANGE TYPE: Modification

CHANGE SUMMARY: Wind Exposure Category A is a legacy category that no longer exists in the IBC and ASCE 7, which is the basis for determination of wind exposure categories. In the 2015 IRC, Exposure Category A is deleted.

In the 2012 IRC, Wind Exposure Category D applied to regions adjacent to open water in non-hurricane-prone regions. Wind Exposure Category D now applies to open water, mud and salt flats, and unbroken ice fields. Exposure Category D also applies in hurricane-prone regions to residences on or near the ocean shore.

2015 CODE: R301.2.1.4 Exposure Category. For each wind direction considered, an exposure category that adequately reflects the characteristics of ground surface irregularities shall be determined for the site at which the building or structure is to be constructed. For a site located in the transition zone between categories, the category resulting in the largest wind forces shall apply. Account shall be taken of variations in ground surface roughness that arises from natural topography and vegetation as well as from constructed features. For a site where multiple detached one- and two-family dwellings, townhouses, or other structures are to be constructed as part of a subdivision, master-planned community, or otherwise designated as a developed area by the authority having jurisdiction, the exposure category for an individual structure shall be based upon the site conditions that will exist at the time when all adjacent structures on the site have been constructed, provided that their construction is expected to begin within one year of the start of construction for the structure for

International Code Council®

Wind Exposure Category C area

International Code Council®

Previously considered in Wind Exposure Category A, large city centers are now classified as Wind Exposure Category B.

which the exposure category is determined. For any given wind direction, the exposure in which a specific building or other structure is sited shall be assessed as being one of the following categories:

~~1. Exposure A. Large city centers with at least 50 percent of the buildings having a height in excess of 70 feet (21 336 mm). Use of this exposure category shall be limited to those areas for which terrain representative of Exposure A prevails in the upwind direction for a distance of at least 0.5 mile (0.8 km) or 10 times the height of the building or other structure, whichever is greater. Possible channeling effects or increased velocity pressures due to the building or structure being located in the wake of adjacent buildings shall be taken into account.~~

1~~2~~. Exposure B. Urban and suburban areas, wooded areas, or other terrain with numerous closely spaced obstructions having the size of single-family dwellings or larger. Exposure B shall be assumed unless the site meets the definition of another type exposure.

2~~3~~. Exposure C. Open terrain with scattered obstructions, including surface undulations or other irregularities, having heights generally less than 30 feet (9,144 mm) extending more than 1,500 feet (457 m) from the building site in any quadrant. This exposure shall also apply to any building located within Exposure B–type terrain where the building is directly adjacent to open areas of Exposure

R301.2.1.4 continues

R301.2.1.4 continued

C–type terrain in any quadrant for a distance of more than 600 feet (183 m). This category includes flat, open country and grasslands.

3~~4~~. Exposure D. Flat, unobstructed areas exposed to wind flowing over open water, smooth mud flats, salt flats, and unbroken ice for a distance of not less than 5,000 feet (1,524 m) ~~1 mile (1.61 km). Shorelines in Exposure D include inland waterways, the Great Lakes, and coastal areas of California, Oregon, Washington and Alaska~~. This exposure shall apply only to those buildings and other structures exposed to the wind coming from over the ~~water~~ unobstructed area. Exposure D extends ~~inland~~ downwind from the ~~shoreline~~ edge of the unobstructed area a distance of 600 feet (183 m) ~~1500 feet (457 m)~~ or 20 ~~10~~ times the height of the building or structure, whichever is greater.

CHANGE SIGNIFICANCE: Wind Exposure Category A is a legacy category that no longer exists in the IBC and ASCE 7, which is the basis for determination of wind exposure categories. Exposure A is deleted in the 2015 IRC. Wind Exposure Category A included residential-height buildings surrounded by taller buildings in an urban environment. Because buildings surrounded by taller buildings may be subjected to increased wind speeds and gusting of winds due to the tunnel effect of taller buildings, the category as a minimal category was dropped from ASCE 7 and the IBC. Buildings in these areas may be required to have wind tunnel testing or have additional factors increasing the basic wind speed applied to them. The category remained in the IRC through the 2012 edition.

Wind Exposure Category D has been updated to match the standard that is the basis for wind exposures, the *Minimum Design Loads for Buildings and Other Structures*, ASCE/SEI 7-10. Due to research in wind speeds at the water surface during hurricanes and other storm events, it is now appropriate to use Exposure D for hurricane-affected coastlines and for large, unusually flat regions that do not have open water nearby. Previously, high winds across the ocean's surface were assumed to create large waves (fetch). Recent research has shown that wave heights directly below hurricanes are dampened, causing the ocean's surface to be relatively smooth. Exposure D is applied when an area is unobstructed and the surface is smooth. This category has always been used for non-hurricane-prone coastlines. With evidence that a hurricane does not significantly roughen the ocean surface, Exposure Category D becomes the more appropriate category along hurricane-prone coastlines as well.

Because Exposure Category D applies to unobstructed areas and smooth surfaces, the category is now used for areas that are salt flats, marshes, and unbroken ice. These areas have little elevation to break up winds before they reach residential construction. Because winds take some time to slow down due to a new obstruction, regions that would otherwise be in Wind Exposure Category B or C that are within 600 feet of the boundary of a body of water or ice, a marsh, or a salt flat also have Exposure Category D requirements applied to them.

This change to the exposure categories brings the IRC in line with the IBC and industry standards. The 2012 IRC definition for Wind Exposure Category D did not match the definition in the 2012 IBC or ASCE 7-10. This code change incorporates the language of ASCE 7-10 Section 26.7.3 into the IRC. For more information on recent high wind research, read ASCE 7-10 commentary section C26.7.

Table R301.2.1.5.1

Modifications for Topographic Wind Effects

CHANGE TYPE: Modification

CHANGE SUMMARY: Table R301.2.1.5.1, Ultimate Design Wind Speed Modification for Topographic Wind Effect, is updated for the change in wind speed values. The table gives minimum ultimate design wind speed values depending upon the slope of the upper portion of the ridge, hill, or escarpment.

2015 CODE:

TABLE R301.2.1.5.1 Ultimate Design Wind Speed Modification for Topographic Wind Effect[a,b]

Ultimate Design Wind Speed from Figure R301.2(4)A	Average Slope of the Top Half of Hill, Ridge, or Escarpment (percent)						
	0.10	0.125	0.15	0.175	0.20	0.23	0.25
	Required Ultimate Design Wind Speed-up, Modified for Topographic Wind Speed Up (mph)						
110	132	137	142	147	152	158	162
115	138	143	148	154	159	165	169
120	144	149	155	160	166	172	176
130	156	162	168	174	179	N/A	N/A
140	168	174	181	N/A	N/A	N/A	N/A
150	180	N/A	N/A	N/A	N/A	N/A	N/A

a. Table applies to a feature height of 500 feet or less and dwellings sited a distance equal or greater than half the feature height.

b. Where the ultimate design wind speed as modified by Table R301.2.1.5.1 equals or exceeds 140 mph, the building shall be considered as "wind design required" in accordance with Section R301.2.1.1.

(Deleted text not shown for clarity.)

International Code Council®

Dwellings on toe of slope

CHANGE SIGNIFICANCE: Table R301.2.1.5.1 gives the required ultimate design wind speed for a building on or above a slope. The value is determined by the slope of the upper portion of a ridge, hill, or escarpment. A new footnote is provided as reference for topographic wind effects making the site a "wind design required" region and requiring use of an alternate standard (ICC-600, WFCM, AISI 230, IBC) for wind design.

The table for modifications due to topographic wind speed effects is based on the concept that wind speed increases as air is compressed when moving upward over a hill or ridge. The process to determine wind speed-up due to topographic effects is complex and requires wind analysis and design. IRC Table R301.2.1.5.1 attempts to simplify this design in a table for slopes of 10 to 25 percent. Note that the table lists percentages as decimal values.

R301.2.4

Floodplain Construction

CHANGE TYPE: Modification

CHANGE SUMMARY: Buildings located in a flood hazard area must comply with the provisions for the most restrictive flood hazard area and may use ASCE 24 for design.

2015 CODE: R301.2.4 Floodplain Construction. Buildings and structures constructed in whole or in part in flood hazard areas (including A or V Zones) as established in Table R301.2(1), and substantial improvement and restoration of substantial damage of buildings and structures in flood hazard areas, shall be designed and constructed in accordance with Section R322. Buildings and structures that are located in more than one flood hazard area shall comply with the provisions associated with the most restrictive flood hazard area. Buildings and structures located in whole or in part in identified floodways shall be designed and constructed in accordance with ASCE 24.

R301.2.4.1 Alternative Provisions. As an alternative to the requirements in Section R322, ~~R322.3 for buildings and structures located in whole or in part in coastal high-hazard areas (V Zones) and Coastal A Zones, if delineated,~~ ASCE 24 is permitted subject to the limitations of this code and the limitations therein.

CHANGE SIGNIFICANCE: ASCE/SEI 24, *Flood Resistant Design and Construction*, provides an alternative design procedure for buildings and structures in flood hazard areas. There are many flood hazard areas where the builder, designer, or building official may deem it appropriate to use an engineered foundation, such as along riverine waterways and some coastal areas (inland of Zone V) where flood depths are significant and

© Delmas Lehman/iStock/Thinkstock

Houses in a floodplain

dwellings would need very tall foundations. Design may be needed in riverine floodplains where flood velocities are very fast as well. ASCE 24 provides assistance for design of these foundations.

Another situation where use of ASCE 24 is appropriate is with dwellings in flood hazard areas on alluvial fans. The IRC does not contain specific provisions for alluvial fans. Specifying ASCE 24 as an alternative allows its use where prescriptive provisions of the IRC do not account for known flood risks.

Design of buildings located in two flood hazard areas is clarified. Where a building is affected by more than one flood hazard, the structure must comply with the more restrictive provisions that take into account flood loads and conditions of the area. For example, a dwelling that straddles a line that separates Zone A from Zone V must comply with the requirements for Zone V. Section R301.2.4 applies to existing dwellings as well as new dwellings. The flood provisions apply to substantial improvement and substantial damage of existing dwellings.

R301.3

Story Height

CHANGE TYPE: Modification

CHANGE SUMMARY: Story height of wood and steel wall framing, insulated concrete, and SIP walls may not exceed 11 feet, 7 inches. Masonry wall height is limited to 13 feet, 7 inches.

2015 CODE: **R301.3 Story Height.** The wind and seismic provisions of this code shall apply to buildings with story heights not exceeding the following:

1. For wood wall framing, the story height shall not exceed 11 feet 7 inches (3531 mm) and the laterally unsupported bearing wall stud height permitted by Table R602.3(5) ~~plus a height of floor framing not to exceed 16 inches (406 mm)~~.

   ~~**Exception:** For wood-framed wall buildings with bracing in accordance with Tables R602.10.3(1) and R602.10.3(3), the wall stud clear height used to determine the maximum permitted story height may be increased to 12 feet (3658 mm) without requiring an engineered design for the building wind and seismic force-resisting systems provided that the length of bracing required by Table R602.10.3(1) is increased by multiplying by a factor of 1.10 and the length of bracing required by Table R602.10.3(3) is increased by multiplying by a factor of 1.20. Wall studs are still subject to the requirements of this section.~~

2. For cold-formed steel wall framing, the story height shall be not more than 11 feet 7 inches (3531 mm) and the ~~a~~ unsupported bearing wall stud height shall be not more than ~~of~~ 10 feet (3048 mm)~~, plus a height of floor framing not to exceed 16 inches (406 mm)~~.

3. For masonry walls, the story height shall be not more than 13 feet 7 inches (4140 mm) and the ~~a maximum~~ bearing wall clear height shall be not greater than ~~of~~ 12 feet (3658 mm) ~~plus a height of floor framing not to exceed 16 inches (406 mm)~~.

   Exception: An additional 8 feet (2438 mm) of bearing wall clear height is permitted for gable end walls.

4. For insulating concrete form walls, the story height shall be not more than 11 feet 7 inches (3531 mm) and the ~~maximum~~ unsupported ~~bearing~~ wall height per story as permitted by Section R611 tables shall not exceed 10 feet (3048 mm) ~~plus a height of floor framing not to exceed 16 inches (406 mm)~~.

5. For structural insulated panel (SIP) walls, the story height shall be not greater than 11 feet 7 inches (3531 mm) and the ~~maximum~~ bearing wall height per story as permitted by Section ~~R613~~ R610 tables shall not exceed 10 feet (3048 mm) ~~plus a height of floor framing not to exceed 16 inches (406 mm)~~.

Individual walls or walls studs shall be permitted to exceed these limits as permitted by Chapter 6 provisions, provided that story heights are not exceeded. ~~Floor framing height shall be permitted to exceed these limits provided the story height does not exceed 11 feet 7 inches (3531 mm).~~ An engineered design shall be provided for the wall or wall framing members

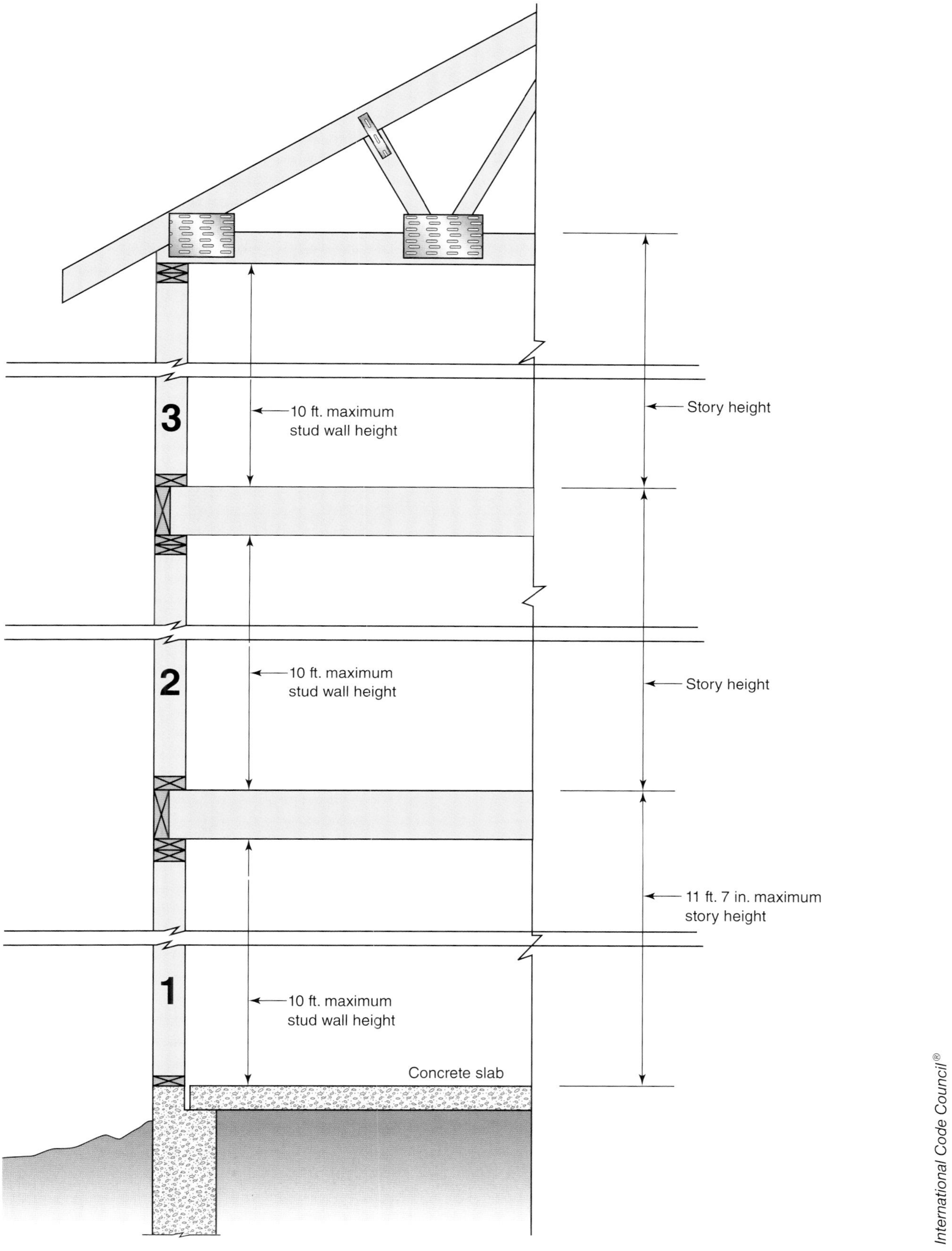

Story height

R301.3 continues

R301.3 continued

~~when they exceed~~ where the limits of Chapter 6 are exceeded. Where the story height limits of this section are exceeded, the design of the building, or the noncompliant portions thereof, to resist wind and seismic loads shall be in accordance with the *International Building Code*.

CHANGE SIGNIFICANCE: This code change revises the story height limits. The 2009 IRC introduced the 11 feet 7 inches story height limit as an alternative to limiting the floor framing height to 16 inches when wall stud heights were 10 feet 0 inches. The exception was added to a paragraph following the five individual limits for wall materials. This has led to conflict with the Chapter 6 provisions limiting stud size and height and with the wall bracing section. This code change moves the story height limit to each of the individual material sections and coordinates the height limit with the material-specific provisions.

The 2012 IRC exception for wood wall studs in Section R301.3 is deleted, as it is redundant with the following provisions of Chapter 6:

1. Table R602.3(5) covers when studs in non-bearing walls can exceed 10 feet.
2. Section R602.3.1 provides limited cases for studs in bearing walls exceeding 10 feet in height.
3. The wall bracing section provides adjustments to wind and seismic bracing amounts for heights up to 12 feet.

This change clarifies the intent that stud heights over 10 feet require engineered design or use of an alternate standard for gravity loads. For determination of wall bracing length, use of the wall bracing provisions permits walls up to 12 feet tall. The wall bracing section does not address structural concerns due to gravity loads resulting from an overall increase in story height. The wall bracing section only applies to in-plane lateral loads. For out-of-plane lateral loads, the limited conditions in Section R602.3.1 allow studs greater than 10 feet in height supporting roof loads.

R302.1

Exterior Walls

CHANGE TYPE: Modification

CHANGE SUMMARY: Unprotected roof overhangs are now permitted to project to within 2 feet of the property line when fireblocking is installed between the top of the wall and the roof sheathing. In most cases, projections are not permitted less than 2 feet from the property line. For dwellings with or without fire sprinkler protection, penetrations of exterior walls do not require fire-resistant protection unless they are located less than 3 feet from the property line.

2015 CODE: R302.1 Exterior Walls. Construction, projections, openings, and penetrations of exterior walls of dwellings and accessory buildings shall comply with Table R302.1(1); or dwellings equipped throughout with an automatic sprinkler system installed in accordance with Section P2904 shall comply with Table R302.1(2).

Exceptions: *(No change to text.)*

R302.1 continues

TABLE R302.1(1) Exterior Walls

Exterior Wall Element		Minimum Fire-Resistance Rating	Minimum Fire Separation Distance
Walls	Fire-resistance rated	1 hour-tested in accordance with ASTM E 119 or UL 263 with exposure from both sides	< 5 feet
	Not fire-resistance rated	0 hours	≥ 5 feet
Projections	Not allowed	N/A	< 2 feet
	Fire-resistance rated	1 hour on the underside[a, b]	≥ 2 feet to < 5 feet
	Not fire-resistance rated	0 hours	≥ 5 feet
Openings in walls	Not allowed	N/A	< 3 feet
	25% maximum of wall area	0 hours	3 feet
	Unlimited	0 hours	5 feet
Penetrations	All	Comply with Section R302.4	< ~~5~~ 3 feet
		None required	~~5~~ 3 feet

For SI: 1 foot = 304.8 mm.
N/A = Not Applicable
a. Roof eave fire-resistance rating shall be permitted to be reduced to 0 hours on the underside of the eave if fireblocking is provided from the wall top plate to the underside of the roof sheathing.
b. Roof eave fire-resistance rating shall be permitted to be reduced to 0 hours on the underside of the eave provided gable vent openings are not installed.

R302.1 continued

TABLE R302.1(2) Exterior Walls—Dwellings with Fire Sprinklers

Exterior Wall Element		Minimum Fire-Resistance Rating	Minimum Fire Separation Distance
Walls	Fire-resistance rated	1 hour-tested in accordance with ASTM E 119 or UL 263 with exposure from the outside	0 feet
	Not fire-resistance rated	0 hours	3 feet[a]
Projections	Not allowed	N/A	< 2 feet
	Fire-resistance rated	1 hour on the underside[b, c]	2 feet[a]
	Not fire-resistance rated	0 hours	3 feet
Openings in walls	Not allowed	N/A	< 3 feet
	Unlimited	0 hours	3 feet[a]
Penetrations	All	Comply with Section R302.4	< 3 feet
		None required	3 feet[a]

For SI: 1 foot = 304.8 mm.
N/A = Not Applicable
a. For residential subdivisions where all dwellings are equipped throughout with an automatic sprinkler system installed in accordance with Section P2904, the fire separation distance for non-rated exterior walls and rated projections shall be permitted to be reduced to zero feet, and unlimited unprotected openings and penetrations shall be permitted, where the adjoining lot provides an open setback yard that is 6 feet or more in width on the opposite side of the property line.
b. Roof eave fire-resistance rating shall be permitted to be reduced to 0 hours on the underside of the eave if fireblocking is provided from the wall top plate to the underside of the roof sheathing.
c. Roof eave fire-resistance rating shall be permitted to be reduced to 0 hours on the underside of the eave provided gable vent openings are not installed.

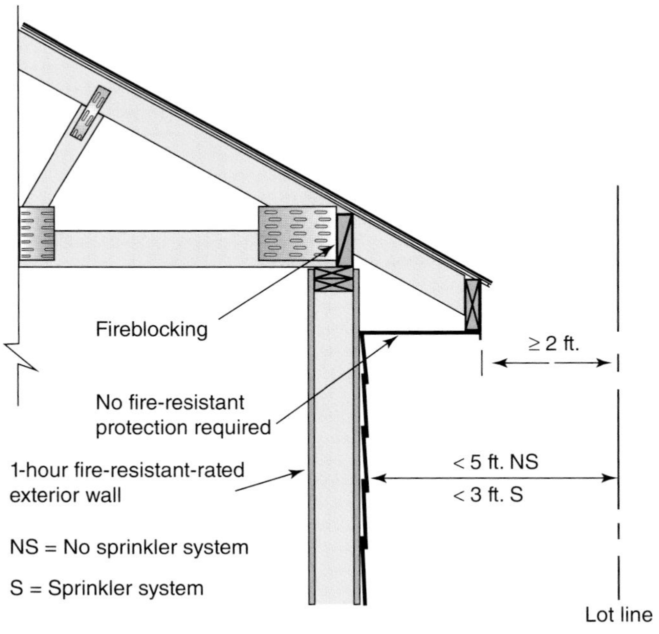

International Code Council®

Fire resistance rating is not required for roof eave projections when fireblocking is installed.

CHANGE SIGNIFICANCE: The code has long recognized the effectiveness of providing space between the exterior wall and the lot line in preventing the spread of fire from a building on one property to a building on another property. Unless the exterior wall is constructed to provide a 1-hour fire-resistance rating in accordance with either ASTM E-119 or UL 263, a minimum fire separation distance is required from the lot line. The consensus as to the minimum distance necessary to provide a sufficient buffer against the spread of fire has changed somewhat over the years, settling on a minimum distance of 5 feet in the 2006 edition of the IRC. Beginning with the 2012 edition, the code reduces the threshold for non-rated walls to 3 feet of separation when the building is protected with an automatic fire sprinkler system. The 5-foot rule still applies to buildings without sprinkler systems. The choice of providing either adequate fire separation distance or fire-resistant-rated construction also extends to openings, penetrations, and projections—some fire resistance measures must be provided when the fire separation distance to the property line falls below the code-prescribed dimension.

Roof eaves constructed without fire-resistant protection are permitted to project to not less than 5 feet from the lot line for buildings without fire sprinkler systems and not less than 3 feet from the lot line for buildings with sprinklers. For eave projections with a separation distance less than those dimensions, the code requires 1-hour fire-resistant protection on the underside of the overhang. The 2015 IRC provides an option to builders to eliminate the soffit protection when fireblocking is installed between the top plate of the exterior fire-resistant-rated wall and the roof sheathing. For a fire originating on the adjacent property, the fireblocking above the wall protects against the spread of fire through the overhang into the attic area. This effectively extends a measure of fire resistance at the exterior wall to the roof line and is considered equivalent protection to a 1-hour-rated soffit. In this case, the unprotected eave projection is viewed as expendable because the barrier to the spread of fire is established at the exterior wall line.

In almost all circumstances, the code does not permit any portion of a roof overhang, with or without fire protection, to be constructed less than 2 feet from a lot line. This point is clarified by the addition of a line in Tables R302.1(1) and R302.1(2) that states that projections are not allowed with a fire separation distance of less than 2 feet. However, there are a couple of exceptions to this rule that have not changed and are still in effect. Exception 4 to the exterior wall provisions of Section R302.1 specifically allows a maximum 4-inch roof eave projection for detached garages located within 2 feet of a lot line. For example, a detached garage that is accessory to the dwelling on the same lot and has an exterior wall located 1 foot from the lot line requires 1-hour fire-resistant-rated construction for that exterior wall. Under the exception, a 4-inch overhang that would project to 8 inches from the lot line is permitted in this case. Most code users have inferred that 1-hour protection is required on the underside of this overhang in accordance with the applicable table and that the exception only applies to the permitted location of the overhang, not the fire-resistance provisions.

The second exception that permits projections less than 2 feet from the lot line first appeared in the 2012 edition of the IRC. Footnote a of Table R302.1(2) allows rated projections with a fire separation distance of 0 feet when other criteria are satisfied. All dwellings in the subdivision require

R302.1 continues

R302.1 continued

TABLE 3-2 Fire Resistance of Roof Overhang Projections

Condition	Minimum Fire Separation Distance			
	5 feet	3 feet	2 feet	0 feet
Dwellings Without Sprinkler System	0 hours	1 hour on underside	1 hour on underside	NP
Dwellings with Sprinkler System	0 hours	0 hours	1 hour on underside	NP
Fireblocking above Top Plate	0 hours	0 hours	0 hours	NP
Sprinklers in All Dwellings and 6-Foot Setback on Adjoining Lot	N/A	N/A	N/A	1 hour on underside
4-Inch Overhang on Detached Garage	N/A	N/A	N/A	1 hour on underside

NP = Not Permitted
N/A = Not Applicable

automatic fire sprinkler systems and buildings on the adjoining property require an open setback yard that is not less than 6 feet. This required setback on the opposite side of the lot line ensures that a minimum 6-foot separation distance is maintained between the exterior walls of the two buildings. With the added protection of a fire sprinkler system, the 6-foot separation is consistent with the provisions for unrated walls and unlimited openings in Table R302.1(2), which requires a 3-foot fire separation distance for each building. Although there are no fire-resistance requirements for the exterior wall under this exception, the 1-hour protection on the underside of the projection is still required. Table 3-2 summarizes the fire separation distance requirements for projections.

In the 2012 edition of the IRC, where wall assemblies are required to be fire-resistance rated, penetrating items require protection to maintain the fire resistance of the wall. For dwellings with automatic fire sprinkler systems, the trigger point for installing a rated wall assembly and penetration protection is a fire separation distance of less than 3 feet. For dwellings without sprinklers, the dimension has been less than 5 feet. However, the IRC has allowed a limited amount of unprotected openings such as windows and doors in exterior walls of unsprinklered dwellings when the fire separation distance was less than 5 feet but not less than 3 feet. In the 2015 IRC, this same allowance is applied to penetrations—fire protection of the penetration is not required unless the exterior wall is less than 3 feet from the lot line. The penetration provisions for exterior walls now match for dwellings with sprinklers and those without. This is considered a reasonable accommodation for small penetrations such as hose bibbs, dryer vent terminations, mechanical draft terminals, and electrical equipment without impairing the effectiveness of the fire-resistant-rated assembly. For penetrations less than 3 feet from the lot line, Section R302.4 prescribes the methods of protection to prevent the passage of flame and hot gases at the penetrations.

R302.2

Townhouse Separation

CHANGE TYPE: Modification

CHANGE SUMMARY: The provisions for separating townhouses with structurally independent fire-resistant-rated walls in accordance with Section R302.1 have been removed in favor of the common wall provisions of Section R302.2. Common walls separating townhouses must now be rated for 2 hours when an automatic fire sprinkler system is not installed in the townhouse dwelling units.

2015 CODE: R302.2 Townhouses. ~~Each townhouse shall be considered a separate building and shall be separated by fire-resistance rated wall assemblies meeting the requirements of Section R302.1 for exterior walls.~~ Common walls separating townhouses shall be assigned a fire resistance rating in accordance with Section R302.2 Item 1 or 2. The common wall shared by two townhouses shall be constructed without plumbing or mechanical equipment, ducts, or vents in the cavity of the common wall. The wall shall be rated for fire exposure from both sides and shall extend to and be tight against exterior walls and the underside of the roof sheathing. Electrical installations shall be in accordance with Chapters 34 through 43. Penetrations of the membrane of common walls for electrical outlet boxes shall be in accordance with Section R302.4.

R302.2 continues

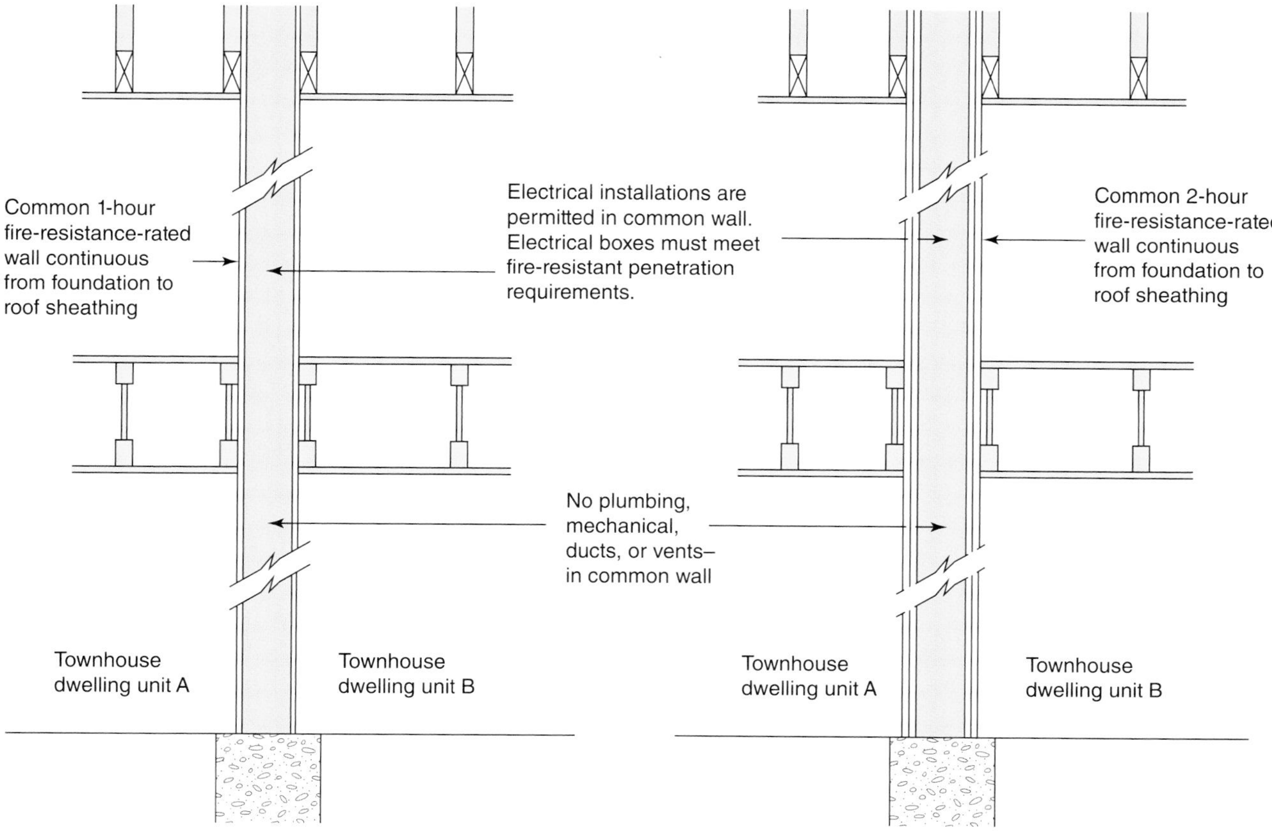

One-hour common wall for townhouses with sprinklers — Two-hour common wall for townhouses without sprinklers

International Code Council®

Common walls separating townhouses

R302.2 continued

~~**Exceptions:**~~

1. Where a fire sprinkler system in accordance with Section P2904 is provided, ~~A~~ the common wall shall be not less than a 1-hour fire-resistance-rated wall assembly tested in accordance with ASTM E 119 or UL 263. ~~is permitted for townhouses if such walls do not contain plumbing or mechanical equipment, ducts or vents in the cavity of the common wall. The wall shall be rated for fire exposure from both sides and shall extend to and be tight against exterior walls and the underside of the roof sheathing. Electrical installations shall be installed in accordance with Chapters 34 through 43. Penetrations of electrical outlet boxes shall be in accordance with Section R302.4.~~
2. Where a fire sprinkler system in accordance with Section P2904 is not provided, the common wall shall be not less than a 2-hour fire-resistance-rated wall assembly tested in accordance with ASTM E 119 or UL 263.

R302.2.4 Structural Independence. Each individual townhouse shall be structurally independent.

Exceptions:

1. Foundations supporting exterior walls or common walls.
2. Structural roof and wall sheathing from each unit fastened to the common wall framing.
3. Nonstructural wall and roof coverings.
4. Flashing at termination of roof covering over common wall.
5. Townhouses separated by a common ~~1-hour fire-resistance-rated~~ wall as provided in Section R302.2, ~~Exception~~ Item 1 or 2.

CHANGE SIGNIFICANCE: In previous editions of the IRC, the general rule required townhouses to be considered separate buildings with each building having a 1-hour fire-resistant-rated wall to separate it from the adjoining townhouse. The 1-hour rating was determined in accordance with Section R302.1 for exterior walls based on the fire separation distance between the individual townhouse units. This resulted in two separate 1-hour-rated walls where townhouses joined. Section R302.2.4 further required that each individual townhouse be structurally independent, meaning that a collapse of the structural wall, floor, ceiling, or roof components of one townhouse in a fire incident would not impair the structural integrity of the adjoining townhouse. As an alternative, the IRC has always provided for constructing a common fire-resistant-rated wall between townhouse units. Because the common wall supports structural floor and roof elements of the townhouse dwelling units on both sides, structural independence is not possible and is not required for the common wall option. But this option has always limited installations in the wall to electrical components. To preserve structural integrity and limit penetrations of the fire-resistant membrane, the code does not permit the installation of plumbing or mechanical equipment, ducts, or vents in the cavity of the common wall.

In the 2015 IRC, the exception for constructing a common wall becomes the rule and the only prescriptive option for separating townhouses. References to the exterior wall provisions in Section R302.1 have been removed. The structural independence requirement of Section R302.2.4 no longer applies because Exception 5 will be in effect for all installations. In practice, this change may not have a significant impact on the way townhouses are constructed. In many geographic regions, the common wall option has been the preferred method for the majority of designers and builders.

Prior to the 2009 edition of the IRC, the common wall for separation of townhouses was required to be a 2-hour fire-resistant-rated wall assembly. With the introduction of mandatory fire sprinkler requirements for all new dwelling units in the 2009 IRC, the rating of the common wall was reduced to 1 hour. The reduced rating reflected a consensus that automatic sprinkler systems would improve fire safety and that the 1-hour rating would provide a reasonable level of passive fire protection. Although the change was based on the sprinkler provisions, the code language did not tie the two together—the sprinkler provisions and the fire resistance of the common wall were independent requirements. Because the basis for the reduced rating was an assumption that sprinklers would be installed, the fire rating of the common wall assembly is now tied to the presence of sprinklers. The 2-hour rating for non-sprinklered buildings has been reinstated in the 2015 IRC. For townhouse dwelling units protected with an automatic sprinkler system, the 1-hour rating is still in effect. The change was prompted by a concern that jurisdictions amending the IRC to remove the fire sprinkler requirements may not be amending the common wall provisions to reflect the 2-hour fire-resistant rating.

R302.13

Fire Protection of Floors

CHANGE TYPE: Clarification

CHANGE SUMMARY: The provisions for fire protection of floors have been relocated from Chapter 5 to the fire-resistant construction provisions of Section R302. New language clarifies that the code does not regulate penetrations or openings in the fire protection membrane.

2015 CODE: **~~R501.3~~ R302.13** **Fire protection of floors.** Floor assemblies~~,~~ that are not required elsewhere in this code to be fire-resistance rated, shall be provided with a ½-inch (12.7 mm) gypsum wallboard membrane, ⅝-inch (16 mm) wood structural panel membrane, or equivalent on the underside of the floor framing member. Penetrations or openings for ducts, vents, electrical outlets, lighting, devices, luminaires, wires, speakers, drainage, piping, and similar openings or penetrations shall be permitted.

Exceptions: *(No change to text.)*

CHANGE SIGNIFICANCE: Fire protection of floors first appeared in Section R501.3 of the 2012 IRC. The provisions call for installation of ½-inch gypsum board, ⅝-inch wood structural panel, or other approved material on the underside of floor assemblies of buildings constructed under the IRC. The application of gypsum wallboard or other approved material intends to provide some protection to the floor system against the effects of fire and delay collapse of the floor. This provision primarily is aimed at light-frame construction consisting of I-joists, manufactured floor trusses, cold-formed steel framing, and other materials and manufactured products considered most susceptible to collapse in a fire. Solid-sawn lumber and structural composite lumber perform fairly well in retaining adequate strength under fire conditions, and floors framed of

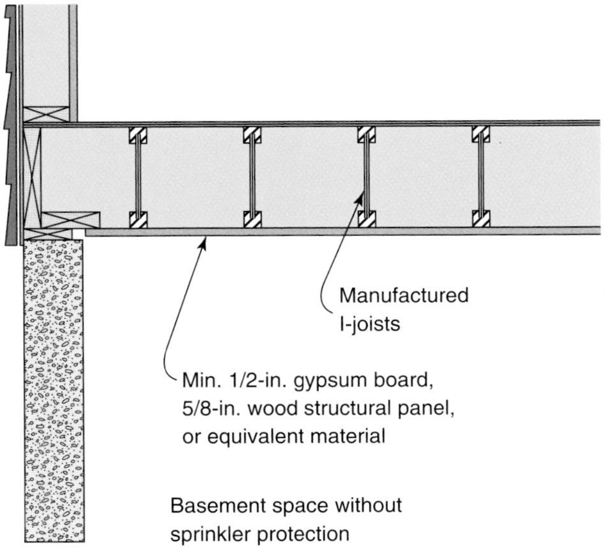

International Code Council®

Fire protection of floors

Open web floor trusses requiring membrane protection on the underside

nominal 2 × 10s or larger of these materials are exempt from these fire protection requirements. Fire protection also is not required if sprinklers are installed to protect the space below the floor assembly.

In the 2009 IRC, there was an effort to organize all of the fire-resistance provisions into a single section to make the code more user-friendly. Because the installation of the code-prescribed membrane intends to provide some limited protection against the effects of fire to the floor system, the requirements have been relocated to the fire-resistant construction provisions of Section R302.

Similar to the fire separation requirements for an attached garage in Section R302.6, the membrane applied to the underside of the floor system does not form a fire-resistant-rated assembly. The membrane acts to shield light-frame floor systems from the heat of a fire originating in the space below the floor. The intent is for the floor system to perform similarly to unprotected 2 × 10 solid-sawn lumber floor joists and to delay structural collapse of the floor system. For that reason, the code does not require any special treatment of joints, penetrations, or openings in the ceiling membrane. For example, the taping of the gypsum board joints is not required and penetrations for electrical boxes and plumbing pipes do not require any firestopping materials. The added language intends to simply clarify that the code does not regulate openings and penetrations in the membrane applied to the underside of the floor system.

R303.7, R303.8

Stairway Illumination

CHANGE TYPE: Clarification

CHANGE SUMMARY: Interior and exterior stairway illumination provisions have been placed in separate sections. Conflicting language has been removed to clarify the requirements.

2015 CODE: ~~**R303.7 Stairway Illumination.** All interior and exterior stairways shall be provided with a means to illuminate the stairs, including the landings and treads. Interior stairways shall be provided with an artificial light source located in the immediate vicinity of each landing of the stairway. For interior stairs the artificial light sources shall be capable of illuminating treads and landings to levels not less than 1 footcandle (11 lux) measured at the center of treads and landings. Exterior stairways shall be provided with an artificial light source located in the immediate vicinity of the top landing of the stairway. Exterior stairways providing access to a basement from the outside grade level shall be provided with an artificial light source located in the immediate vicinity of the bottom landing of the stairway.~~

> ~~**Exception:** An artificial light source is not required at the top and bottom landing, provided an artificial light source is located directly over each stairway section.~~

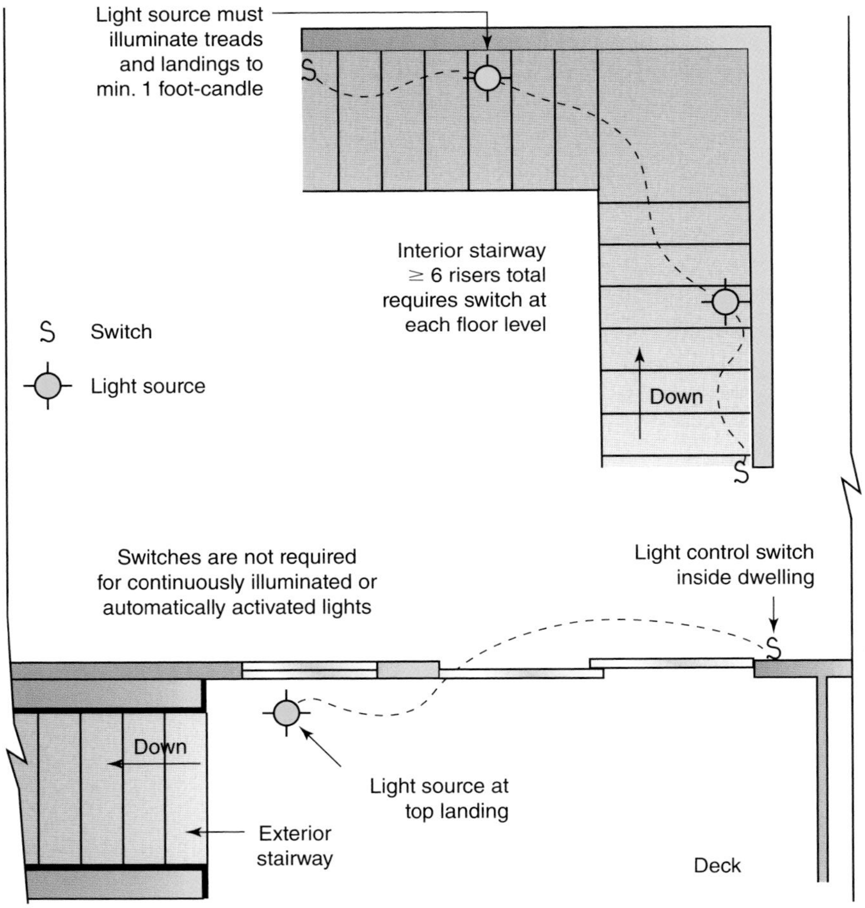

International Code Council®

Stairway illumination

~~**R303.7.1 Light Activation.** Where lighting outlets are installed in interior stairways, there shall be a wall switch at each floor level to control the lighting outlet where the stairway has six or more risers. The illumination of exterior stairways shall be controlled from inside the dwelling unit.~~

> ~~**Exception:** Lights that are continuously illuminated or automatically controlled.~~

R303.7 Interior Stairway Illumination. Interior stairways shall be provided with an artificial light source to illuminate the landings and treads. The light source shall be capable of illuminating treads and landings to levels of not less than 1 foot-candle (11 lux) as measured at the center of treads and landings. There shall be a wall switch at each floor level to control the light source where the stairway has six or more risers.

> **Exception:** A switch is not required where remote, central, or automatic control of lighting is provided.

R303.8 Exterior Stairway Illumination. Exterior stairways shall be provided with an artificial light source located at the top landing of the stairway. Exterior stairways providing access to a basement from the outdoor grade level shall be provided with an artificial light source located at the bottom landing of the stairway.

CHANGE SIGNIFICANCE: Editorial changes to the stairway illumination provisions clarify their application. The requirements for interior and exterior illumination have been placed in separate sections. Interior stairways require illumination of treads and landings. The code no longer prescribes the location of the light source for interior stairways, but allows design flexibility in satisfying the minimum illumination level at the walking surface along the entire stairway. Exterior stairways are treated differently and the code does not prescribe a minimum illumination level. In this case the code requires a light source located at the top landing. In addition, bottom landings require a light source if they provide access to the basement from grade level. The IRC electrical provisions do not address exterior stairway illumination but do require a wall-switch-controlled lighting outlet on the exterior side of each outdoor egress door, including those serving garages.

Previously, the stairway illumination section began with the general statement that "all interior and exterior stairways shall be provided with a means to illuminate the stairs, including the landings and treads." That language did not align with the specific location requirements later in the section that required a light source at only the top landing of exterior stairs, and in some cases the bottom landing. The conflicting language has been removed.

R304.1

Minimum Habitable Room Area

CHANGE TYPE: Modification

CHANGE SUMMARY: The requirement for one habitable room with a minimum floor area of 120 square feet has been removed from the code.

2015 CODE: **R304.1 Minimum Area.** ~~Every dwelling unit shall have at least one habitable room that shall have not less than 120 square feet (11 m^2) of gross floor area.~~

~~**R304.2 Other Rooms.**~~ ~~Other~~ Habitable rooms shall have a floor area of not less than 70 square feet (6.5 m^2).

Exception: Kitchens.

CHANGE SIGNIFICANCE: The IRC sets minimum requirements for a healthy interior living environment, including provisions for room size, ceiling height, light, ventilation, and heating. The code has long provided a minimum room area of 120 square feet for at least one habitable room with all other habitable rooms having a floor area not less than 70 square feet. Most modern homes have rooms that exceed those dimensions, but the intent has been to at least provide a small 12-foot by 10-foot living room with one or more bedrooms measuring approximately 7 feet by 10 feet. The requirement for one habitable room with a minimum floor area of 120 square feet has been removed from the code. The 70-square-foot minimum area now applies to all habitable rooms as the smallest acceptable size for occupants to move about and use the habitable space as intended. The minimum area of 120 square feet was not based on scientific analysis or on identified safety hazards but was generally accepted by code users and in the marketplace. Recently, however, proponents of minimalist living have advocated smaller dwellings to reduce environmental impact and provide for lower living costs through reduced mortgage and maintenance expenses. These dwellings are intended to allow for a minimalist lifestyle that doesn't demand large volumes of living space. Extreme examples of

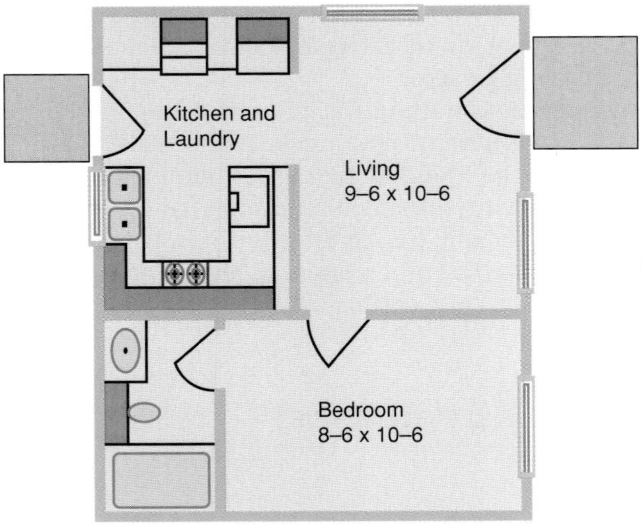

International Code Council®

Small dwelling complying with minimum area requirements

these minimalist dwellings are often termed "tiny houses." Proponents of this change reasoned that consumers make a purposeful and informed decision as to the appropriateness of the housing they choose to live in and that the code should not place arbitrary restrictions on room size that have no demonstrable life-safety benefit. Although the change will not impact typical residential construction, it will accommodate alternatives for very small dwellings that would previously not be allowed under the IRC. It may also encourage greater acceptance of and compliance with the residential code by those pursuing a minimalist lifestyle.

R305

Ceiling Height

CHANGE TYPE: Modification

CHANGE SUMMARY: The minimum ceiling height for bathrooms, toilet rooms, and laundry rooms has been reduced to 6 feet 8 inches. The exception for allowing beams, girders, ducts, or other obstructions to project to within 6 feet, 4 inches of the finished floor has been expanded to include basements with habitable space.

2015 CODE:

SECTION R305
CEILING HEIGHT

R305.1 Minimum Height. Habitable space, hallways, ~~bathrooms, toilet rooms, laundry rooms~~ and portions of basements containing these spaces shall have a ceiling height of not less than 7 feet (2134 mm). Bathrooms, toilet rooms, and laundry rooms shall have a ceiling height of not less than 6 feet 8 inches (2032 mm).

Exceptions:

1. For rooms with sloped ceilings, the required floor area of the room must have a ceiling height of not less than 5 feet (1524 mm) and not less than 50 percent of the required floor

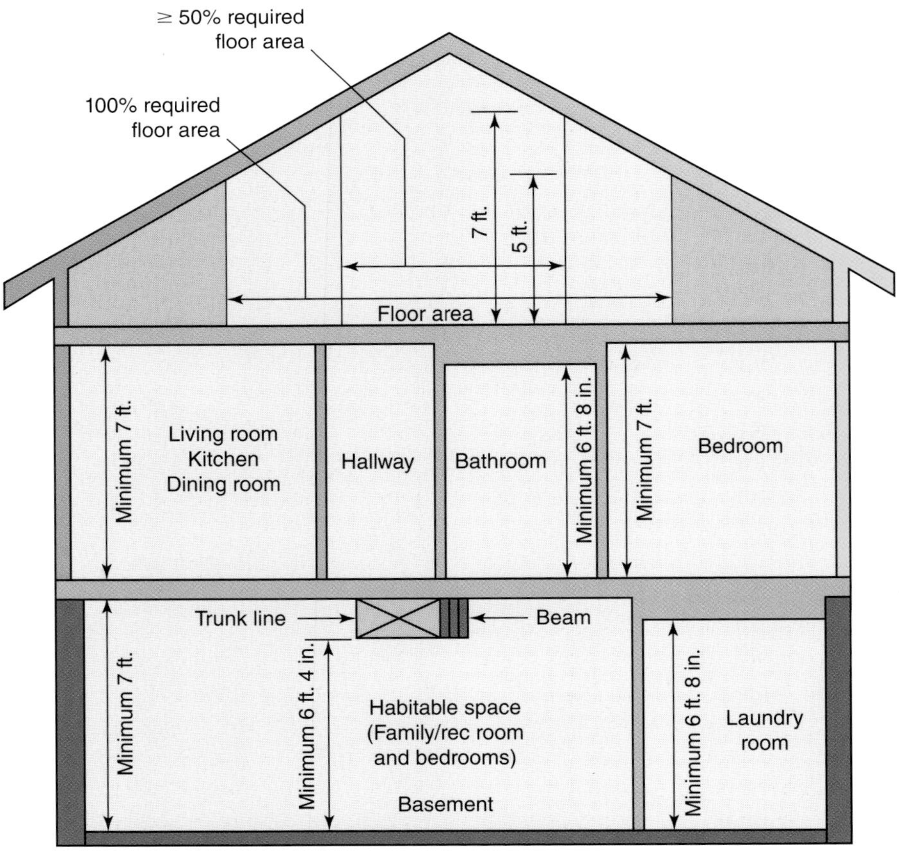

International Code Council®

Ceiling height

area shall have a ceiling height of not less than 7 feet (2134 mm). ~~at least 50 percent of the required floor area of the room must have a ceiling height of at least 7 feet (2134 mm) and no portion of the required floor area may have a ceiling height of less than 5 feet (1524 mm).~~

2. ~~Bathrooms shall have a minimum ceiling height of 6 feet 8 inches (2032 mm) at the center of the front clearance area for fixtures as shown in Figure R307.1.~~ The ceiling height above bathroom and toilet room fixtures shall be such that the fixture is capable of being used for its intended purpose. A shower or tub equipped with a showerhead shall have a ~~minimum~~ ceiling height of not less than 6 feet 8 inches (2032 mm) above an ~~minimum~~ area not less than 30 inches (762 mm) by 30 inches (762 mm) at the showerhead.
3. Beams, girders, ducts, or other obstructions in basements containing habitable space shall be permitted to project to within 6 feet 4 inches (1931 mm) of the finished floor.

R305.1.1 Basements. Portions of basements that do not contain habitable space~~,~~ or hallways~~, bathrooms, toilet rooms and laundry rooms~~ shall have a ceiling height of not less than 6 feet 8 inches (2032 mm).

> **Exception:** At ~~B~~beams, girders, ducts, or other obstructions, the ceiling height shall be not less than ~~may project to within~~ 6 feet 4 inches (1931 mm) ~~of~~ from the finished floor.

CHANGE SIGNIFICANCE: The exceptions to the minimum 7-foot ceiling height have been expanded in relation to bathrooms, toilet rooms, and basements. Laundry rooms have been added to the list of exceptions allowing a lower ceiling height. In previous editions, the code has recognized that the areas in front of plumbing fixtures and in showers may have ceiling heights of 6 feet, 8 inches without impairing the function of the space. The intent is that bathrooms are not habitable space and a lower ceiling height does not cause any safety or health hazard, or inconvenience to the occupants. Proponents of the change reasoned that there was no justification to limit the exception to the area around plumbing fixtures. With that in mind, the code now permits the entire bathroom or toilet room to have a lower ceiling height of 6 feet, 8 inches. For other than showers, the code does not set a ceiling height requirement above bathroom and toilet room fixtures provided the fixtures can be used. Because laundry rooms also are not habitable space and have a temporary use similar to bathrooms, they are now included in the exception for the lower ceiling height. Most modern homes exceed the minimum ceiling height requirements and the provisions for 7-foot and 6-foot, 8-inch ceilings more often come into play during remodeling of existing homes.

The code has long given special consideration to unfinished basements, those without habitable space, hallways, and bathrooms, for example. The 2015 IRC maintains the provisions for a reduced ceiling height of 6 feet, 8 inches for basements without habitable space or hallways. There is no longer a need to mention bathrooms, toilet rooms, and laundry rooms when considering basement ceiling height because these rooms now also permit a ceiling height of 6 feet, 8 inches. Basements

R305 continues

R305 continued

often have support beams and ductwork below the floor system above, and the code provides for a minimum ceiling height of 6 feet, 4 inches below these and other similar obstructions. In previous editions, the IRC limited the exception for projections to basements without habitable space. This can present problems when the basement is finished off to include habitable space such as a family room or recreation room and the ceiling height under these projections no longer is in compliance with the code. For this reason, the code now allows a 6-foot, 4-inch height below beams, girders, ducts, or other obstructions in basements containing habitable space. With this language added, the designer can establish the ceiling height of an unfinished basement at 7 feet, while setting the beam height at 6 feet, 4 inches above the finished floor, thereby allowing for the basement to be converted to habitable space in the future.

R308.4.2

Glazing Adjacent to Doors

CHANGE TYPE: Modification

CHANGE SUMMARY: Glazing installed perpendicular to a door in a closed position and within 24 inches of the door only requires safety glazing if it is on the hinge side of an in-swinging door.

2015 CODE: R308.4.2 Glazing Adjacent <u>to</u> Doors. Glazing in an individual fixed or operable panel adjacent to a door <u>shall be considered to be a hazardous location</u> ~~where the nearest vertical edge of the glazing is within a 24-inch (610 mm) arc of either vertical edge of the door in a closed~~

R308.4.2 continues

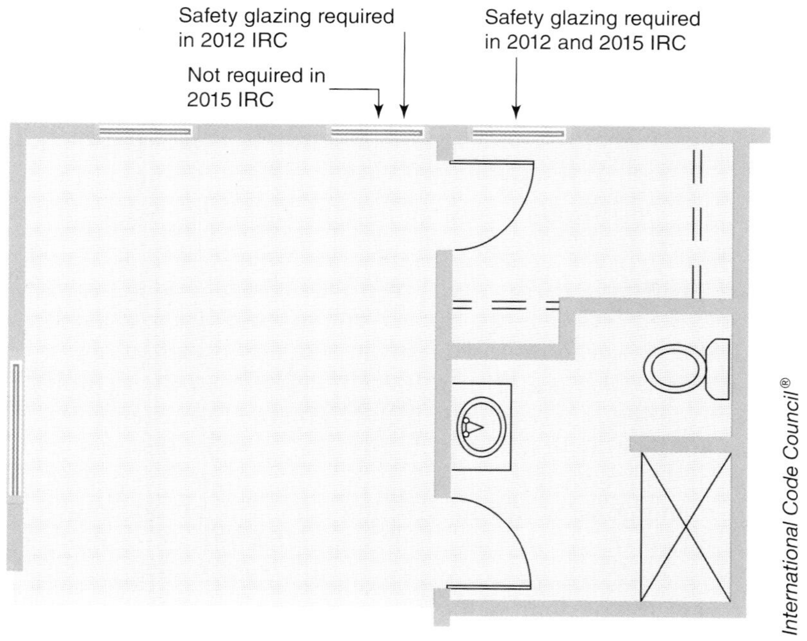

Glazing in windows adjacent and perpendicular to door

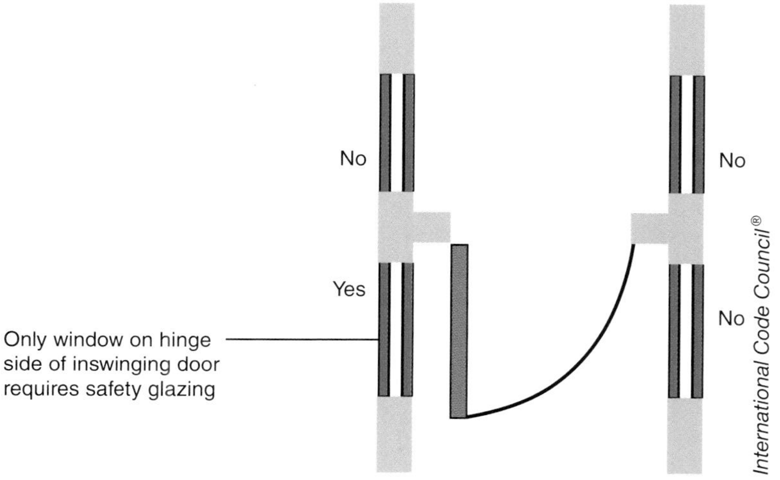

Windows adjacent and perpendicular to door

R308.4.2 continued

~~position and~~ where the bottom exposed edge of the glazing is less than 60 inches (1524 mm) above the floor or walking surface ~~shall be considered a hazardous location~~ and it meets either of the following conditions:

1. Where the glazing is within 24 inches (610 mm) of either side of the door in the plane of the door in a closed position.
2. Where the glazing is on a wall perpendicular to the plane of the door in a closed position and within 24 inches (610 mm) of the hinge side of an in-swinging door.

Exceptions:

1. Decorative glazing.
2. ~~When~~ Where there is an intervening wall or other permanent barrier between the door and the glazing.
3. ~~Glazing in walls on the latch side of and perpendicular to the plane of the door in a closed position~~

~~4~~3. Where access through the door is to a closet or storage area 3 feet (914 mm) or less in depth. Glazing in this application shall comply with Section R308.4.3.

~~5~~4. Glazing that is adjacent to the fixed panel of patio doors.

CHANGE SIGNIFICANCE: Traditionally, the code has provided that glazing installed less than 24 inches from a door in the closed position required safety glazing unless the lowest edge of the glazing was at least 60 inches above the floor. This most often applies to door sidelights and windows installed in the same wall and therefore in the same plane as the door. The application of the code in this case is straightforward and easily understood. The hazard was more difficult to identify for windows installed perpendicular to the plane of the door. The general rule for safety glazing applied to glazing installed within 24 inches of the door, but Exception 3 exempted glazing on the latch side of and perpendicular to a door, regardless of door swing. To most code users, this meant that an adjacent window installed perpendicular to the door and on the hinge side required safety glazing, even if the door swing was away from the window. With the door swinging away from the glazing and the glazing installed parallel to the direction of travel (perpendicular to the door), it was difficult to explain the hazard and justify the requirement for safety glazing. The revised language identifies the hazard of someone being pushed into the glazing when the door swings open. Therefore, for glazing installed perpendicular to a door, the code now identifies only one position as a hazardous location—where the window is located on the hinge side and the door swings in the direction of the glazing. To prevent injury to a person being pushed into or through the window in this position, safety glazing is required.

R308.4.5

Glazing and Wet Surfaces

CHANGE TYPE: Modification

CHANGE SUMMARY: The exception from the safety glazing requirements for glazing that is 60 inches or greater from the water's edge of a bathtub, hot tub, spa, whirlpool, or swimming pool has been expanded to include glazing that is an equivalent distance from the edge of a shower, sauna, or steam room.

2015 CODE: R308.4.5 Glazing and Wet Surfaces. Glazing in walls, enclosures, or fences containing or facing hot tubs, spas, whirlpools, saunas, steam rooms, bathtubs, showers, and indoor or outdoor swimming pools where the bottom exposed edge of the glazing is less than 60 inches (1524 mm) measured vertically above any standing or walking surface shall be considered to be a hazardous location. This shall apply to single glazing and all panes in multiple glazing.

> **Exception:** Glazing that is more than 60 inches (1524 mm), measured horizontally and in a straight line, from the water's edge of a bathtub, hot tub, spa, whirlpool, or swimming pool or from the edge of a shower, sauna, or steam room.

CHANGE SIGNIFICANCE: In the reorganization of the safety glazing provisions in the last two code cycles, the language for determining a hazardous location for glazing adjacent to swimming pools and hot tubs was adapted to the provisions for bathtubs and showers and similar locations. In the 2012 IRC, the provisions for bathtubs, showers, and swimming pools were combined into one section titled "Glazing and Wet Surfaces." The change clarified that the code was regulating the area inside as well as outside and adjacent to bathtubs and showers as a slipping hazard requiring safety glazing. In regard to glazing, the hazardous location for swimming pool decks has traditionally been defined as a location less than 60 inches horizontally from the water's edge. For glazing installed 60 inches or more from the water's edge, safety glazing has not been required. This exception for a 60-inch horizontal distance was applied to bathtubs and whirlpool tubs. Inadvertently, showers were omitted from

R308.4.5 continues

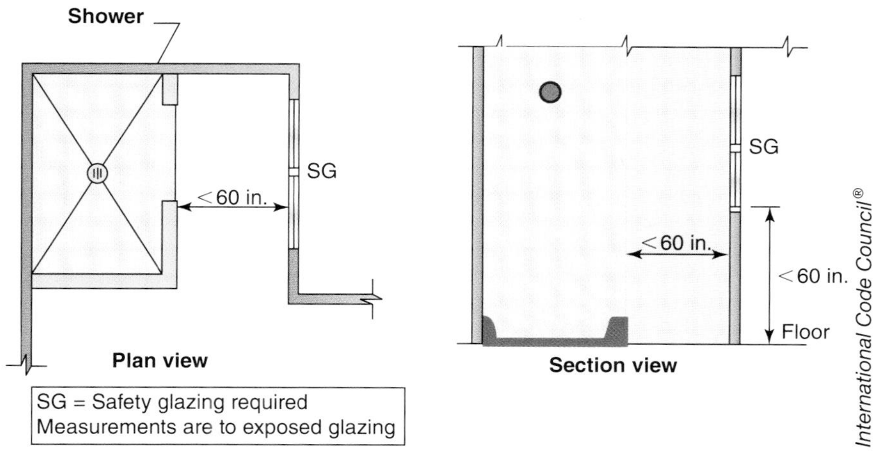

Glazing facing shower

R308.4.5 continued

the exception, although the intent was the same—to exempt glazing that was at least 60 inches away from a shower. In defining a hazardous location as the wet surface adjacent to a laundry list of fixtures, it is reasonable to expect that the exception for a safe distance away from the hazard should match that list in the main section. The code now recognizes that glazing installed 60 inches or greater from the edge of a shower, sauna, or steam room does not require safety glazing, the same distance that applies when measuring from the water's edge of a bathtub, hot tub, spa, whirlpool, or swimming pool.

R308.4.7

Glazing Adjacent to the Bottom Stair Landing

CHANGE TYPE: Clarification

CHANGE SUMMARY: Glazing adjacent to the bottom stair landing is now defined as the area in front of the plane of the bottom tread.

2015 CODE: R308.4.7 Glazing Adjacent to the Bottom Stair Landing. Glazing adjacent to the landing at the bottom of a stairway where the glazing is less than 36 inches (914 mm) above the landing and within a 60-inch~~es~~ (1524 mm) ~~horizontally of~~ horizontal arc less than 180 degrees from the bottom tread nosing shall be considered to be a hazardous location.

> **Exception:** The glazing is protected by a guard complying with Section R312 and the plane of the glass is more than 18 inches (457 mm) from the guard.

CHANGE SIGNIFICANCE: Based on a concern that the code might be misapplied to require safety glazing in an area that was behind the horizontal plane of the nose of the bottom tread, the new language intends to better define the area adjacent to the bottom stair landing that is considered a hazardous location for glazing. As an occupant walks down a stair, the hazard of falling into and breaking glazing occurs in the bottom landing area in front of and to either side of the direction of travel beyond the bottom tread. It is unlikely that a person would fall into glazing that was placed behind the plane of the bottom tread. Section R308.4.6 regulates glazing at the sides of stairs.

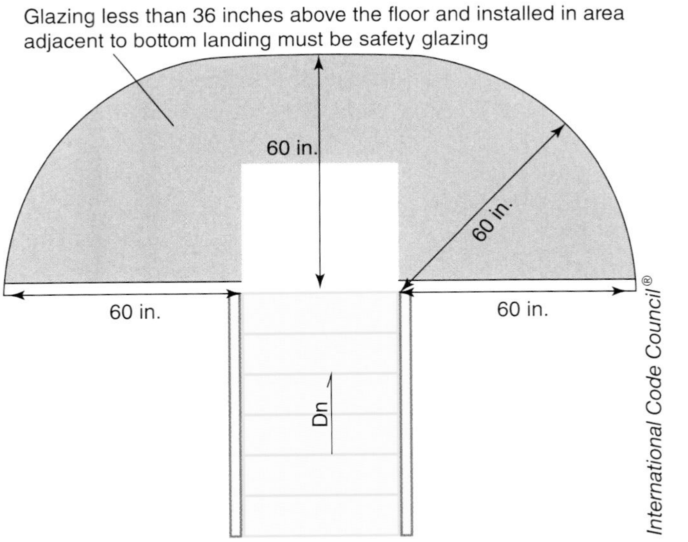

Glazing adjacent to bottom stair landing

R310

Emergency Escape and Rescue Openings

CHANGE TYPE: Clarification

CHANGE SUMMARY: The emergency escape and rescue openings provisions have been reorganized. Separate provisions spell out the requirements for windows and doors used for emergency escape and rescue.

2015 CODE:

SECTION R310
EMERGENCY ESCAPE AND RESCUE OPENINGS

R310.1 Emergency Escape and Rescue Opening Required. Basements, habitable attics, and every sleeping room shall have not less than one operable emergency escape and rescue opening. Where basements contain one or more sleeping rooms, an emergency escape and rescue opening shall be required in each sleeping room. Emergency escape and rescue openings shall open directly into a public way, or to a yard or court that opens to a public way.

> **Exception:** Storm shelters and basements used only to house mechanical equipment not exceeding a total floor area of 200 square feet (18.58 m^2)

R310.1.1 Operational Constraints and Opening Control Devices. Emergency escape and rescue openings shall be operational from the inside of the room without the use of keys, tools or special knowledge. Window opening control devices complying with ASTM F 2090 shall be permitted for use on windows serving as a required emergency escape and rescue opening.

R310.2 Emergency Escape and Rescue Openings. Emergency escape and rescue openings shall have minimum dimensions as specified in this section.

R310.2.1 Minimum Opening Area. All emergency and escape rescue openings shall have a net clear opening of not less than 5.7 square feet (0.530 m^2). The net clear opening dimensions required by this section shall be obtained by the normal operation of the emergency escape and rescue opening from the inside. The net clear height opening shall be not less than 24 inches (610 mm) and the net clear width shall be not less than 20 inches (508 mm).

> **Exception:** Grade floor or below-grade openings shall have a net clear opening of not less than 5 square feet (0.465 m^2).

R310.2.2 Window Sill Height. Where a window is provided as the emergency escape and rescue opening, it shall have a sill height of not more than 44 inches (1118 mm) above the floor; where the sill height is below grade, it shall be provided with a window well in accordance with Section R310.2.3.

R310.2.3 Window Wells. The horizontal area of the window well shall be not less than 9 square feet (0.9 m^2), with a horizontal projection and width of not less than 36 inches (914 mm). The area of the window well shall allow the emergency escape and rescue opening to be fully opened.

Exception: The ladder or steps required by Section R310.2.3.1 shall be permitted to encroach not more than 6 inches (152 mm) into the required dimensions of the window well.

R310.2.3.1 Ladder and Steps. Window wells with a vertical depth greater than 44 inches (1118 m) shall be equipped with a permanently affixed ladder or steps usable with the window in the fully open position. Ladders or steps required by this section shall not be required to comply with Sections R311.7 and R311.8. Ladders or rungs shall have an inside

R310 continues

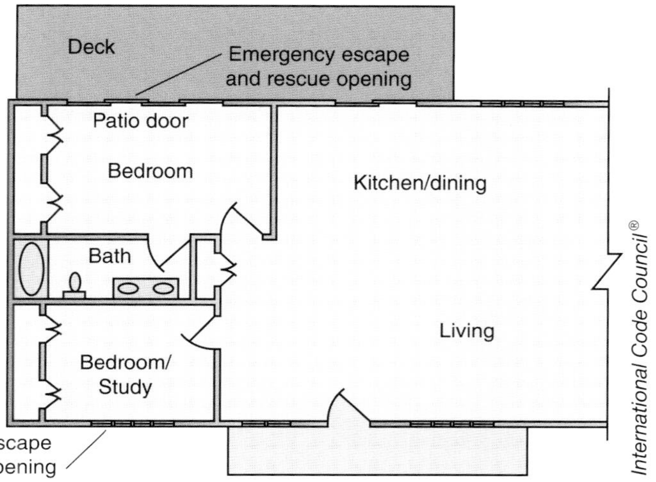

Typical use of patio doors and windows as emergency escape and rescue openings

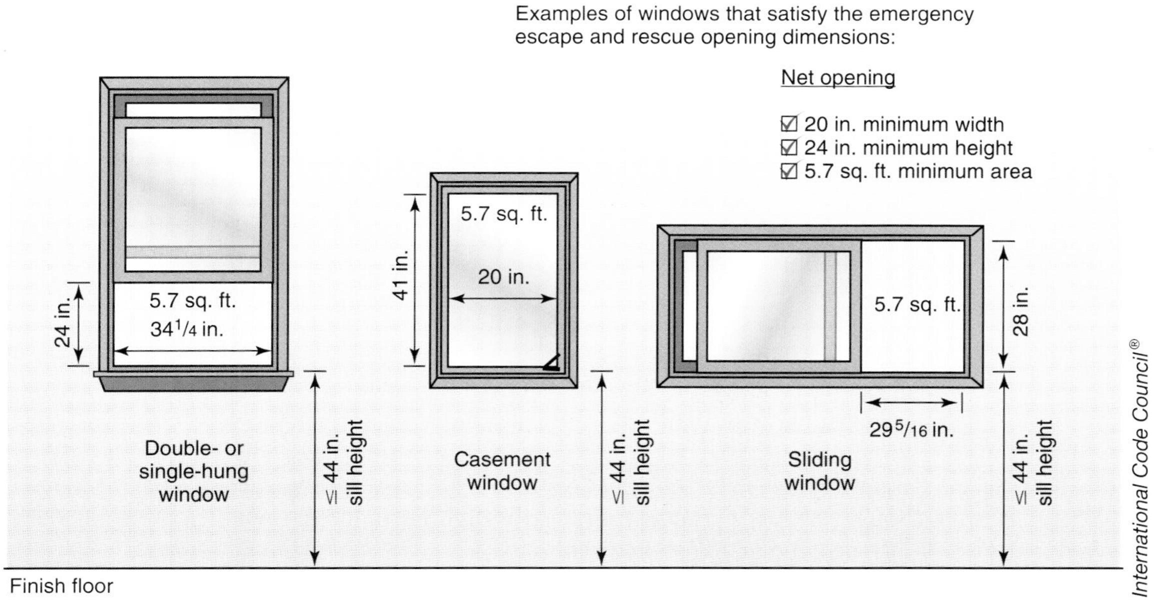

Emergency escape and rescue windows

R310 continued

width of not less than 12 inches (305 mm), shall project not less than 3 inches (76 mm) from the wall and shall be spaced not more than 18 inches (457 mm) on center vertically for the full height of the window well.

R310.2.3.2 Drainage. Window wells shall be designed for proper drainage by connecting to the building's foundation drainage system required by Section R405.1 or by an approved alternative method.

> **Exception:** A drainage system for window wells is not required when the foundation is on well-drained soil or sand-gravel mixture soils according to the United Soil Classification System, Group I Soils, as detailed in Table R405.1.

R310.2.4 Emergency Escape and Rescue Openings under Decks and Porches. Emergency escape and rescue openings shall be permitted to be installed under decks and porches provided the location of the deck allows the emergency escape and rescue openings to be fully opened and provides a path not less than 36 inches (914 mm) in height to a yard or court.

R310.3 Emergency Escape and Rescue Doors. Where a door is provided as the required emergency escape and rescue opening, it shall be permitted to be a side hinged door or a slider. Where the opening is below the adjacent ground elevation, it shall be provided with a bulkhead enclosure.

R310.3.1 Minimum Door Opening Size. The minimum net clear height opening for any door that serves as an emergency and escape rescue opening shall be in accordance with Section R310.2.1.

R310.3.2 Bulkhead Enclosures. Bulkhead enclosures shall provide direct access from the basement. The bulkhead enclosure shall provide the minimum net clear opening equal to the door in the fully open position.

R310.3.2.1 Drainage. Bulkhead enclosures shall be designed for proper drainage by connecting to the building's foundation drainage system required by Section R405.1 or by an approved alternative method.

> **Exception:** A drainage system for bulkhead enclosures is not required when the foundation is on well-drained soil or sand-gravel mixture soils according to the United Soil Classification System, Group I Soils, as detailed in Table R405.1.

R310.4 Bars, Grilles, Covers, and Screens. Bars, grilles, covers, screens or similar devices are permitted to be placed over emergency escape and rescue openings, bulkhead enclosures, or window wells that serve such openings, provided the minimum net clear opening size complies with Sections R310.1.1 to R310.2.3, and such devices shall be releasable or removable from the inside without the use of a key, tool, special knowledge, or force greater than that which is required for normal operation of the escape and rescue opening.

(*Deleted text is not shown for brevity and clarity.*)

CHANGE SIGNIFICANCE: There are no intended technical changes in the reorganization of the emergency escape and rescue provisions in Sections R310.1 through R310.4. (New provisions for existing buildings are addressed in the next change in this publication.) Specific language related to doors used for emergency escape and rescue openings has been added, but there are no changes to the opening dimension requirements. Because exterior doors are typically much larger than the required dimensions for the net opening, they have always been acceptable for satisfying the emergency escape and rescue opening requirements. The reorganization addresses windows and doors separately to avoid any confusion.

Section R310.1.1 clarifies that window opening control devices complying with ASTM F 2090 are approved for use on windows serving as a required emergency escape and rescue opening. Window opening control devices are one option for satisfying the window fall protection requirements of Section R312.2. The device limits the operation of the window such that the net opening does not permit a 4-inch sphere to pass through but has a quick-release mechanism that is approved for emergency escape and rescue. Window opening control devices must comply with ASTM F 2090, *Specification for Window Fall Prevention Devices—with Emergency Escape (Egress) Release Mechanisms*.

R310.5, R310.6

Emergency Escape and Rescue Openings for Additions, Alterations and Repairs

CHANGE TYPE: Clarification

CHANGE SUMMARY: The basement of a dwelling addition does not require an emergency escape and rescue opening if there is access to a basement that does have an emergency escape and rescue opening. Remodeling of an existing basement does not trigger the emergency escape and rescue opening requirements unless a new bedroom is created.

2015 CODE: **R310.5 Dwelling Additions.** Where dwelling additions occur that contain sleeping rooms, an emergency escape and rescue opening shall be provided in each new sleeping room. Where dwelling additions occur that have basements, an emergency escape and rescue opening shall be provided in the new basement.

Exceptions:

1. An emergency escape and rescue opening is not required in a new basement that contains a sleeping room with an emergency escape and rescue opening.
2. An emergency escape and rescue opening is not required in a new basement where there is an emergency escape and rescue opening in an existing basement that is accessible from the new basement.

R310.6 Alterations or Repairs of Existing Basements. An emergency escape and rescue opening is not required where existing basements undergo alterations or repairs.

Exception: New sleeping rooms created in an existing basement shall be provided with emergency escape and rescue openings in accordance with Section R310.1.

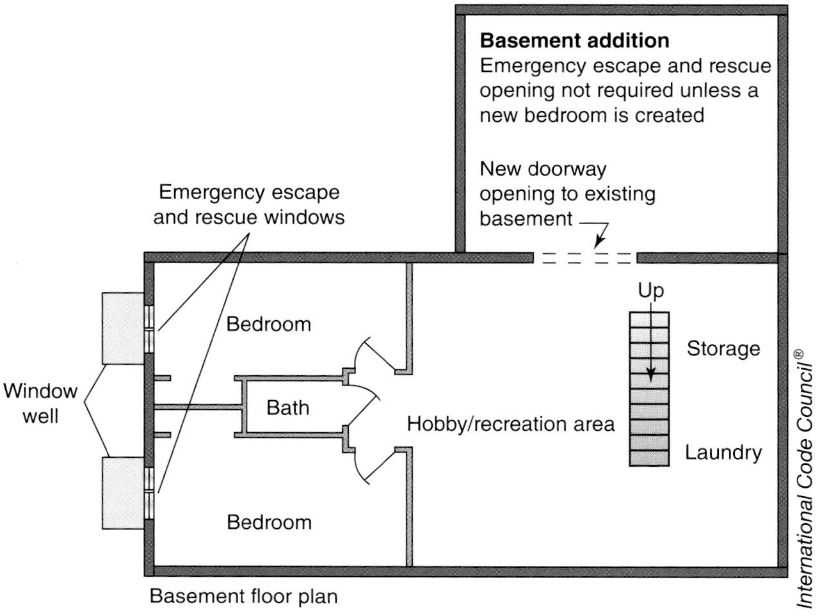

A basement addition does not require an emergency escape and rescue opening if access is provided to the existing basement.

CHANGE SIGNIFICANCE: The new sections addressing emergency escape and rescue openings in existing buildings intend to clarify the correct application of the code during remodeling and construction of dwelling additions. Remodeling of a basement does not trigger a requirement to install an emergency escape and rescue opening. Only the creation of a new bedroom requires installation of an emergency escape and rescue opening. Constructing a new basement is subject to the applicable requirements in the code and an emergency escape and rescue opening is required. The exception recognizes that the appropriate level of safety is achieved if there is an existing emergency escape and rescue opening in the existing basement and access is provided to the existing basement.

R311.1

Means of Egress

CHANGE TYPE: Clarification

CHANGE SUMMARY: The required egress door of a dwelling unit must open directly into a public way or to a yard or court that opens to a public way.

2015 CODE: **R311.1 Means of Egress.** ~~All d~~Dwellings shall be provided with a means of egress ~~as provided~~ in accordance with this section. The means of egress shall provide a continuous and unobstructed path of vertical and horizontal egress travel from all portions of the dwelling to the ~~exterior of the dwelling at the~~ required egress door without requiring travel through a garage. The required egress door shall open directly into a public way or to a yard or court that opens to a public way.

CHANGE SIGNIFICANCE: Proponents of this change reasoned that the means of egress should not have less restrictive requirements than those for emergency escape and rescue openings, which are required to open to a public way or to a yard or court that leads to a public way. The new language is consistent with Section R310.1. In practice, almost all new homes exit to a yard that has access to a public sidewalk or street, and this clarification to the code is not likely to impact construction.

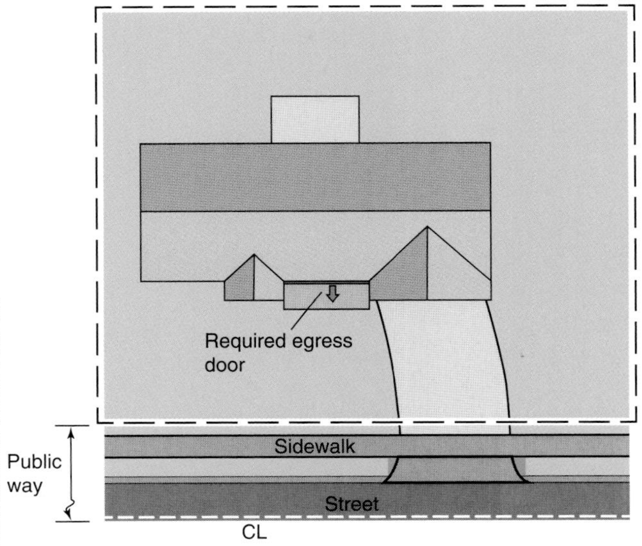

International Code Council®

The required egress door must have access to a public way.

© Lindasj22/Shutterstock.com.

Dwelling with front-yard access to a public way

R311.7.3, R311.7.5.1

Stair Risers

CHANGE TYPE: Modification

CHANGE SUMMARY: The total vertical rise in a stairway without an intermediate landing has increased from 144 inches to 147 inches. The provision for allowing open risers has been clarified. It is based on the distance above grade or the floor below, not on the total rise of the stair. A new exception clarifies that open risers are permitted on spiral stairways.

2015 CODE: R311.7.3 Vertical Rise. A flight of stairs shall not have a vertical rise larger than ~~12 feet (3658 mm)~~ <u>147 inches (3734 mm)</u> between floor levels or landings.

R311.7.5.1 Risers. The ~~maximum~~ riser height shall be <u>not more than</u> 7¾ inches (196 mm). The riser shall be measured vertically between leading edges of the adjacent treads. The greatest riser height within any flight of stairs shall not exceed the smallest by more than ⅜ inch (9.5 mm). Risers shall be vertical or sloped from the underside of the nosing of the tread above at an angle not more than 30 degrees (0.51 rad) from the vertical. Open risers are permitted provided that the opening<u>s</u> ~~between treads~~ <u>located more than 30 inches (762 mm), as measured vertically, to the floor or grade below</u> do~~es~~ not permit the passage of a 4-inch-diameter (102 mm) sphere.

Exception<u>s</u>:

<u>1.</u> The opening between adjacent treads is not limited on ~~stairs with a total rise of 30 inches (762 mm) or less~~ <u>spiral stairways.</u>

<u>2. The riser height of spiral stairways shall be in accordance with Section R311.7.10.1.</u>

CHANGE SIGNIFICANCE: The code limits the total rise between landings in a stair. Traditionally, that vertical distance has been 12 feet before a landing is required for the user to pause and rest when climbing a high stair.

R311.7.3, R311.7.5.1 continues

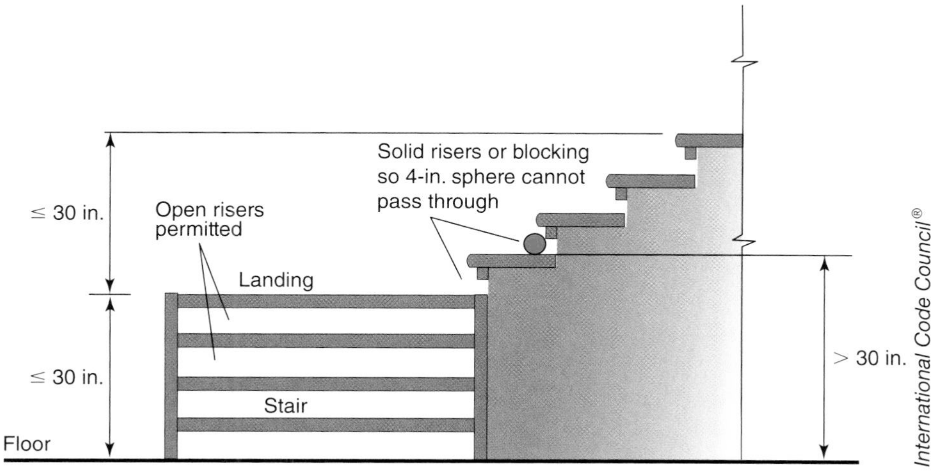

Open risers are only permitted when they are 30 inches or less above the floor.

R311.7.3, R311.7.5.1 continued

This is typically not an issue in conventional home construction because the story height—the distance from one floor surface to the floor above—does not usually exceed 12 feet. However, with open floor plans, long-span engineered wood floor systems of greater depth, and high ceilings, story height can exceed 12 feet and an intermediate landing is required. In the 2015 IRC, the 12-foot limit has been increased by 3 inches to 147 inches. The elevation of 147 inches is a multiple of the maximum riser height of 7¾ inches (19 risers × 7.75 inches = 147.25 inches). This minor change of just 3 inches in the total rise of the flight will in some cases eliminate the cost of incorporating a landing and the additional space required, reducing construction costs. This represents no discernible difference consequential to the user. The change to the 147-inch dimension provides a direct relationship between the total vertical rise and the maximum riser height requirements.

The change to the open riser exception clarifies that the fall hazard is based on the distance to the grade or floor below, not the total rise of an individual flight of stairs. The previous exception allows unrestricted openings in risers if the stair flight has a 30-inch total rise. Flights stacked in a well could each have a total rise of 30 inches and an exposure to a much greater fall distance to the next level or flight below. The exception has been removed and the mandatory language found in Section R312, "Guards and Window Fall Protection," is placed in the main section. Stairs located more than 30 inches above the floor or grade below require solid risers or must be constructed such that a 4-inch-diameter sphere cannot pass through. The code permits larger openings between treads of stairs located 30 inches or less above the surface below.

Spiral stairways typically are constructed with open risers. Although the *International Building Code* (IBC) states that solid risers are not required for spiral stairways, the IRC has not previously addressed this riser provision. An exception has been added to Section R311.7.5.1 to clarify that there are no restrictions for openings between the treads of spiral stairways.

CHANGE TYPE: Modification

R311.7.10.1

Spiral Stairways

CHANGE SUMMARY: The code adds a definition of spiral stairway that omits any requirement for a center post to allow for design flexibility. The code now limits the size of spiral stairways by restricting the radius at the walkline to a dimension not greater than 24½ inches. The method of measurement for tread depth now matches the winder provisions and measures at the intersection of the walkline and the tread nosings rather than perpendicular to the leading edge of the tread.

International Code Council®

Spiral stairway

2015 CODE: R311.7.10.1 Spiral Stairways. Spiral stairways are permitted, provided that the ~~minimum~~ clear width at and below the handrail ~~shall be~~ is not less than 26 inches (660 mm) ~~with~~ and the walkline radius is not greater than 24½ inches (622 mm). ~~e~~Each tread ~~having~~ shall have a ~~7½-inch (190 mm) minimum~~ tread depth ~~at 12 inches (914 mm) from the narrower edge~~ of not less than 6¾ inches (171 mm) at the walkline. All treads shall be identical, and the rise shall be not more than 9½ inches (241 mm). ~~A minimum~~ Headroom shall be not less than 6 feet, 6 inches (1982 mm) ~~shall be provided~~.

SECTION R202
DEFINITIONS

STAIRWAY, SPIRAL. A stairway with a plan view of closed circular form and uniform section-shaped treads radiating from a minimum-diameter circle.

CHANGE SIGNIFICANCE: Spiral stairways provide a space-saving alternative and are permitted to serve any portion of a dwelling as part of the means of egress. Due to their narrow, winding design, they are considered safe for use with less headroom, taller risers, and treads that are narrower at the walkline than conventional stairs. In previous editions of the code, spiral stairways were not restricted in size. The revised provisions define a reasonable limit of the radius at the walkline for the design of spiral stairways, while still maintaining the exceptions for headroom, riser height, and tread depth when compared to conventional stairs. Stairs beyond the limit stated would be considered curved stairs. This change correlates with the new IRC definition of spiral stairway, which omits any reference to a supporting column as found in the IBC.

The other change to this section is largely editorial. Treads within spiral stairways meet the definition of winder treads and are sometimes interpreted to be measured for tread depth in the same fashion. This change simply adjusts the spiral stair tread depth in conformance with the method of measuring for winder tread depth at the intersections of the walkline with the nosings instead of the prior method, which was square to the leading edge. The effective tread depth remains unchanged. The intent of the change in measuring methods, which occurred in the 2009 edition of the IRC, was to provide for consistent tread depth measurements conforming with stair design methodology, not to change or increase tread depth. The long-accepted 7½-inch tread depth was based

R311.7.10.1 continues

R311.7.10.1 continued on the typical spiral layout with 13 treads per revolution or 27.692 degrees per tread. A 7½-inch measurement made square to the leading edge of the tread is equal to a 6¹³⁄₁₆-inch dimension when measured at the intersections of the walkline and nosings. For the ease of applying the requirements, the required tread depth is rounded to 6¾ inches. This change intends to allow long-accepted manufacturing, material, and design standards to continue to meet the requirement and does not change the effective depth of the tread.

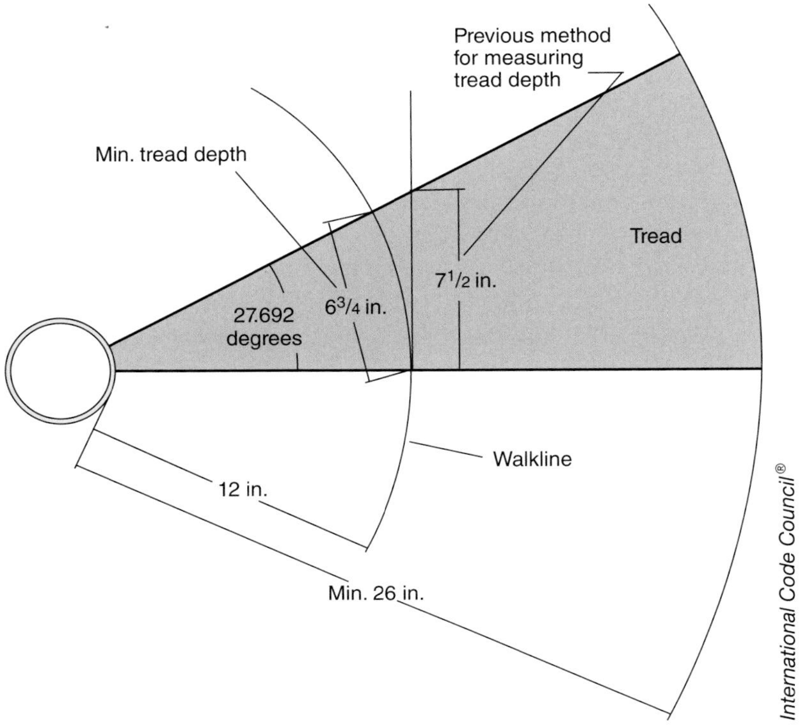

Measuring the tread of spiral stairs at the intersection of the tread nosings and the walkline

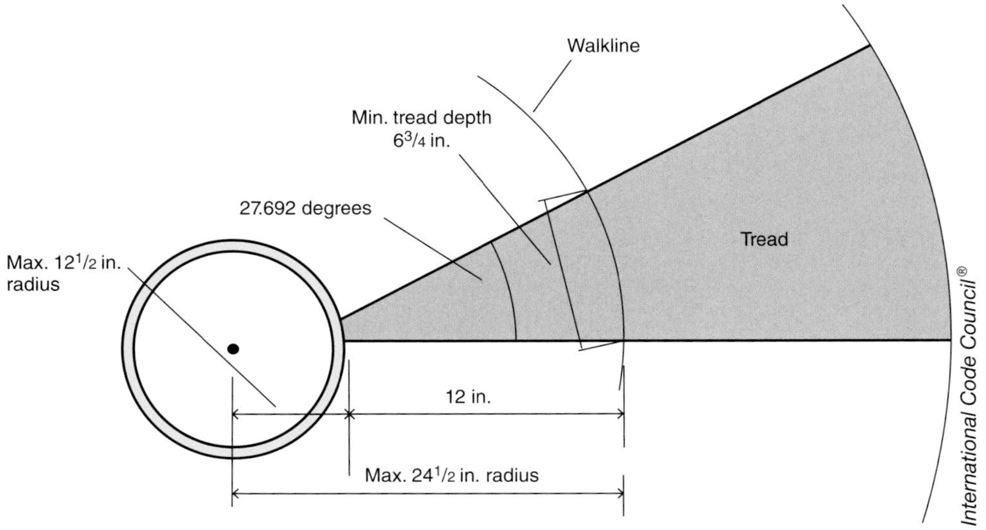

Maximum walkline radius for a spiral stairway

R311.7.11, R311.7.12

Alternating Tread Devices and Ship Ladders

CHANGE TYPE: Addition

CHANGE SUMMARY: Alternating tread devices and ship ladders have been added to the stair provisions. Neither device is approved for use as a means of egress.

2015 CODE:

SECTION R202
DEFINITIONS

ALTERNATING TREAD DEVICE. A device that has a series of steps between 50 and 70 degrees (0.87 and 1.22 rad) from horizontal, usually attached to a center support rail in an alternating manner so that the user does not have both feet on the same level at the same time.

R311.7.11 Alternating Tread Devices. Alternating tread devices shall not be used as an element of a means of egress. Alternating tread devices shall be permitted provided the required means of egress stairway or ramp serves the same space at each adjoining level or where a means of egress is not required. The clear width at and below the handrails shall be not less than 20 inches (508 mm).

R311.7.11, R311.7.12 continues

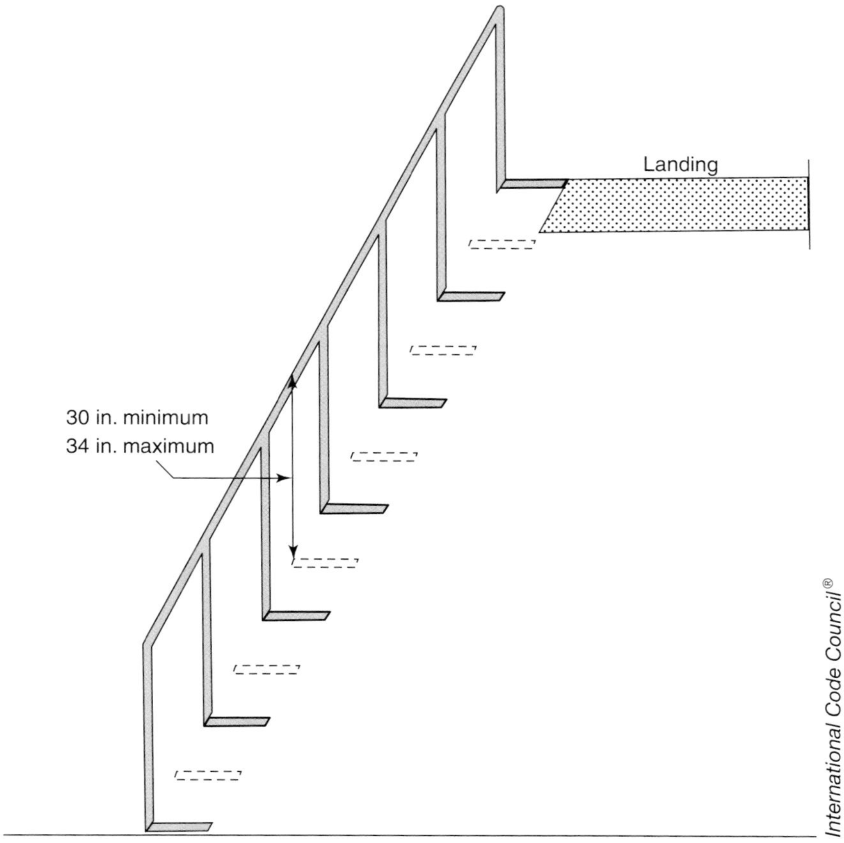

Alternating tread device

R311.7.11, R311.7.12 continued

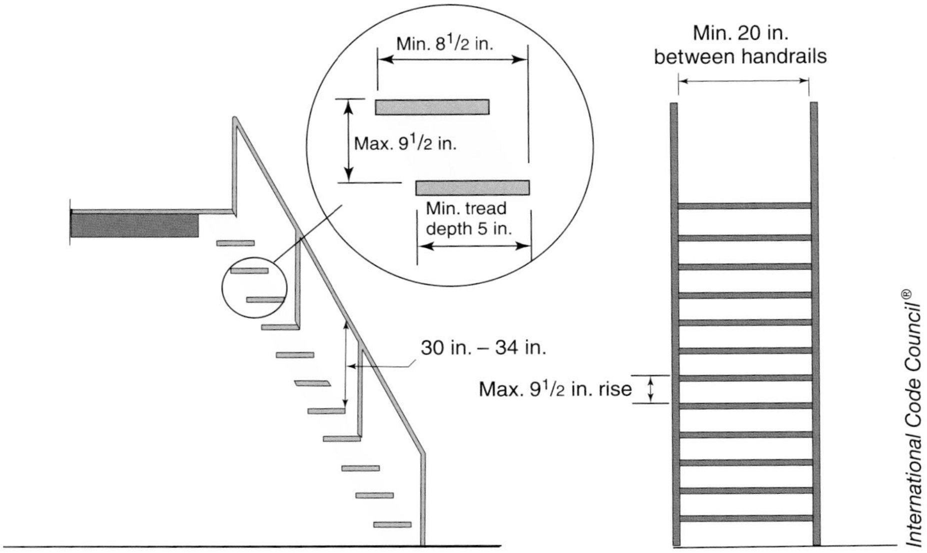

Ship ladder

R311.7.11.1 Treads of Alternating Tread Devices. Alternating tread devices shall have a tread depth of not less than 5 inches (127 mm), a projected tread depth of not less than 8½ inches (216 mm), a tread width of not less than 7 inches (178 mm), and a riser height of not more than 9½ inches (241 mm). The tread depth shall be measured horizontally between the vertical planes of the foremost projections of adjacent treads. The riser height shall be measured vertically between the leading edges of adjacent treads. The riser height and tread depth provided shall result in an angle of ascent from the horizontal of between 50 and 70 degrees (0.87 and 1.22 rad). The initial tread of the device shall begin at the same elevation as the platform, landing, or floor surface.

R311.7.11.2 Handrails of Alternating Tread Devices. Handrails shall be provided on both sides of alternating tread devices and shall comply with R311.7.8.2 thru R311.7.8.4. Handrail height shall be uniform, not less than 30 inches (762 mm) and not more than 34 inches (864 mm).

R311.7.12 Ship Ladders. Ship ladders shall not be used as an element of a means of egress. Ship ladders shall be permitted provided a required means of egress stairway or ramp serves the same space at each adjoining level or where a means of egress is not required. The clear width at and below the handrails shall be not less than 20 inches.

R311.7.12.1 Treads of Ship Ladders. Treads shall have a tread depth of not less than 5 inches (127 mm). The tread shall be projected such that the total of the tread depth plus the nosing projection is not less than 8½ inches (216 mm). The riser height shall be not more than 9½ inches (241 mm).

R311.7.12.2 Handrails of Ship Ladders. Handrails shall be provided on both sides of ship ladders and shall comply with Sections R311.7.8.2 thru R311.7.8.4. Handrail height shall be uniform, not less than 30 inches (762 mm) and not more than 34 inches (864 mm).

CHANGE SIGNIFICANCE: Alternating tread devices and ship ladders have been used in residential applications but have previously not appeared in the IRC. The new provisions adopt the dimensions and other specifications from the IBC to provide guidance when they are used. An alternating tread device or ship ladder cannot be used as an element of a means of egress, and can only be used when a means of egress is not required or when the required means of egress stairway or ramp is provided to serve the same space. Proponents held that these types of stairs will become more common and that introducing them into the code provides needed guidelines and allows for more design flexibility.

R311.8

Ramps

CHANGE TYPE: Modification

CHANGE SUMMARY: Ramps that do not serve the required egress door are now permitted to have a slope not greater than 1 unit vertical in 8 units horizontal.

2015 CODE: R311.8 Ramps.

R311.8.1 Maximum Slope. Ramps serving the egress door required by Section R311.2 shall have a ~~maximum~~ slope of not more than 1 unit vertical in 12 units horizontal (8.3 percent slope). All other ramps shall have a maximum slope of 1 unit vertical to 8 units horizontal (12.5 percent slope).

> **Exception:** Where it is technically infeasible to comply because of site constraints, ramps shall have a ~~maximum~~ slope of not more than 1 unit vertical in 8 units horizontal (12.5 percent slope).

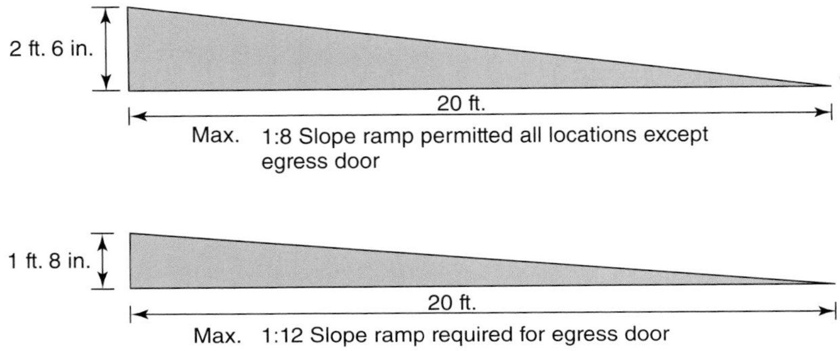

International Code Council®

Maximum ramp slopes

© Daryl Brooks/Shutterstock.com.

Ramp serving the required egress door

CHANGE SIGNIFICANCE: Prior to the 2006 edition of the IRC, the code permitted ramps with a maximum slope of 1 unit vertical in 8 units horizontal (1:8 or 12.5 percent slope). The maximum slope was decreased to 1:12 in the 2006 IRC to provide accessibility for physically disabled persons and to promote designs that allow an aging population to stay in their homes longer, now often referred to as "aging in place." Such design features intend to increase safety, accessibility, and independence for older adults in their own homes. For the most part, IRC buildings are not required to be accessible and ramps are not required. Ramps that are steeper than 1:12 are difficult or impossible to use for those in wheelchairs or using canes or walkers. The IRC has maintained an exception to allow a 1:8 slope ramp where it is not feasible to install a 1:12 slope ramp—for example, if the house is constructed on a sloped site or is located close to the property line. Consensus for the 2015 IRC is to allow the steeper 1:8 ramp in all situations, interior and exterior, unless the ramp serves the one required egress door. Proponents for the change to a less stringent rule reasoned that where ramps are constructed to serve the required egress door, requiring a 1:12 maximum slope is a reasonable accommodation for accessibility and the elderly. They also pointed to a similar provision in the IBC, which allows a 1:8 slope for pedestrian ramps not used as part of a means of egress. Shallower slope ramps take up more space, and the intent of the change is to provide more design flexibility for residential buildings constructed under the IRC. Construction of a ramp is an option in the IRC and is not mandatory. Likewise, building ramps with a lesser slope than the maximum allowed by the IRC also remains an option.

R312.1.2
Guard Height

Example of a guard with fixed seating

CHANGE TYPE: Modification

CHANGE SUMMARY: The provision requiring that the guard height be measured from the surface of adjacent fixed seating has been removed from the code.

2015 CODE: R312.1.2 Height. Required guards at open-sided walking surfaces, including stairs, porches, balconies, or landings, shall be not less than 36 inches (914 mm) ~~high~~ in height as measured vertically above the adjacent walking surface~~, adjacent fixed seating~~ or the line connecting the leading edges of the treads.

Exceptions:

1. Guards on the open sides of stairs shall have a height not less than 34 inches (864 mm) measured vertically from a line connecting the leading edges of the treads.
2. Where the top of the guard ~~also~~ serves as a handrail on the open sides of stairs, the top of the guard shall be not less than 34 inches (864 mm) and not more than 38 inches (965 mm) as measured vertically from a line connecting the leading edges of the treads.

CHANGE SIGNIFICANCE: The requirement to extend a guard 36 inches above the surface of fixed seating that is adjacent to a required guard, which appeared in the 2009 and 2012 IRC, has been removed from the code. A similar requirement appeared in the 2009 IBC but was deleted from the 2012 IBC. The provision was initially placed in the code due

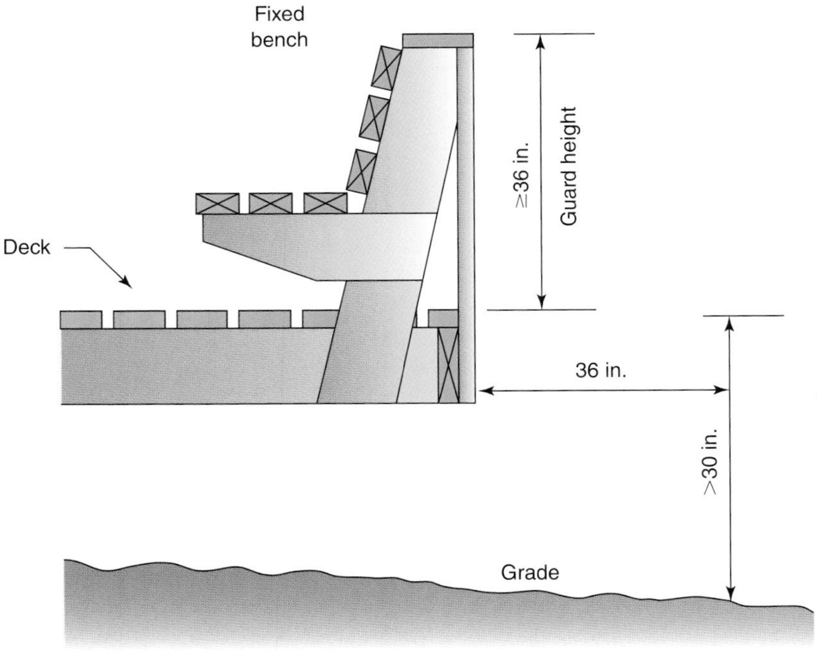

Measuring guard height at fixed seating

to the concern of children climbing on fixed seating and falling over the usual guard that had a height of 36 inches above the walking surface of the deck. With the removal of the provision, the IRC and IBC are now consistent with each other in this area. Proponents of this change reasoned that there was no technical justification to raise the height of the guard at the back of fixed seating when it was placed in the code and that the requirement was overly restrictive. There also were concerns that there has been no definition of fixed seating. Consensus held that fixed seating, similar to movable furniture and other objects found adjacent to guards on a deck, should not be regulated as a walking surface.

R312.2.1

Window Fall Protection

CHANGE TYPE: Clarification

CHANGE SUMMARY: The window fall prevention provisions have been revised to clarify the meaning, remove redundant language, and achieve consistency with the IBC provisions.

2015 CODE: R312.2.1 Window Sills. In dwelling units, where the top of the sill ~~opening~~ of an operable window opening is located less than 24 inches (610 mm) above the finished floor and greater ~~more~~ than 72 inches (1829 mm) above the finished grade or other surface below on the exterior of the building, the operable window shall comply with one of the following: ~~the lowest part of the clear opening of the window shall be a minimum of 24 inches (610 mm) above the finished floor of the room in which the window is located. Operable sections of windows shall not permit openings that allow passage of a 4-inch-diameter (102 mm) sphere where such openings are located within 24 inches (610 mm) of the finished floor.~~

~~Exceptions:~~

1. Operable windows ~~whose~~ with openings that will not allow a 4-inch-diameter (102 mm) sphere to pass through the opening ~~when~~ where the opening is in its largest opened position.
2. ~~Openings~~ Operable windows that are provided with window fall prevention devices that comply with ASTM F 2090.
3. Operable windows that are provided with window opening control devices that comply with Section R312.2.2.

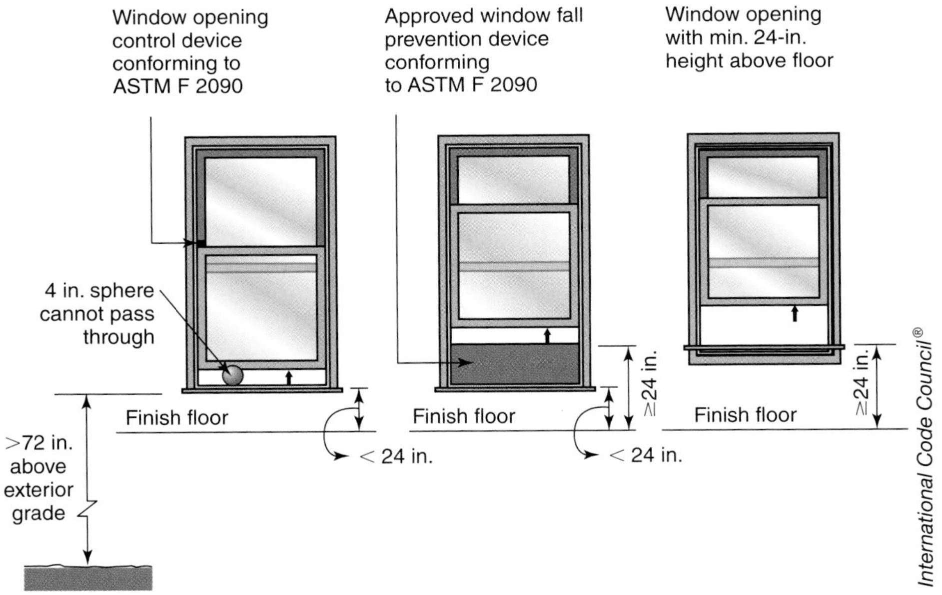

Three methods for complying with the window fall prevention provisions

R312.2.2 Window Opening Control Devices. Window opening control devices shall comply with ASTM F 2090. The window opening control device, after operation to release the control device allowing the window to fully open, shall not reduce the ~~minimum~~ net clear opening area of the window unit to less than the area required by Section R310.2.1.

CHANGE SIGNIFICANCE: The provisions for fall protection intend to reduce the number of injuries to children resulting from falls through windows. This section has been reorganized to clarify the application. Previously, the code required that the lowest portion of an operable window opening had to be at least 24 inches above the floor when located more than 72 inches above the finished grade or other surface below. The code then listed three exceptions, the first of which repeated language from the main section regarding the 4-inch-sphere rule for the window opening. The rules are the same, but the code now takes another approach in explaining the requirements. The language in the main section is changed to say that when the window opening does not meet the 24-inch rule for sill height, then one of the three options must be chosen for compliance with the code. These options are now listed as conditions rather than exceptions.

R314

Smoke Alarms

CHANGE TYPE: Modification

CHANGE SUMMARY: Battery-operated smoke alarms are permitted for satisfying the smoke alarm power requirements when alterations, repairs, and additions occur. Household fire alarm systems no longer require monitoring by an approved supervising station. New provisions address smoke alarms installed near bathrooms and cooking appliances.

2015 CODE: **R314.1 General.** Smoke alarms shall comply with NFPA 72 and Section R314.

R314.1.1 Listings. Smoke alarms shall be listed in accordance with UL 217. Combination smoke and carbon monoxide alarms shall be listed in accordance with UL 217 and UL 2034.

R314.2 Where Required. Smoke alarms shall be provided in accordance with this section.

R314.2.1 New Construction. Smoke alarms shall be provided in dwelling units.

~~R314.3.1~~ R314.2.2 Alterations, Repairs, and Additions. Where alterations, repairs, or additions requiring a permit occur, or when one or more sleeping rooms are added or created in existing dwellings, the individual dwelling unit shall be equipped with smoke alarms located as required for new dwellings.

Exceptions:

1. Work involving the exterior surfaces of dwellings, such as the replacement of roofing or siding, or the addition or replacement of windows or doors, or the addition of a porch or deck, are exempt from the requirements of this section.

Smoke alarm

2. Installation, alteration, or repairs of plumbing or mechanical systems are exempt from the requirements of this section.

R314.3 Location. Smoke alarms shall be installed in the following locations:

1. In each sleeping room.
2. Outside each separate sleeping area in the immediate vicinity of the bedrooms.
3. On each additional story of the dwelling, including basements and habitable attics ~~but~~ and not including crawl spaces and uninhabitable attics. In dwellings or dwelling units with split levels and without an intervening door between the adjacent levels, a smoke alarm installed on the upper level shall suffice for the adjacent lower level provided that the lower level is less than one full story below the upper level.
4. Smoke alarms shall be installed not less than 3 feet (914 mm) horizontally from the door or opening of a bathroom that contains a bathtub or shower unless this would prevent placement of a smoke alarm required by Section R314.3.

R314.3.1 Installation Near Cooking Appliances. Smoke alarms shall not be installed in the following locations unless this would prevent placement of a smoke alarm in a location required by Section R314.3.

1. Ionization smoke alarms shall not be installed less than 20 feet (6096 mm) horizontally from a permanently installed cooking appliance.
2. Ionization smoke alarms with an alarm-silencing switch shall not be installed less than 10 feet (3048 mm) horizontally from a permanently installed cooking appliance.

R314 continues

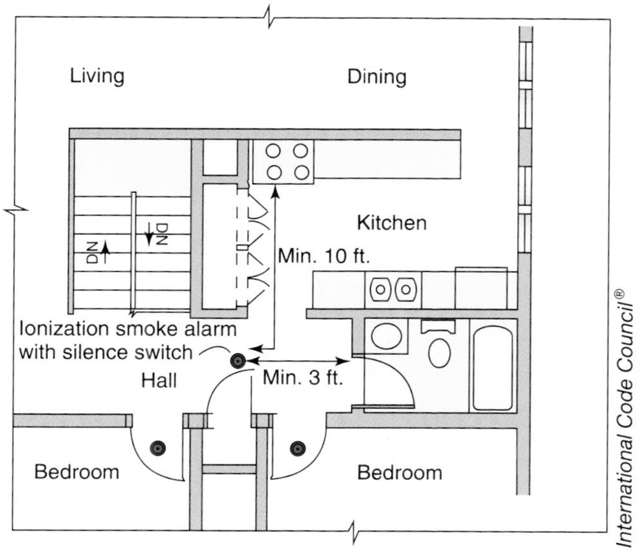

Smoke alarm distances from bathrooms and cooking appliances

R314 continued

3. Photoelectric smoke alarms shall not be installed less than 6 feet (1828 mm) horizontally from a permanently installed cooking appliance.

~~R314.5~~ R314.4 Interconnection. Where more than one smoke alarm is required to be installed within an individual dwelling unit in accordance with Section R314.3, the alarm devices shall be interconnected in such a manner that the actuation of one alarm will activate all of the alarms in the individual dwelling unit. Physical interconnection of smoke alarms shall not be required where listed wireless alarms are installed and all alarms sound upon activation of one alarm.

Exception: Interconnection of smoke alarms in existing areas shall not be required where alterations or repairs do not result in removal of interior wall or ceiling finishes exposing the structure, unless there is an attic, crawl space, or basement available which could provide access for interconnection without the removal of interior finishes.

R314.5 Combination Alarms. Combination smoke and carbon monoxide alarms shall be permitted to be used in lieu of smoke alarms.

~~R314.4~~ R314.6 Power Source. Smoke alarms shall receive their primary power from the building wiring where such wiring is served from a commercial source, and where primary power is interrupted, shall receive power from a battery. Wiring shall be permanent and without a disconnecting switch other than those required for overcurrent protection.

Exceptions:

1. Smoke alarms shall be permitted to be battery operated when installed in buildings without commercial power.
2. Smoke alarms installed in accordance with Section R314.2.2 shall be permitted to be battery powered. ~~Hard wiring of smoke alarms in existing areas shall not be required where the alterations or repairs do not result in the removal of interior wall or ceiling finishes exposing the structure, unless there is an attic, crawl space or basement available which could provide access for hard wiring without the removal of interior finishes.~~

R314.7 Fire Alarm Systems. Fire alarm systems shall be permitted to be used in lieu of smoke alarms and shall comply with Sections R314.7.1 through R314.7.4.

~~R314.2 Smoke Detection Systems.~~ R314.7.1 General. Fire alarm systems shall comply with the provisions of this code and the household fire warning equipment provisions of NFPA 72. Smoke detectors shall be listed in accordance with UL 268. ~~Household fire alarm systems installed in accordance with NFPA 72 that include smoke alarms, or a combination of smoke detector and audible notification device installed as required by this section for smoke alarms, shall be permitted. The household fire~~

~~alarm system shall~~ ~~provide the same level of smoke detection and alarm as required by this section for smoke alarms.~~

R314.7.2 Location. Smoke detectors shall be installed in the locations specified in Section R314.3.

~~**R314.2 Smoke Detection Systems.**~~ **R314.7.3 Permanent Fixture.** Where a household fire alarm system is installed ~~using a combination of smoke detector and audible notification device(s)~~, it shall become a permanent fixture of the occupancy ~~and~~, owned by the homeowner. ~~The system shall be monitored by an approved supervising station and be maintained in accordance with NFPA 72.~~

R314.7.4 Combination Detectors. Combination smoke/carbon monoxide detectors shall be permitted to be installed in fire alarm systems in lieu of smoke detectors, provided they are listed in accordance with UL 268 and UL 2075.

(Some 2012 IRC provisions not shown for brevity and clarity.)

CHANGE SIGNIFICANCE: The smoke alarm provisions in Section R314 have been reorganized in a user-friendly sequential order to clarify their application. For example, the household fire alarm system provisions have been placed in a separate Section R314.7 following all of the smoke alarm provisions in Sections R314.1 through R314.6. In addition, new charging sections have been added to clarify the scope and make the provisions easier to locate.

New provisions specifically permit the installation of combination smoke and carbon monoxide alarms complying with the applicable standards. Combination alarms are commonly installed outside of bedroom areas in residential construction as an acceptable method for satisfying both smoke alarm and carbon monoxide alarm provisions in the IRC, and this change simply recognizes a method that is already in practice.

The changes to the smoke alarm provisions in Section R314 are largely editorial and in most cases do not intend to create any technical changes. However, there is a minor change in the language regarding power requirements for smoke alarms installed in existing buildings. Smoke alarms are one of the few requirements that are retroactive in the IRC. Interior remodeling work and room additions that require permits do trigger the installation of smoke alarms in the same locations as are required for new dwellings. Previous editions of the code have generally required these smoke alarms to also meet the power and interconnection requirements. An exception has recognized that it is not always feasible to install the additional wiring necessary to bring electricity to the devices or to connect the devices so that when one alarm sounds all alarms in the dwelling activate. Therefore, hard wiring of smoke alarms in existing areas was not required if the alterations or repairs did not result in the removal of interior wall or ceiling finishes exposing the structure. The code further stated that if there existed an attic, crawl space, or basement that could provide access for hard wiring without the removal of interior finishes, then connection to the dwelling unit electrical system was required. Otherwise, the code permitted the installation of battery-operated smoke alarms in

R314 continues

R314 continued these existing areas. This has always been a judgment call on the part of the building official, and many jurisdictions have developed procedures or guidelines for determining if it is feasible to bring power to new smoke alarms in existing buildings. Although installing a battery-operated smoke alarm is relatively easy and inexpensive, installing electrical wiring in an existing building can be very costly. The new language in the 2015 IRC does not address the feasibility of connecting to the electrical system in existing buildings. Exception 2 of Section R314.6 says that smoke alarms installed in accordance with Section R314.2.2 for alterations, repairs, and additions are permitted to be battery powered. The change to more prescriptive language will simplify the administration of the code and encourage consistency in the application during remodeling of existing buildings without imposing excessive costs, while at the same time providing an acceptable level of safety with the installation of battery-operated smoke alarms in all of the locations required for new buildings.

The provisions for interconnecting smoke alarms in existing areas have not changed from the 2012 edition. The exception in Section R314.4 requires interconnection of smoke alarms in existing areas where interior wall or ceiling finishes are removed or where there is an attic, crawl space, or basement available that could provide access for interconnection without the removal of interior finishes. As an alternative, the code specifically allows wireless interconnection of smoke alarms in lieu of physical interconnection.

Another new approach in the 2015 IRC intends to reduce nuisance alarms by requiring minimum separation distances between smoke alarms and cooking appliances, and between smoke alarms and bathrooms. The new requirements are similar to those in NFPA 72, *National Fire Alarm Code*, which is a referenced standard in the smoke alarm provisions of Section R314. The code now requires a minimum separation of 3 feet from bathrooms because steam and water vapor produced by bathtubs and showers can trigger operation of the smoke alarm. The minimum separation requirements from permanently installed cooking appliances vary based on the type of smoke alarm installed. Ionization smoke alarms generally require a separation distance of 20 feet, but that distance may be reduced to 10 feet if the smoke alarm has an alarm-silencing switch. Photoelectric smoke alarms are less susceptible to activation by smoke and cooking vapors and are permitted to be located as close as 6 feet from a permanently installed cooking appliance. The intent is to regulate separation distance from built-in cook tops and ovens as well as stand-alone kitchen ranges. The word "permanent" intends to exclude movable countertop cooking appliances from the separation requirements.

For installation in proximity to both bathrooms and cooking appliances, exceptions permit installation less than the prescribed separation distances if such installation is required by the location requirements of Section R314.3. For example, in a small house the kitchen, bathroom, and bedroom may be grouped closely together. The code would require installation of a smoke alarm in the hallway outside the bedroom even though the location does not meet the separation requirements from the bathroom or cooking appliance. The primary concern is safety by providing early warning of a fire for the occupants, particularly if they are sleeping, and nuisance alarms are of secondary importance. In most cases builders follow the manufacturer's installation instructions and industry-accepted practices to provide adequate separation from cooking appliances and bathrooms, and to avoid costly callbacks from unhappy customers.

As an alternative to the individual smoke alarm requirements, the code permits the installation of a household fire alarm system installed in accordance with NFPA 72. These fire alarm systems rely on separate detection devices installed in the same required locations as smoke alarms and separate annunciating devices installed in various locations of the home in accordance with the design. These systems become a permanent fixture of the occupancy and are owned by the homeowner as prescribed in the code. This provision intends to avoid systems that are leased to the homeowner by an alarm company, and could subsequently be removed by the alarm company if the homeowner discontinued service, leaving the home with no smoke detection and notification protection. The permanent fixture and ownership provisions remain in the 2015 IRC. However, there has been confusion regarding the requirement for systems to be monitored by an approved supervising station, and this requirement was considered difficult to enforce. Proponents reasoned that a system that provides local alarm notification satisfies the intent of the code to provide early warning to occupants and that it was difficult to justify the extra costs associated with monitoring by a supervising station. The code does not prohibit monitoring, but it is now an option rather than a requirement. In addition, the reference in Section R314.2 of the 2012 IRC to systems being maintained in accordance with NFPA 72 has been removed because it was considered outside the scope and intent of the IRC.

R315

Carbon Monoxide Alarms

Carbon monoxide alarm

CHANGE TYPE: Modification

CHANGE SUMMARY: Carbon monoxide alarms now require connection to the house wiring system with battery backup. Exterior work such as roofing, siding, windows, doors, and deck and porch additions no longer trigger the carbon monoxide alarm provisions for existing buildings. An attached garage is one criterion for requiring carbon monoxide alarms, but only if the garage has an opening into the dwelling. A carbon monoxide alarm is required in bedrooms when there is a fuel-fired appliance in the bedroom or adjoining bathroom. Carbon monoxide detection systems only require detectors installed in the locations prescribed by the code and not those locations described in NFPA 720.

2015 CODE: **R315.1 General.** Carbon monoxide alarms shall comply with Section R315.

R315.1.1 Listings. Carbon monoxide alarms shall be listed in accordance with UL 2034. Combination carbon monoxide/smoke alarms shall be listed in accordance with UL 2034 and UL 217.

R315.2 Where Required. Carbon monoxide alarms shall be provided in accordance with Sections R315.2.1 and R315.2.2.

~~R315.1 Carbon Monoxide Alarms.~~ R315.2.1 New Construction. For new construction, ~~an approved~~ carbon monoxide alarms shall be ~~installed outside of each separate sleeping area in the immediate vicinity of the bedrooms in dwelling units within which fuel-fired appliances are installed and in dwelling units that have attached garages.~~ provided in dwelling units where either or both of the following conditions exist.

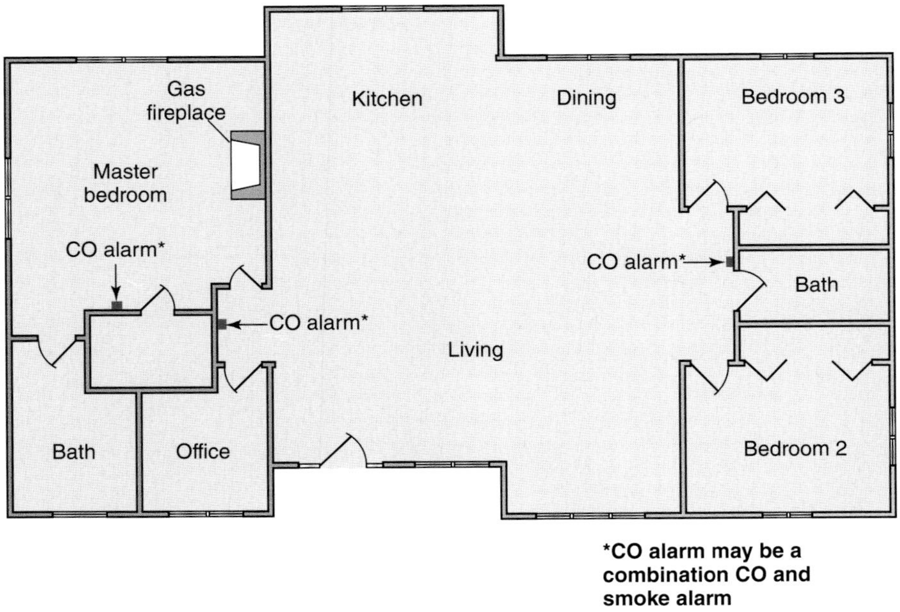

Carbon monoxide alarms required outside each separate sleeping area and in bedrooms containing fuel-fired appliances

1. The dwelling unit contains a fuel-fired appliance.
2. The dwelling unit has an attached garage with an opening that communicates with the dwelling unit.

~~R315.3 Where Required in Existing Dwellings.~~ R315.2.2 Alterations, Repairs, and Additions. Where ~~work~~ alterations, repairs, or additions requiring a permit occurs ~~in existing dwellings that have attached garages or in existing dwellings within which fuel-fired appliances exist~~, or where one or more sleeping rooms are added or created in existing dwellings, ~~carbon monoxide alarms shall be provided in accordance with Section R315.1.~~the individual dwelling unit shall be equipped with carbon monoxide alarms located as required for new dwellings.

Exceptions:

1. Work involving the exterior surfaces of dwellings, such as the replacement of roofing or siding, or the addition or replacement of windows or doors, or the addition of a porch or deck, are exempt from the requirements of this section.
2. Installation, alteration, or repairs of plumbing or mechanical systems are exempt from the requirements of this section.

R315.3 Location. Carbon monoxide alarms in dwelling units shall be installed outside of each separate sleeping area in the immediate vicinity of the bedrooms. Where a fuel-burning appliance is located within a bedroom or its attached bathroom, a carbon monoxide alarm shall be installed within the bedroom.

R315 continues

Exterior work, such as a deck addition or re-siding, does not trigger the carbon monoxide provisions.

R315 continued

R315.4 Combination Alarms. Combination carbon monoxide and smoke alarms shall be permitted to be used in lieu of carbon monoxide alarms.

R315.5 Power Source. Carbon monoxide alarms shall receive their primary power from the building wiring where such wiring is served from a commercial source, and where primary power is interrupted, shall receive power from a battery. Wiring shall be permanent and without a disconnecting switch other than those required for overcurrent protection.

Exceptions:

1. Carbon monoxide alarms shall be permitted to be battery operated where installed in buildings without commercial power.
2. Carbon monoxide alarms installed in accordance with Section R315.2.2 shall be permitted to be battery powered.

R315.6 Carbon Monoxide Detection Systems. Carbon monoxide detection systems shall be permitted to be used in lieu of carbon monoxide alarms and shall comply with Sections R315.6.1 to R315.6.4.

R315.6.1 General. Household carbon monoxide detection systems shall comply with NFPA 720. Carbon monoxide detectors shall be listed in accordance with UL 2075.

R315.6.2 Location. Carbon monoxide detectors shall be installed in the locations specified in Section R315.3. These locations supersede the locations specified in NFPA 720.

R315.6.3 Permanent Fixture. Where a household carbon monoxide detection system is installed, it shall become a permanent fixture of the occupancy and owned by the homeowner. ~~and shall be monitored by an approved supervising station.~~

R315.6.4 Combination Detectors. Combination carbon monoxide/smoke detectors shall be permitted to be installed in carbon monoxide detection systems in lieu of carbon monoxide detectors, provided they are listed in accordance with UL 2075 and UL 268.

(Some 2012 IRC provisions are not shown for brevity and clarity.)

CHANGE SIGNIFICANCE: The carbon monoxide alarm provisions in Section R315 have been reorganized to clarify their application. For example, the carbon monoxide detection system provisions have been placed in a separate Section R315.6 following all of the carbon monoxide alarm provisions in Sections R315.1 through R315.5. In addition, new charging sections have been added to clarify the scope and make the provisions easier to locate. Several technical changes have also occurred. The carbon monoxide alarm provisions have been rewritten to generally align with the smoke alarm provisions. For example, connection to the house wiring system with battery backup is now required for carbon monoxide alarms installed in new dwellings. Previously, the code only referenced compliance with UL 2075, which permits battery-operated, plug-in, or hard-wired alarms. Battery-operated carbon monoxide alarms are still permitted for satisfying the requirements in existing buildings, matching the revised provisions for smoke alarms. When work requiring a permit occurs, alarms must be installed in the locations prescribed by the code, but hard wiring is not required. New to the 2015 IRC, exterior work requiring a permit, such as roofing, siding, windows, doors, porches, and decks, does not trigger the retroactive carbon monoxide alarm requirements. This language mirrors the exemption in the smoke alarm provisions. Unlike the smoke alarm provisions, when two or more carbon monoxide alarms are required, interconnection of the alarms is not required.

The carbon monoxide alarm provisions are only in effect when the dwelling contains fuel-fired appliances or has an attached garage. A malfunctioning fuel-fired appliance, such as a gas-fired furnace, water heater, or fireplace, is the most common cause of carbon monoxide poisoning in homes. Automobile exhaust migrating into the home from an attached garage is the other hazard addressed by the code requirements. Because the hazard of carbon monoxide gas from a garage depends on an opening between the garage and the dwelling unit, typically a door in the common wall, the code now only addresses garages with openings into the dwelling. Attached garages that do not communicate with the house do not trigger the carbon monoxide alarm requirements.

The code requires carbon monoxide alarms to be installed outside of each separate sleeping area in the immediate vicinity of the bedrooms to protect people when they are most vulnerable to the effects of carbon monoxide poisoning—when they are sleeping or not fully alert. The 2015 IRC requires an additional alarm to be located in the bedroom when a fuel-fired appliance is installed in the bedroom or the adjoining bathroom. The IRC allows fuel-burning appliances to be installed in bedrooms and bathrooms, but this is not a common practice. When one is installed, the new requirement intends to provide early warning to protect occupants who sleep with their bedroom door closed from a potential source of carbon monoxide poisoning within the space.

New provisions specifically permit the installation of combination carbon monoxide and smoke alarms complying with the applicable standards. Combination alarms are commonly installed outside of bedroom areas in residential construction as an acceptable method for satisfying both the smoke alarm and the carbon monoxide alarm provisions in the IRC, and this change simply recognizes a method that is already in practice.

As an alternative to the individual carbon monoxide alarm requirements, the code permits the installation of a carbon monoxide detection system installed in accordance with NFPA 720. These systems have separate detection devices installed in the same required locations as carbon monoxide alarms. Carbon monoxide detection systems become a permanent fixture of the dwelling unit and are owned by the homeowner as prescribed in the code. This provision intends to avoid systems that are leased to the homeowner by an alarm company, and could subsequently be removed by the alarm company if the homeowner discontinued service, leaving the home with no protection for detecting rising levels of carbon monoxide gas. The permanent fixture and ownership provisions remain in the 2015 IRC. However, there has been confusion regarding the requirement for systems to be monitored by an approved supervising station, and this requirement was considered difficult to enforce. Proponents reasoned that a system that provides local alarm

R315 continues

R315 continued

notification satisfies the intent of the code to provide early warning to occupants of increased levels of carbon monoxide gas and that it was difficult to justify the extra costs associated with monitoring by a supervising station. The code does not prohibit monitoring, but it is now an option rather than a requirement. Language has also been added to the detection system requirements to clarify that the detectors only need to be installed in locations specified in Section R315.3, outside of each separate sleeping area in the immediate vicinity of the bedrooms, not in all of the locations specified in NFPA 720.

R322.1, R322.2

Flood Hazards

CHANGE TYPE: Modification

CHANGE SUMMARY: Section R322.1 is modified to emphasize that the provision applies to existing buildings in flood hazard areas where 50 percent or more of the structure has damage and requires restoration. Section R322.2 limits the minimum elevation allowed for dwellings in flood hazard areas and defines a Coastal A Zone.

2015 CODE: R322.1 General. Buildings and structures constructed in whole or in part in flood hazard areas, including A or V Zones and Coastal A Zones as established in Table R301.2(1), and substantial improvement and restoration of substantial damage of buildings and structures in flood hazard areas, shall be designed and constructed in accordance with the provisions contained in this section. Buildings and structures that are located in more than one flood hazard area shall comply with the provisions associated with the most restrictive flood hazard area. Buildings and structures located in whole or in part in identified floodways shall be designed and constructed in accordance with ASCE 24.

R322.1.1 Alternative Provisions. As an alternative to the requirements in Section R322, ~~R322.3 for buildings and structures located in whole or in part in coastal high-hazard areas (V Zones) and Coastal A Zones, if delineated,~~ ASCE 24 is permitted subject to the limitations of this code and the limitations therein.

R322.2 Flood Hazard Areas (Including A Zones). Areas that have been determined to be prone to flooding and that are not subject to high-velocity wave action shall be designated as flood hazard areas. Flood hazard areas that have been delineated as subject to wave heights between 1.5 feet and 3 feet or otherwise designated by the jurisdiction shall be designated as Coastal A Zones and are subject to the requirements in Section R322.3. Buildings and structures constructed in whole or in part in flood hazard areas shall be designed and constructed in accordance with Sections R322.2.1 through R322.2.3.

R322.1, R322.2 continues

International Code Council®

Residential construction in a coastal A zone

R322.1, R322.2 continued

R322.2.1 Elevation requirements.

1. Buildings and structures in flood hazard areas ~~not~~ including flood hazard areas designated as Coastal A Zones shall have the lowest floors elevated to or above the base flood elevation plus 1 foot (305 mm) or the design flood elevation.

~~2. Buildings and structures in flood hazard areas designated as Coastal A Zones shall have the lowest floors elevated to or above the base flood elevation plus 1 foot (305 mm), or to the design flood elevation, whichever is higher.~~

2.~~3~~ In areas of shallow flooding (AO Zones), buildings and structures shall have the lowest floor (including basement) elevated to a height of not less than ~~at least as high above~~ the highest adjacent grade as the depth number specified in feet (mm) on the FIRM plus 1 foot (305 mm) or not less than 3 feet (15 mm) ~~or at least 2 feet (610 mm)~~ if a depth number is not specified.

3.4 Basement floors that are below grade on all sides shall be elevated to or above base flood elevation plus 1 foot (305 mm), or the design flood elevation whichever is higher.

Exception: Enclosed areas below the design flood elevation, including basements with floors that are not below grade on all sides, shall meet the requirements of Section R322.2.2.

(Some provisions not shown for brevity and clarity.)

CHANGE SIGNIFICANCE: The changes to R322.1 clarify that, as stated in R102.7.1, because the IRC applies to work on existing dwellings, the flood provisions apply to substantial improvement and substantial damage of existing dwellings. This has been a source of confusion in previous editions of the IRC.

Buildings located where they are affected by more than one flood hazard must comply with the more restrictive provisions that take into account flood loads and conditions of the area. For example, a dwelling that straddles a line that separates Zone A from Zone V must comply with the requirements for Zone V.

ASCE/SEI 24, *Flood Resistant Design and Construction,* provides design requirements for buildings and structures in flood hazard areas as an alternative to the IRC. There are flood hazard areas where the builder, designer, or building official may choose to not use or allow use of prescriptive foundations, such as along riverine waterways and some coastal areas (inland of Zone V) where flood depths are significant and dwellings would need very tall foundations. Design may also be required in riverine floodplains where flood velocities are very fast. ASCE 24 may be used for design of the foundation.

Another situation where use of ASCE 24 is appropriate is with dwellings in flood hazard areas on alluvial fans. Because the IRC does not have specific provisions for alluvial fans, ASCE 24 may be used as an alternative where prescriptive provisions of the IRC do not account for known flood risks.

The Coastal A Zone (CAZ) was added to the 2009 edition of IRC Section R322.2, with a requirement that if an area subject to waves between 1.5 feet and 3 feet is delineated, then the area is designated a

Coastal A Zone and the lowest floors shall be at least 1 foot above the design flood elevation. Dwellings in Coastal A Zones must also comply with all requirements for dwellings in Zone A.

The inland boundary of the coastal high-hazard area (Zone V) is drawn by FEMA where breaking wave heights are expected to drop below 3 feet during base flood conditions. The requirements for foundations of dwellings that are located just landward of the Zone V boundary assume that hydrodynamic loads associated with waves—even waves that are 2.9 feet tall—are not significant and that conventional foundations such as perimeter walls can resist these loads and associated erosion and local scour.

Post-disaster investigations after severe coastal storms led to recommendations of application of coastal high-hazard area (Zone V) requirements to areas inland of the Zone V/Zone A boundary—in the area subject to waves between 1.5 feet and 3 feet—the area now referred to as "Coastal A Zone." The total land area likely to be designated as CAZ is small. Less than 3 percent of all mapped flood hazard areas are Zone V, and the inland edge of the Coastal A Zone is a relatively short distance inland from the Zone V boundary. This inland boundary is called the Limit of Moderate Wave Action (LiMWA) on FIRMs.

R322.3

Coastal High-Hazard Areas

CHANGE TYPE: Modification

CHANGE SUMMARY: Coastal A Zones are defined and an exception for foundation types in Coastal A Zones is added.

2015 CODE: R322.3 Coastal High-Hazard Areas (Including V Zones and Coastal A Zones, Where Designated). Areas that have been determined to be subject to wave heights in excess of 3 feet (914 mm) or subject to high-velocity wave action or wave-induced erosion shall be designated as coastal high-hazard areas. Flood hazard areas that have been delineated as subject to wave heights between 1½ feet (457 mm) and 3 feet (914 mm) or otherwise designated by the jurisdiction shall be designated as Coastal A Zones. Buildings and structures constructed in whole or in part in coastal high-hazard areas and in Coastal A Zones, where designated, shall be designed and constructed in accordance with Sections R322.3.1 through R322.3.7.

R322.3.3 Foundations. All buildings and structures erected in coastal high-hazard areas and Coastal A Zones, shall be supported on pilings or columns and shall be adequately anchored to such pilings or columns. The space below the elevated building shall be either free of obstruction or, if enclosed with walls, the walls shall meet the requirements of Section R322.3.4. Piling shall have adequate soil penetrations to resist the combined wave and wind loads (lateral and uplift). Water loading values used shall be those associated with the design flood. Wind loading values shall be those required by this code. Pile embedment shall include consideration of decreased resistance capacity caused by scour of soil strata surrounding the piling. Pile systems design and installation shall be certified in accordance with Section R322.3.6. Spread footing, mat, raft, or other foundations that support columns shall not be permitted where soil investigations that are required in accordance with Section R401.4 indicate that soil material under the spread footing, mat, raft, or other foundation is subject to scour or erosion from wave–velocity flow

International Code Council®

Pier and beam construction in a coastal high-hazard area

conditions. If permitted, spread footing, mat, raft, or other foundations that support columns shall be designed in accordance with ASCE 24. Slabs, pools, pool decks, and walkways shall be located and constructed to be structurally independent of buildings and structures and their foundations to prevent transfer of flood loads to the buildings and structures during conditions of flooding, scour, or erosion from wave-velocity flow conditions, unless the buildings and structures and their foundation are designed to resist the additional flood load.

> **Exception:** In Coastal A Zones, stem wall foundations supporting a floor system above and backfilled with soil or gravel to the underside of the floor system shall be permitted provided the foundations are designed to account for wave action, debris impact, erosion, and local scour. Where soils are susceptible to erosion and local scour, stem wall foundations shall have deep footings to account for the loss of soil.

(Some provisions in section not included for brevity and clarity.)

CHANGE SIGNIFICANCE: Dwellings in areas designated as "Coastal A Zones" must meet the requirements of Section 322.3 for dwellings in coastal high-hazard areas (Zone V), including open foundations (pilings or columns), but they may have filled stemwalls as foundations.

The Coastal A Zone (CAZ) has been in ASCE/SEI 7, *Minimum Design Loads for Buildings and Other Structures*, since the late 1990s and in ASCE/SEI 24, *Flood Resistant Design and Construction*, since its initial publication in 1998. Recognition of CAZ was added to the 2009 edition of the *International Residential Code* (IRC) in Section R322.2. CAZs had only one requirement: if an area subject to waves between 1.5 feet and 3 feet was delineated, then the area was designated a Coastal A Zone. The lowest floors were required to be at least 1 foot above the design flood elevation. Otherwise, the 2009 and 2012 IRC required dwellings in Coastal A Zones to comply with the requirements for Zone A.

The inland boundary of the coastal high-hazard area (Zone V) is drawn by FEMA where breaking wave heights are expected to drop below 3 feet during base flood conditions. The requirements for foundations of dwellings that are located just landward of the Zone V boundary are assumed to be primarily affected by the waves. Waves, even waves that are 2.9 feet tall, are not significant. It had been assumed that conventional foundations such as perimeter walls could resist the wave loads and associated erosion and local scour.

Post-disaster investigations after recent severe coastal storms have shown that in the area subject to waves between 1.5 feet and 3 feet, the area now referred to as "Coastal A Zone," significant damage may occur. FEMA reports have recommended implementing requirements for Zone V in Coastal A Zones. All coastal flood studies by FEMA now include analyses of moderate wave action and FIRMs show the Limit of Moderate Wave Action (LiMWA). An area defined as experiencing LiMWA is determined by a number of factors, including fetch (length of open water over which wind blows to generate waves), orientation of the shoreline to prevalent direction of wind and waves, land elevation relative to water depths, and the presence of dunes, buildings, and other elements of the landscape that have the effect of breaking up waves. Many reaches of

R322.3 continues

R322.3 continued

shoreline subject to tidal flooding do not have conditions that produce moderate wave action, in which case the FIRM does not show a LiMWA.

The total land area that is likely to be designated as CAZ is small. Less than 3 percent of all mapped flood hazard areas are Zone V and the LiMWA is a relatively short distance inland from the Zone V boundary. Some communities currently augment the minimum NFIP requirements because of observed wave damage to conventional, closed foundations in this area of shallow wave action.

Observations after Superstorm Sandy continue to reinforce the damage potential in areas just inland of the Zone V boundary. Given that open foundations (piles and columns) perform well under velocity and wave conditions, dwellings in Coastal A Zones should meet the same requirements as dwellings in coastal high-hazard areas, except foundations of filled stemwalls that account for the potential for scour and erosion are allowed.

CHANGE TYPE: Addition

CHANGE SUMMARY: New provisions in Section R325 place limitations on the construction of mezzanines related to ceiling height and openness consistent with the *International Building Code* (IBC).

R325

Mezzanines

2015 CODE:

A mezzanine is not considered a story but must not exceed one-third of the floor area of the room where it is located.

SECTION R202
DEFINITIONS

MEZZANINE~~, LOFT~~. An intermediate level or levels between the floor and ceiling of any story ~~with an aggregate floor area of not more than one-third of the area of the room or space in which the level or levels are located~~.

R301.2.2.3.1 Height Limitations. Wood-framed buildings shall be limited to three stories above grade plane or the limits given in Table R602.10.3(3). Cold-formed, steel-framed buildings shall be limited to less than or equal to three stories above grade plane in accordance with AISI S230. Mezzanines as defined in Section R202 that comply with Section R325 shall not be considered as stories. Structural insulated panel buildings shall be limited to two stories above grade plane.

SECTION R325
MEZZANINES

R325.1 General. Mezzanines shall comply with Section R325.

R325.2 Mezzanines. The clear height above and below mezzanine floor construction shall be not less than 7 feet (2134 mm).

R325.3 Area Limitation. The aggregate area of a mezzanine or mezzanines shall be not greater than one-third of the floor area of the room or space in which they are located. The enclosed portion of a room shall not be included in a determination of the floor area of the room in which the mezzanine is located.

R325.4 Means of Egress. The means of egress for mezzanines shall comply with the applicable provisions of Section R311.

R325.5 Openness. Mezzanines shall be open and unobstructed to the room in which they are located except for walls not more than 42 inches (1067 mm) in height, columns, and posts.

Exceptions:

1. Mezzanines or portions thereof are not required to be open to the room in which they are located, provided that the aggregate floor area of the enclosed space is not greater than 10 percent of the mezzanine area.
2. In buildings that are no more than two stories above grade plane and equipped throughout with an automatic sprinkler system in accordance with NFPA 13R or NFPA 13D,

R325 continues

R325 continued

a mezzanine having two or more means of egress shall not be required to be open to the room in which the mezzanine is located.

CHANGE SIGNIFICANCE: In previous editions of the code, mezzanines and lofts were defined as intermediate levels with an aggregate floor area of not more than one-third of the area of the room or space in which they were located. In the height limitations provisions of Section R301.2.2.3 for buildings located in Seismic Design Categories D_0, D_1, and D_2, mezzanines are not considered as stories. The intent is that mezzanines are not considered stories in Seismic Design Categories A, B, and C either, considering that the hazard for taller buildings is less in geographic locations where anticipated seismicity is lower. The only advantage for identifying a floor level as a mezzanine rather than a story under the IRC is to construct taller buildings. The scope of the IRC limits dwellings and townhouses to three stories above grade plane. Construction of a mezzanine could add a usable intermediate level or levels in addition to the three stories. Unlike the IBC, the IRC does not place limits on floor areas.

A new section in the 2015 IRC establishes provisions for mezzanines consistent with the IBC provisions. The limitation for mezzanine size to not exceed one-third of the floor area of the room or space in which it is located has been moved out of the definition into the new section. The code now stipulates a minimum ceiling height of 7 feet for mezzanines that is consistent with the ceiling height provisions for habitable rooms and hallways in Section R305. Mezzanines are generally required to be open to the space in which they are located, but the code provides for a limited area to be enclosed. With the installation of sprinklers and two exits, mezzanines in two-story buildings are permitted to be enclosed. In most cases, floor levels in two-story buildings will be identified as stories rather than mezzanines and the mezzanine provisions will not apply. The term "loft" has been deleted from the definition because there are no provisions in the code for lofts.

PART **3**

Building Construction

Chapters 4 through 10

- **Chapter 4** Foundations
- **Chapter 5** Floors
- **Chapter 6** Wall Construction
- **Chapter 7** Wall Covering
- **Chapter 8** Roof-Ceiling Construction
- **Chapter 9** Roof Assemblies
- **Chapter 10** Chimneys and Fireplaces
  No changes addressed

Chapters 4 through 10 address the prescriptive methods for building foundations, floor construction, wall construction, wall coverings, roof construction, roof assemblies, chimneys, and fireplaces. Concrete, masonry, and wood foundations; retaining walls; supporting soil properties; surface drainage; and foundation dampproofing and drainage are found in Chapter 4. Chapters 5, 6, and 8 contain the construction provisions for floors, walls, and roofs, respectively, with most of the provisions addressing light-frame construction. Chapter 7 addresses interior finishes, such as drywall and plaster installations, and exterior wall coverings, including water-resistive barriers, flashings, siding, and veneer, to provide a durable weather-resistant exterior. Chapter 9 covers the various waterproof roof assemblies, including roofing underlayment, roof eave ice barrier, flashings, asphalt shingles, and other roof coverings. Site-built masonry fireplaces and chimneys as well as prefabricated fireplaces and chimneys, including their weather-tight roof terminations, are addressed in the provisions of Chapter 10. ■

R403.1.1

Minimum Footing Size

R403.1.2, R602.10.9.1

Continuous Footings in Seismic Design Categories D_0, D_1, and D_2

R403.1.3

Footing and Stem Wall Reinforcing in Seismic Design Categories D_0, D_1, and D_2

R403.1.6

Foundation Anchorage

R404.1.4.1

Masonry Foundation Walls in SDC D_0, D_1, and D_2

R404.4

Retaining Walls

TABLES R502.3.1(1), R502.3.1(2)

Floor Joist Spans for Common Lumber Species

R502.10

Framing of Floor Openings

R507.1, R507.4

Decking

R507.2

Deck Ledger Connection to Band Joist

R507.2.4

Alternative Deck Lateral Load Connection

R507.5, R507.6, R507.7

Deck Joists and Beams

R507.8

Deck Posts

TABLE R602.3(1)

Fastening Schedule—Roof Requirements

TABLE R602.3(1)

Fastening Schedule—Wall Requirements

TABLE R602.3(1)

Fastening Schedule—Floor Requirements

R602.3.1

Stud Size, Height, and Spacing

R602.7

Headers

TABLE R602.10.3(1)

Bracing Requirements Based on Wind Speed

TABLE R602.10.5

Contributing Length of Method CS-PF Braced Wall Panels

R602.10.6.2

Method PFH: Portal Frame with Hold-Downs

R602.10.11

Cripple Wall Bracing

R602.12

Simplified Wall Bracing

R603.9.5

Structural Sheathing over Steel Framing for Stone and Masonry Veneer

R606

Masonry Walls

R606.3.5

Grouting Requirements for Masonry Construction

R610.7

Drilling and Notching in Structural Insulated Panels

R703.3

Siding Material Thickness and Attachment

R703.5

Wood, Hardboard, and Wood Structural Panel Siding

R703.6

Wood Shakes and Shingles on Exterior Walls

R703.9

Exterior Insulation and Finish Systems (EIFS)

R703.11.1

Vinyl Siding Attachment

R703.13, R703.14

Insulated Vinyl Siding and Polypropylene Siding

R703.15, R703.16, R703.17

Cladding Attachment over Foam Sheathing

TABLES R802.4, R802.5

Ceiling Joist and Rafter Tables

R806.1

Attic Ventilation

TABLE R806.5

Insulation for Condensation Control in Unvented Attics

R905.1.1

Underlayment

R905.7.5

Wood Shingle Application

R905.8.6

Wood Shake Application

R905.16

Photovoltaic Shingles

R907

Rooftop-Mounted Photovoltaic Systems

R403.1.1

Minimum Footing Size

CHANGE TYPE: Modification

CHANGE SUMMARY: This code change divides minimum footing size and thickness into three expanded tables based on the type of construction being supported: light frame, light frame with veneer, and concrete or masonry. The values are also based on the type of foundation: slab on grade, crawl space, or basement.

2015 CODE: **R403.1.1 Minimum Size.** The minimum ~~sizes~~ width, W, and thickness, T, for concrete ~~and masonry~~ footings shall be ~~as set forth~~ in accordance with Tables R403.1(1) through R403.1(3) and Figure R403.1(1) or R403.1.3, as applicable. The footing width~~, W,~~ shall be based on the load-bearing value of the soil in accordance with Table R401.4.1. ~~Spread footings shall be at least 6 inches (152 mm) in thickness, T.~~ Footing projections, P, shall be ~~at least~~ not less than 2 inches (51 mm) and shall not exceed the thickness of the footing. Footing thickness and projection for fireplaces shall be in accordance with Section R1001.2. The size of footings supporting piers and columns shall be based on the tributary load and allowable soil pressure in accordance with Table R401.4.1. Footings for wood foundations shall be in accordance with the details set forth in Section R403.2, and Figures R403.1(2) and R403.1(3).

R403.1.1 continues

TABLE R403.1(1) Minimum Width and Thickness for Concrete Footings for Light Frame Construction (inches)[a, b]

Snow Load or Roof Live Load	Story and Type of Structure with Light Frame	Load-Bearing Value of Soil (psf)					
		1500	2000	2500	3000	3500	4000
20 psf	1 story - slab on grade	12 × 6	12 × 6	12 × 6	12 × 6	12 × 6	12 × 6
	1 story - with crawl space	12 × 6	12 × 6	12 × 6	12 × 6	12 × 6	12 × 6
	1 story - plus basement	18 × 6	14 × 6	12 × 6	12 × 6	12 × 6	12 × 6
	2 story - slab on grade	12 × 6	12 × 6	12 × 6	12 × 6	12 × 6	12 × 6
	2 story - with crawl space	16 × 6	12 × 6	12 × 6	12 × 6	12 × 6	12 × 6
	2 story - plus basement	22 × 6	16 × 6	13 × 6	12 × 6	12 × 6	12 × 6
	3 story - slab on grade	14 × 6	12 × 6	12 × 6	12 × 6	12 × 6	12 × 6
	3 story - with crawl space	19 × 6	14 × 6	12 × 6	12 × 6	12 × 6	12 × 6
	3 story - plus basement	25 × 8	19 × 6	15 × 6	13 × 6	12 × 6	12 × 6
30 psf	1 story - slab on grade	12 × 6	12 × 6	12 × 6	12 × 6	12 × 6	12 × 6
	1 story - with crawl space	13 × 6	12 × 6	12 × 6	12 × 6	12 × 6	12 × 6
	1 story - plus basement	19 × 6	14 × 6	12 × 6	12 × 6	12 × 6	12 × 6
	2 story - slab on grade	12 × 6	12 × 6	12 × 6	12 × 6	12 × 6	12 × 6
	2 story - with crawl space	17 × 6	13 × 6	12 × 6	12 × 6	12 × 6	12 × 6
	2 story - plus basement	23 × 6	17 × 6	14 × 6	12 × 6	12 × 6	12 × 6
	3 story - slab on grade	15 × 6	12 × 6	12 × 6	12 × 6	12 × 6	12 × 6
	3 story - with crawl space	20 × 6	15 × 6	12 × 6	12 × 6	12 × 6	12 × 6
	3 story - plus basement	26 × 8	20 × 6	16 × 6	13 × 6	12 × 6	12 × 6

(continues)

R403.1.1 continued

TABLE R403.1(1) ***(Continued)***

Snow Load or Roof Live Load	Story and Type of Structure with Light Frame	Load-Bearing Value of Soil (psf)					
		1500	2000	2500	3000	3500	4000
50 psf	1 story - slab on grade	12 × 6	12 × 6	12 × 6	12 × 6	12 × 6	12 × 6
	1 story - with crawl space	16 × 6	12 × 6	12 × 6	12 × 6	12 × 6	12 × 6
	1 story - plus basement	21 × 6	16 × 6	13 × 6	12 × 6	12 × 6	12 × 6
	2 story - slab on grade	14 × 6	12 × 6	12 × 6	12 × 6	12 × 6	12 × 6
	2 story - with crawl space	19 × 6	14 × 6	12 × 6	12 × 6	12 × 6	12 × 6
	2 story - plus basement	25 × 7	19 × 6	15 × 6	12 × 6	12 × 6	12 × 6
	3 story - slab on grade	17 × 6	13 × 6	12 × 6	12 × 6	12 × 6	12 × 6
	3 story - with crawl space	22 × 6	17 × 6	13 × 6	12 × 6	12 × 6	12 × 6
	3 story - plus basement	28 × 9	21 × 6	17 × 6	14 × 6	12 × 6	12 × 6
70 psf	1 story - slab on grade	12 × 6	12 × 6	12 × 6	12 × 6	12 × 6	12 × 6
	1 story - with crawl space	18 × 6	13 × 6	12 × 6	12 × 6	12 × 6	12 × 6
	1 story - plus basement	24 × 7	18 × 6	14 × 6	12 × 6	12 × 6	12 × 6
	2 story - slab on grade	16 × 6	12 × 6	12 × 6	12 × 6	12 × 6	12 × 6
	2 story - with crawl space	21 × 6	16 × 6	13 × 6	12 × 6	12 × 6	12 × 6
	2 story - plus basement	27 × 9	20 × 6	16 × 6	14 × 6	12 × 6	12 × 6
	3 story - slab on grade	19 × 6	14 × 6	12 × 6	12 × 6	12 × 6	12 × 6
	3 story - with crawl space	25 × 7	18 × 6	15 × 6	12 × 6	12 × 6	12 × 6
	3 story - plus basement	30 × 10	23 × 6	18 × 6	15 × 6	13 × 6	12 × 6

a. Interpolation allowed. Extrapolation is not allowed.
b. Based on 32 foot wide house with load-bearing center wall that carries half of the tributary attic, and floor framing. For every 2 feet of adjustment to the width of the house add or subtract 2 inches of footing width and 1 inch of footing thickness (but not less than 6 inches thick).

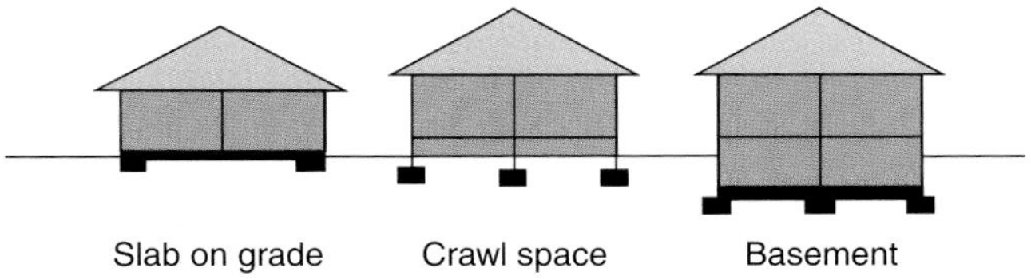

(Additional tables not shown for brevity.)

CHANGE SIGNIFICANCE: Due to concern that the 2012 IRC Table R403.1, covering minimum width of footings, was too conservative for concrete footings, the table for minimum footing size has been expanded into three tables. Loading conditions are clarified to more accurately determine the footing size required. The new minimum footing may be smaller, the same size, or larger than the 2012 IRC minimum footing size. Often the new minimum footing is slightly larger than the 2012 minimum for light-frame construction. For buildings with veneer, the minimum footing may be smaller.

The new tables are Table R403.1(1), light-frame construction; Table R403.1(2), light-frame construction with veneer; and Table R403.1(3), cast-in-place concrete or masonry construction. See the 2015 IRC for

Footing–stem wall

Footing–basement wall

Tables R403.1(2) and R403.1(3). Note that the new tables apply to concrete footings only. The tables do not address grouted or solid masonry, crushed stone footings, or wood foundations.

Although not specifically stated in the code, for the 2012 IRC Table R403.1, footing size and depth were based upon the following assumptions:

- Snow load of 50 psf
- 20 feet of tributary roof area
- 16 feet of tributary floor area
- 10-foot first-floor height
- 8-foot second- and third-floor heights

The 2015 IRC minimum footing size tables are based on similar but not identical factors. The following assumptions are made:

- Snow or roof live load of 20, 30, 50 or 70 psf (the maximum allowed prescriptively by the IRC in accordance with Section R301.2.3)
- 18 feet of tributary roof area
- 16 feet of tributary floor area
- 8-foot third floor height
- 9-foot second floor height
- 10-foot first floor height
- 3-foot crawlspace wall height
- 10-foot basement wall height, 10-inch basement wall thickness, basement wall material weight of 125 pcf

Footnote a allows interpolation of soil and snow load conditions. Footnote b accounts for an increase or decrease in building width, allowing a change in the footing width and thickness.

R403.1.1 continues

R403.1.1 continued

Examples—Minimum Required Footing

Two-story house with slab on grade foundation:

Light-frame construction
Soil-bearing strength = 1500 psf
Roof Live Load = 20 psf
32 ft wide building with interior load-bearing wall

Minimum Footing Width		
2012	**2015**	Smaller footing width allowed
15×6	12×6	

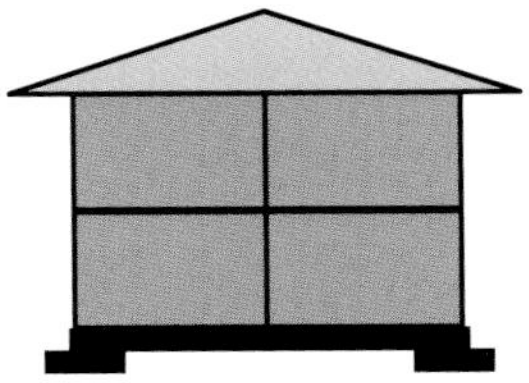

One-story house with stem wall foundation (crawl space):

Light-frame construction
Soil-bearing strength = 1500 psf
Snow Load = 30 psf
28 ft. wide building with interior load-bearing wall (see footnote b)
Footnote b allows buildings with roof widths smaller than 32 ft. to subtract 2 in. from the footing width for every 2 ft. of width less than 32 ft.

Minimum Footing Width		
2012	**2015**	Same size footing required
12×6	13″ – 2×2″ < 12″ 12×6	

4-inch brick veneer over light-frame construction
Soil-bearing strength = 1500 psf
Snow Load = 30 psf
32 ft. wide building with interior load-bearing wall

Minimum Footing Width		
2012	**2015**	Larger footing width required
12×6	16×6	

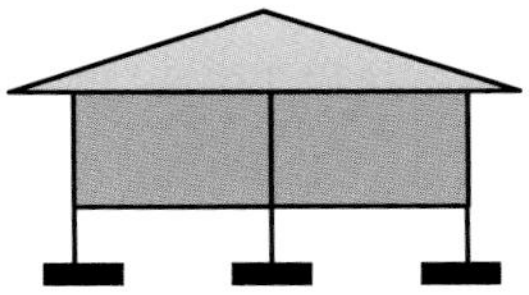

Two-story house with basement foundation:

Light-frame construction
Soil-bearing strength = 2000 psf
Snow Load = 30 psf
28 ft. wide building with interior load-bearing wall (see footnote b)
Footnote b allows buildings with roof widths smaller than 32 ft. to subtract 2 in. from the footing width for every 2 ft. of width less than 32 ft.

Minimum Footing Width		
2012	**2015**	Larger footing width required
12×6	17″ – 2×2″ = 13″ 13×6	

4-inch brick veneer over light-frame construction
Soil-bearing strength = 2000 psf
Snow Load = 30 psf
32 ft. wide building with interior load-bearing wall

Minimum Footing Width		
2012	**2015**	Larger footing width required
16×6	21×6	

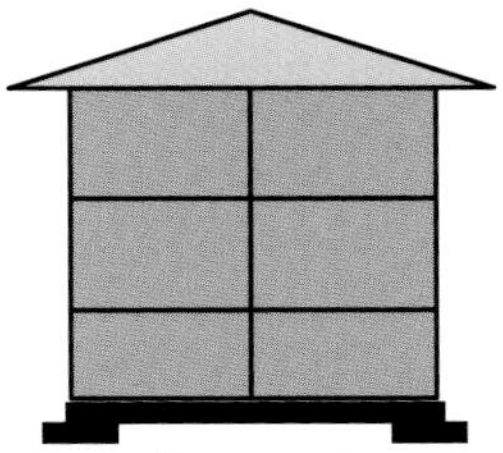

Minimum footing size examples

As displayed in the examples, the tables are based on the type of foundation. Categories include:

1. One, two, and three stories built on a slab on grade (without a first-floor load),
2. One, two, and three stories built over a crawl space (with a first-floor load and foundation wall/footing load), and
3. One, two, and three stories built with basement (with a first-floor load and basement wall load—previously, the table was silent on how to handle the extra load from a masonry or concrete basement wall).

In the new tables, footing size increases for homes with a crawl space or basement.

The width of the footing is provided based on loads described above and the minimum footing projection. The tables are based on the following load case:

Total load (TL) equal to dead load (D) plus 75% of the snow or roof live load (S, L_R)

$$TL = D + .75(S \text{ or } L_R)$$

In combining the two loads, a reduction is allowed in the live load.

The minimum footing thickness is 6 inches, the minimum footing width is 12 inches. Calculation of footing size may result in a smaller footing but the code requires a minimum 12 × 6 or 12 inches wide and 6 inches deep footing.

R403.1.2, R602.10.9.1

Continuous Footings in Seismic Design Categories D_0, D_1, and D_2

CHANGE TYPE: Clarification

CHANGE SUMMARY: This code change clarifies the continuous footing requirement in Section R403.1.2 and moves requirements in Section R602.10.9.1 to the foundation chapter.

2015 CODE: R403.1.2 Continuous Footing in Seismic Design Categories D_0, D_1, and D_2. ~~The braced wall panels at~~ Exterior walls of buildings located in Seismic Design Categories D_0, D_1, and D_2 shall be supported by continuous solid or fully grouted masonry or concrete footings. Other footing materials or systems shall be designed in accordance with accepted engineering practice. All required interior braced wall panels in buildings located in Seismic Design Categories D_0, D_1, and D_2 with plan dimensions greater than 50 feet (15 240 mm) shall be supported by continuous solid or fully grouted masonry or concrete footings in accordance with Section R403.1.3.4, except for two-story buildings in Seismic Design Category D_2, in which all braced wall panels, interior and exterior, shall be supported on continuous foundations.

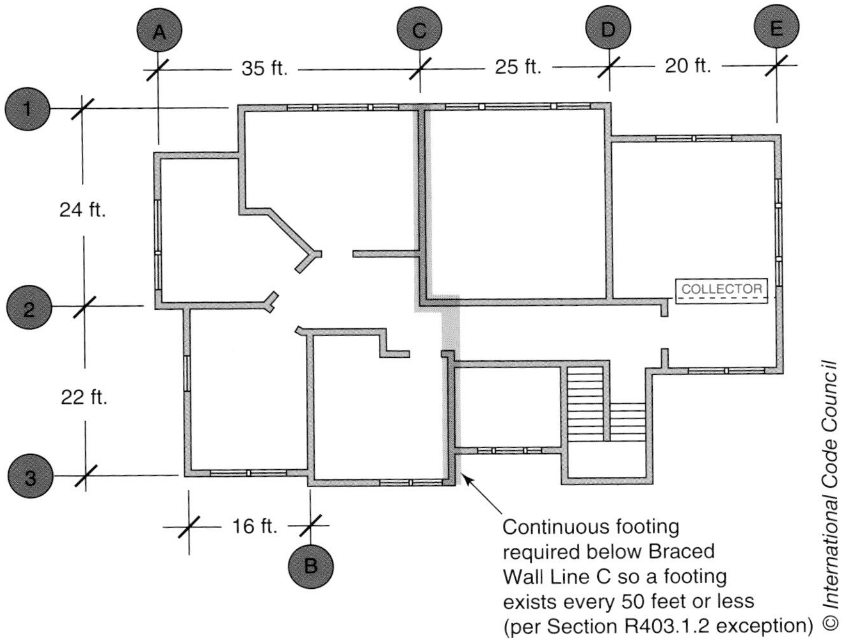

Continuous footing

Exception: Two-story buildings shall be permitted to have interior braced wall panels supported on continuous foundations at intervals not exceeding 50 feet (15 240 mm) provided that:

1. The height of cripple walls does not exceed 4 feet (1219 mm).
2. First-floor braced wall panels are supported on doubled floor joists, continuous blocking, or floor beams.
3. The distance between bracing lines does not exceed twice the building width measured parallel to the braced wall line.

R602.10.9.1 Braced Wall Panel Support for Seismic Design ~~Category~~ Categories D_0, D_1, and D_2. ~~In one-story buildings located in Seismic Design Category D_2, braced wall panels shall be supported on continuous foundations at intervals not exceeding 50 feet (15 240 mm). In two-story buildings located in Seismic Design Category D_2, all braced wall panels shall be supported on continuous foundations.~~ In Seismic Design Categories D_0, D_1, and D_2 braced wall panel footings shall be as specified in Section R403.1.2.

~~**Exception:** Two-story buildings shall be permitted to have interior braced wall panels supported on continuous foundations at intervals not exceeding 50 feet (15 240 mm) provided that:~~

1. ~~The height of cripple walls does not exceed 4 feet (1219 mm).~~
2. ~~First-floor braced wall panels are supported on doubled floor joists, continuous blocking or floor beams.~~
3. ~~The distance between bracing lines does not exceed twice the building width measured parallel to the braced wall line.~~

CHANGE SIGNIFICANCE: In previous editions of the IRC, provisions in Chapters 4 and 6 for continuous footings and foundations were widely separated and confusing. In Section R602.10.9.1, there was no specific guidance on what to do in Seismic Design Categories (SDC) D_0 and D_1 below interior braced wall panels. This code change clarifies the foundation requirements under braced wall panels in all high-seismic regions. In previous IRC editions, the requirements for SDC D_2 in the wall bracing section added to requirements in Section R403.1.2.

All requirements for footings in high-seismic regions are now located in Section R403.1.2 within the foundation chapter. Section R403.1.2 requires continuous footings for the cases listed in Table 4-1.

R403.1.2, R602.10.9.1 continues

R403.1.2, R602.10.9.1 continued

TABLE 4-1 Continuous Footing Requirements in High-Seismic Regions

SDC	# of Stories	Wall Location	LENGTH OF PLAN DIMENSIONS: Both dimensions ≤50 ft.	LENGTH OF PLAN DIMENSIONS: One or more dimensions >50 ft.
D_0, D_1, D_2	One Story	Exterior Walls	Continuous footing required	Continuous footing required
		Interior Walls	No continuous footings required	Continuous footings required below all interior braced wall panels
D_0, D_1	Two Story	Exterior Walls	Continuous footing required	Continuous footing required
		Interior Walls	No continuous footings required	Continuous footings are required below all interior braced wall panels unless the Section R403.1.2 exception is used. Then interior wall lines with braced wall panels may be supported according to the exception with footings below interior braced wall panels spaced a maximum of 50 ft or less.
D_2	Two Story	Exterior Walls	Continuous footing required	Continuous footing required
		Interior Walls	Continuous footings are required below all interior braced wall panels unless the Section R403.1.2 exception is used.	Continuous footings are required below all interior braced wall panels unless the Section R403.1.2 exception is used. Then interior wall lines with braced wall panels may be supported according to the exception with footings below interior braced wall panels spaced a maximum of 50 ft. or less.

R403.1.3

Footing and Stem Wall Reinforcing in Seismic Design Categories D_0, D_1, and D_2

CHANGE TYPE: Clarification

CHANGE SUMMARY: Updated figures and code provisions in Section R403.1.3 now clearly define minimum required reinforcement in footings and stem walls located in Seismic Design Categories (SDC) D_0, D_1, and D_2.

2015 CODE: R403.1.3 ~~Seismic Reinforcing~~ Footing and Stem Wall Reinforcing in Seismic Design Categories D_0, D_1 and D_2. Concrete footings located in Seismic Design Categories D_0, D_1 and D_2, as established in Table R301.2(1), shall have minimum reinforcement in accordance with this section and Figure R403.1.3. ~~Bottom r~~Reinforcement shall be ~~located~~ installed with support and cover in accordance with Section R403.1.3.5. ~~a minimum of 3 inches (76 mm) clear from the bottom of the footing.~~

R403.1.3.1 Concrete Stem Walls with Concrete Footings. In Seismic Design Categories D_0, D_1 and D_2 where a construction joint is created between a concrete footing and a concrete stem wall, a minimum of one No. 4 vertical bar shall be installed at not more than 4 feet (1219 mm) on center. The vertical bar shall have a standard hook and extend to the bottom of the footing and shall have support and cover as specified in Section R403.1.3.5.3 and extend a minimum of 14 inches (357 mm) into the stem wall. Standard hooks shall comply with Section ~~R611.5.4.5~~ R608.5.4.5. A minimum of one No. 4 horizontal bar shall be installed within 12 inches (305 mm) of the top of the stem wall and one No. 4 horizontal bar shall be located 3 to 4 inches (76 mm to 102 mm) from the bottom of the footing. ~~The vertical bar shall extend to 3 inches (76 mm) clear of the bottom of the footing have a standard hook and extend a minimum of 14 inches (357 mm) into the stem wall.~~

R403.1.3.2 Masonry Stem Walls with Concrete Footings. In Seismic Design Categories D_0, D_1 and D_2 where a masonry stem wall is supported on a concrete footing~~ and stem wall~~, a minimum of one No. 4 vertical bar shall be installed at not more than 4 feet (1219 mm) on center. The vertical bar shall have a standard hook and extend to the bottom of the footing and shall have support and cover as specified in Section R403.1.3.5.3 and extend a minimum of 14 inches (357 mm) into the stem wall. Standard hooks shall comply with Section ~~R611.5.4.5~~ R608.5.4.5. A minimum of one No. 4 horizontal bar shall be installed within 12 inches (305 mm) of the top of the wall and one No. 4 horizontal bar shall be located 3 to 4 inches (76 mm to 102 mm) from the bottom of the footing. Masonry stem walls shall be solid grouted. ~~The vertical bar shall extend to 3 inches (76 mm) clear of the bottom of the footing and have a standard hook. In Seismic Design Categories D_0, D_1 and D_2 masonry stem walls without solid grout and vertical reinforcing are not permitted.~~

> **Exception:** ~~In detached one- and two-family dwellings which are three stories or less in height and constructed with stud bearing walls, isolated plain concrete footings, supporting columns or pedestals are permitted.~~

R403.1.3 continues

R403.1.3 continued

~~**R403.1.3.1 Foundations with Stemwalls.** Foundations with stem walls shall have installed a minimum of one No. 4 bar within 12 inches (305 mm) of the top of the wall and one No. 4 bar located 3 inches (76 mm) to 4 inches (102 mm) from the bottom of the footing.~~

~~R403.1.3.2~~ R403.1.3.3 Slabs-on-Ground with Turned-Down Footings. In Seismic Design Categories D_0, D_1 and D_2, ~~S~~slabs on ground cast monolithically with turned down footings shall have a minimum of one No. 4 bar at the top and the bottom of the footing or one No. 5 bar or two No. 4 bars in the middle third of the footing depth.

> ~~**Exception:** For slabs-on-ground cast monolithically with the footing, locating one No. 5 bar or two No. 4 bars in the middle third of the footing depth shall be permitted as an alternative to placement at the footing top and bottom.~~

Where the slab is not cast monolithically with the footing, No. 3 or larger vertical dowels with standard hooks on each end shall be ~~provided~~ installed at not more than 4 feet (1219 mm) on center in accordance with Figure ~~R403.1.3.2~~ R403.1.3, Detail 2. Standard hooks shall comply with Section ~~R611.5.4.5~~ R608.5.4.5.

~~R403.1.4.2 Seismic Conditions~~ R403.1.3.4 Interior Bearing and Braced Wall Panel Footings in Seismic Design Categories D_0, D_1, and D_2. In Seismic Design Categories D_0, D_1 and D_2, interior footings supporting bearing walls or braced wall panels, ~~bracing walls~~ and cast monolithically with a slab on grade, shall extend to a depth of not less than 12 inches (305 mm) below the top of the slab.

R403.1.3.5 Reinforcement. Footing and stem wall reinforcement shall comply with Sections R403.1.3.5.1 through R403.1.3.5.4.

R403.1.3.5.1 Steel Reinforcement. Steel reinforcement shall comply with the requirements of ASTM A615, A706, or A996. ASTM A996 bars produced from rail steel shall be Type R. The minimum yield strength of reinforcing steel shall be 40,000 psi (Grade 40) (276 MPa).

R403.1.3.5.2 Location of Reinforcement in Wall. The center of vertical reinforcement in stem walls shall be located at the centerline of the wall. Horizontal and vertical reinforcement shall be located in footings and stem walls to provide the minimum cover required by Section R403.1.3.5.3.

~~R404.1.2.3.7.4~~ R403.1.3.5.3 Support and Cover. Reinforcement shall be secured in the proper location in the forms with tie wire or other bar support system to prevent displacement during the concrete placement operation. Steel reinforcement in concrete cast against the earth shall have a minimum cover of 3 inches (75 mm). Minimum cover for reinforcement in concrete cast in removable forms that will be exposed to the earth or weather shall be 1½ inches (38 mm) for No. 5 bars and smaller, and 2 inches (50 mm) for No. 6 bars and larger. For concrete cast in removable forms that will not be exposed to the earth or weather, and for concrete cast in stay-in-place forms, minimum cover shall be ¾ inch (19 mm).

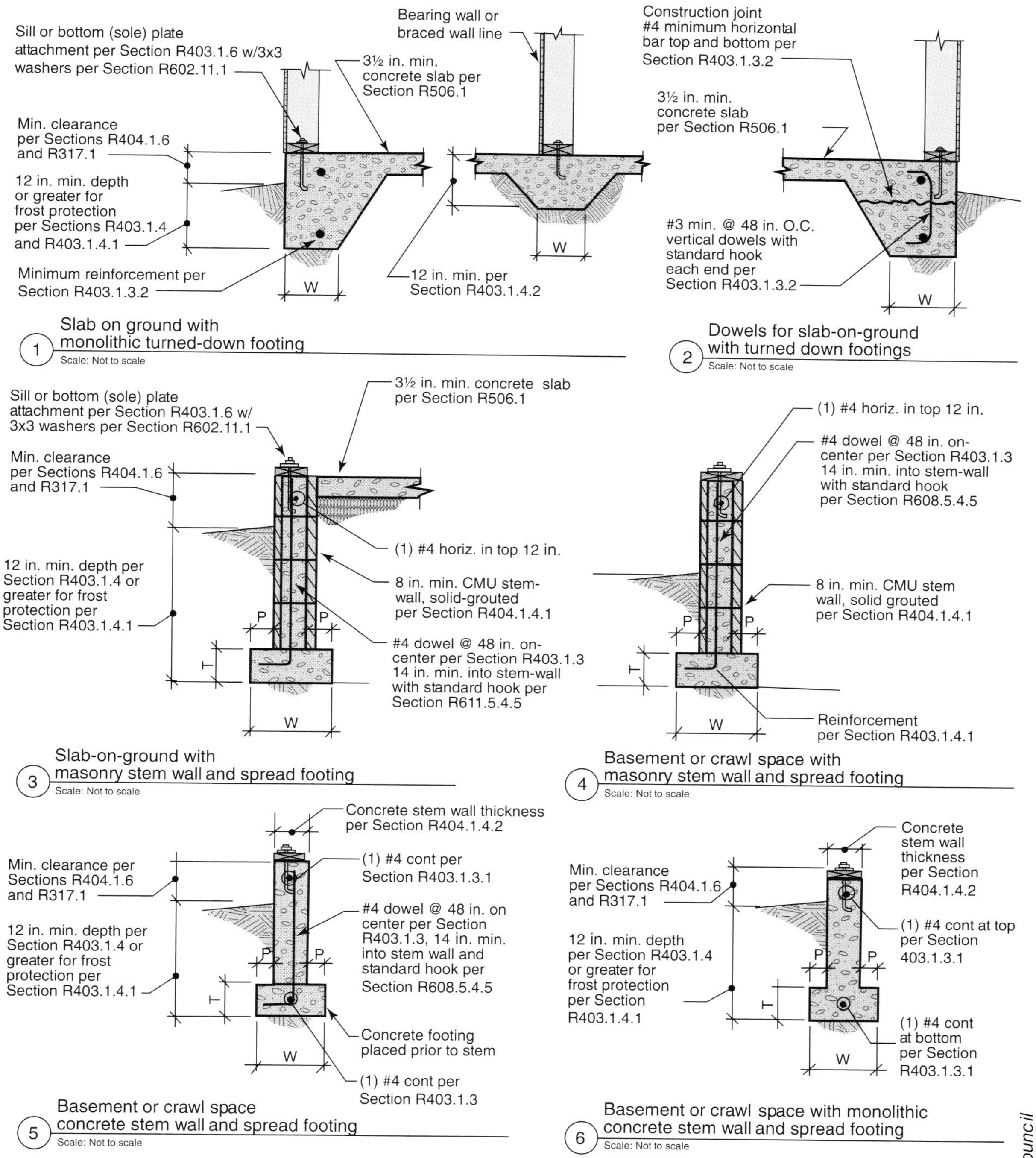

W = Width of Footing, T = Thickness of Footing and P = Projection per Section R 403.1.1.

a. See Section R404.3 for sill requirements.
b. See Section R403.1.6 for sill attachment.
c. See Section R506.2.3 for vapor barrier requirements.
d. See Section R403.1 for base.
e. See Section R408 for under-floor ventilation and access requirements.
f. See Section R403.1.3.5 for reinforcement requirements.

Figure R403.1.3 Reinforced Concrete Footings and Masonry and Concrete Stemwalls in SDC D_0, D_1 and D_2[a,b,c,d,e,f]

R403.1.3 continues

R403.1.3 continued

R403.1.3.5.4 Lap Splices. Vertical and horizontal reinforcement shall be the longest lengths practical. Where splices are necessary in reinforcement, the length of lap splice shall be in accordance with Table R608.5.4(1) and Figure R608.5.4(1). The maximum gap between noncontact parallel bars at a lap splice shall not exceed the smaller of one-fifth the required lap length and 6 inches (152 mm). See Figure R608.5.4(1).

R403.1.3.6 Isolated Concrete Footings. In detached one- and two-family dwellings which are three stories or less in height and constructed with stud bearing walls, isolated plain concrete footings, supporting columns, or pedestals are permitted.

CHANGE SIGNIFICANCE: Revisions of the title and language in Section R403.1.3 are done for clarity. A note is added that references new Section R403.1.3.5 clarifying material and installation requirements for reinforcement. Existing language describing concrete stem walls and masonry stem walls on concrete footings is separated into two sections: Section R403.1.3.1, Concrete stem walls, and Section R403.1.3.2, Masonry stem walls.

Section R403.1.3.3 for slabs on ground is updated to clarify that the section addresses turned-down footings cast monolithically with the slab. Reinforcement installed in the middle third of the footing is moved into the section instead of being an exception.

Section R403.1.3.5.3 Support and Cover is a new section, moved from 2012 IRC Section R404.1.2.3.7.4, covering all concrete clearance and support.

The footing figures in Section R403.1 are revised and updated. The graphic quality of the figures is improved and additional information helpful to the code user is added. For the first time, a set of figures shows minimum footing size and applicable reinforcement requirements for SDC D_0, D_1, and D_2 in Figure R403.1.3.

CHANGE TYPE: Modification

CHANGE SUMMARY: Anchor bolts are now required to be placed in the middle third of the sill plate. Approved anchors may be used instead of ½-inch anchor bolts.

R403.1.6

Foundation Anchorage

2015 CODE: R403.1.6 Foundation Anchorage. Wood sill ~~Sill~~ plates and wood walls supported directly on continuous foundations shall be anchored to the foundation in accordance with this section.

Cold-formed steel framing shall be anchored directly to the foundation or fastened to wood sill plates anchored to the foundation. Anchorage of cold-formed steel framing and sill plates supporting cold-formed steel framing shall be in accordance with this section and Sections R505.3.1 or R603.3.1.

Wood sole plates at all exterior walls on monolithic slabs, wood sole plates of braced wall panels at building interiors on monolithic slabs and all wood sill plates shall be anchored to the foundation with minimum ½ inch (12.7 mm) diameter anchor bolts spaced a maximum of 6 feet (1829 mm) on center or *approved* anchors or anchor straps spaced as required to provide equivalent anchorage to ½-inch-diameter (12.7 mm) anchor bolts. Bolts shall ~~be at least ½ inch (12.7 mm) in diameter and shall~~ extend a minimum of 7 inches (178 mm) into concrete or grouted cells of concrete masonry units. The bolts shall be located in the middle third of the width of the plate. A nut and washer shall be tightened on each anchor bolt. There shall be a minimum of two bolts per plate section with one bolt located not more than 12 inches (305 mm) or less than seven bolt diameters from each end of the plate section. Interior bearing wall sole plates on monolithic slab foundation that are not part of a braced wall panel

R403.1.6 continues

© International Code Council

Anchor bolt placement - centered in sill plate

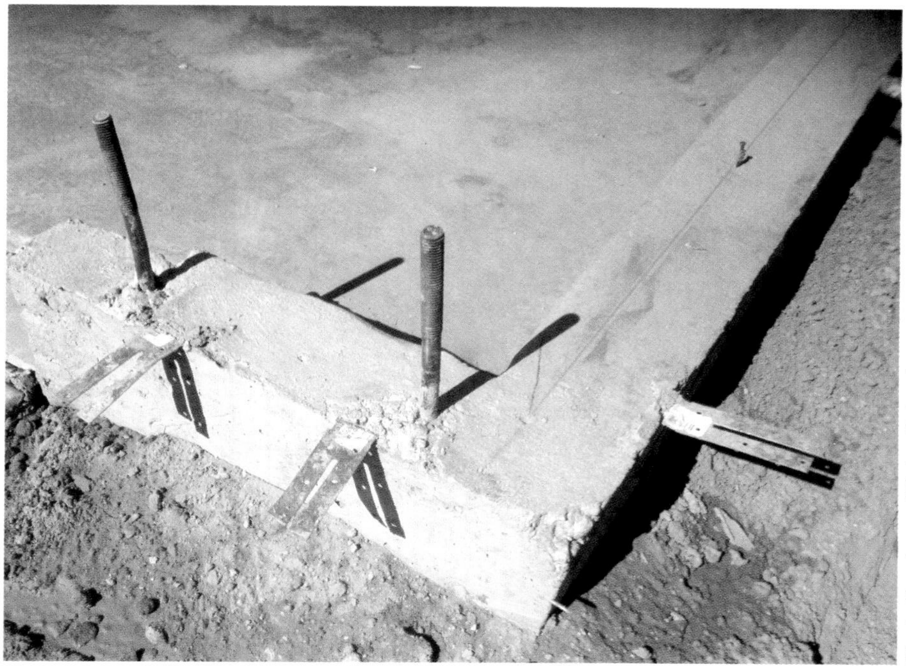

Photo Courtesy of Peter Kulczyk

Alternate anchorage

R403.1.6 continued shall be positively anchored with approved fasteners. Sill plates and sole plates shall be protected against decay and termites where required by Sections R317 and R318. ~~Cold-formed steel framing systems shall be fastened to wood sill plates or anchored directly to the foundation as required in Section R505.3.1 or R603.3.1.~~

Exceptions:

~~1. Foundation anchorage, spaced as required to provide equivalent anchorage to ½-inch-diameter (12.7 mm) anchor bolts.~~

1.~~2.~~ Walls 24 inches (610 mm) total length or shorter connecting offset *braced wall panels* shall be anchored to the foundation with a minimum of one anchor bolt located in the center third of the plate section and shall be attached to adjacent *braced wall panels* at corners, as shown in Item 9 of Table R602.3(1).

2.~~3.~~ Connection of walls 12 inches (305 mm) total length or shorter connecting offset *braced wall panels* to the foundation without anchor bolts shall be permitted. The wall shall be attached to adjacent *braced wall panels* at corners, as shown in Item 9 of Table R602.3(1).

CHANGE SIGNIFICANCE: It is common to see an anchor bolt placed near the edge of a wood sole plate. The general industry standard is for the bolt to be located at least two bolt diameters from the plate's edge, but there have

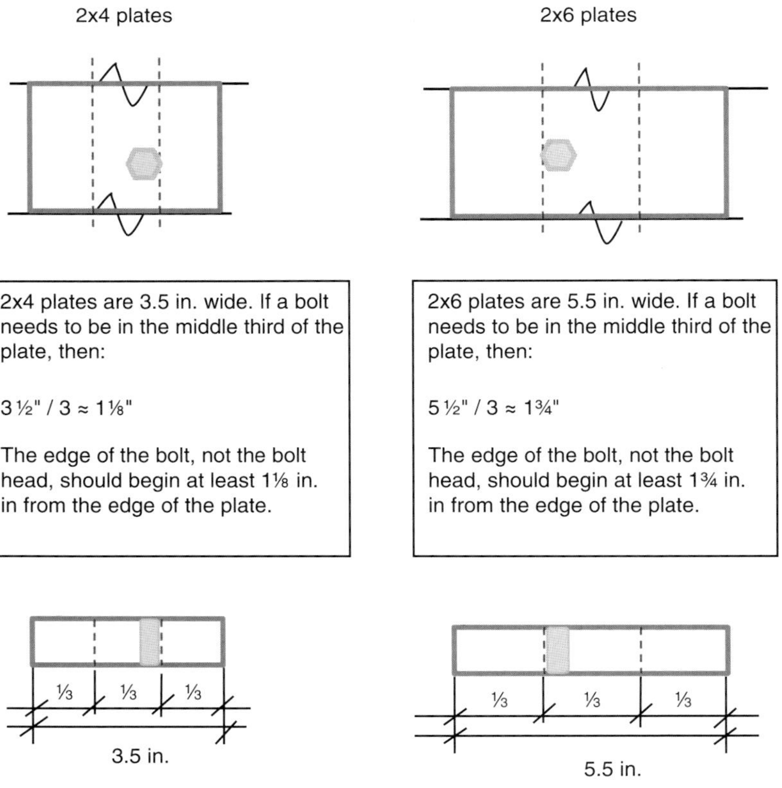
2x4 plates
2x6 plates
2x4 plates are 3.5 in. wide. If a bolt needs to be in the middle third of the plate, then:
3½" / 3 ≈ 1⅛"
The edge of the bolt, not the bolt head, should begin at least 1⅛ in. in from the edge of the plate.
2x6 plates are 5.5 in. wide. If a bolt needs to be in the middle third of the plate, then:
5½" / 3 ≈ 1¾"
The edge of the bolt, not the bolt head, should begin at least 1¾ in. in from the edge of the plate.
⅓
⅓
⅓
3.5 in.
⅓
⅓
⅓
5.5 in.

Minimum edge distance

been no provisions in the IRC to govern edge distance. Requirements of the 2012 IRC included two bolts per plate, within 12 inches of the end of the plate, and spaced no more than 6 feet apart. Adding a requirement for placement of a bolt within the middle third of the wood plate width allows some flexibility while providing for a minimum edge distance.

Testing has demonstrated that a bolt loses anchoring capacity when placed closer than 1¾ inches from the plate's edge. Manufacturers of anchor bolts require a minimum plate edge distance of 1¾ inches in their installation and technical documents. This code change places an anchor bolt at least 1⅛ inches from the edge of a 2 × 4 sill plate. With 2 × 6 construction, the bolt is a minimum of 1¾ inches from the plate edge.

Foundation anchorage requirements for alternate foundation anchor systems providing equivalent capacity to ½-inch anchor bolts spaced at 6 feet on center (or as otherwise required by the code or design) are moved from the exceptions into the main text. Revised language is similar to the 2012 *International Building Code* (IBC) Section 2308.6. The provision allows use of anchors such as foundation anchors (mudsill anchors), wedge anchors, expansion anchors, adhesive anchors, and other alternatives approved by the building official as alternates to cast-in-place anchor bolts within Section R403.1.6.

Anchorage requirements for cold-formed steel framing systems have been separated from the requirements for wood. The new provision points to appropriate cold-formed steel provisions in Chapters 5 (Floors) and 6 (Walls). In addition, language is revised to clarify that both provisions of Section R403.1.6 and the applicable provisions of Section R505.3.1 (for cold-formed steel floor framing) and Section R603.3.1 (for cold-formed steel wall framing) need to be followed. The change adds a pointer to anchor bolt spacing and embedment requirements specific to cold-formed steel.

R404.1.4.1

Masonry Foundation Walls in SDC D_0, D_1, and D_2

Masonry foundation wall

CHANGE TYPE: Modification

CHANGE SUMMARY: Minimum vertical reinforcement in masonry stem walls has been increased from No. 3 bars to No. 4 bars spaced a maximum of 4 feet on center in grouted cells.

2015 CODE: R404.1.4.1 Masonry Foundation Walls. In~~ addition to the requirements of Table R404.1.1(1) plain masonry foundation walls~~ buildings assigned to Seismic Design Category D_0, D_1 or D_2, as established in Table R301.2(1), masonry foundation walls shall comply with this section. In addition to the requirements of Table R404.1.1(1), plain masonry foundation walls shall comply with the following.

1. Wall height shall not exceed 8 feet (2438 mm).
2. Unbalanced backfill height shall not exceed 4 feet (1219 mm).
3. Minimum nominal thickness for plain masonry foundation walls shall be 8 inches (203 mm).
4. Masonry stem walls shall have a minimum vertical reinforcement of one No. ~~3~~ 4 (No. ~~10~~ 13) bar located a maximum of 4 feet (1219 mm) on center in grouted cells. Vertical reinforcement shall be tied to the horizontal reinforcement in the footings.

Foundation walls~~ in buildings assigned to Seismic Design Category D_0, D_1 or D_2, as established in Table R301.2(1)~~, supporting more than 4 feet (1219 mm) of unbalanced backfill or exceeding 8 feet (2438 mm) in height shall be constructed in accordance with Table R404.1.1(2), R404.1.1(3), or R404.1.1(4). Masonry foundation walls shall have two No. 4 (No. 13) horizontal bars located in the upper 12 inches (305 mm) of the wall.

CHANGE SIGNIFICANCE: This code change clarifies requirements for masonry and concrete foundation walls by following the same format in each section. For masonry, the minimum vertical reinforcement is increased from one No. 3 bar to one No. 4 bar for seismic reinforcement in SDC D_0, D_1, and D_2.

In Section 1.18.4.4 of TMS 402/ACI 530/ASCE 5, *Building Code Requirements for Masonry Structures and Commentary*, the adopted standard for masonry design, vertical reinforcement is required to be at least a No. 4 bar spaced a maximum of 48 inches on center. The IRC now agrees with the standard referenced throughout the masonry sections of the IRC.

Table R404.1.1(1) is updated to differentiate between solid and grouted masonry in plain masonry walls for minimum wall thickness. In general, with good soils, minimum wall thickness for plain masonry walls with solid masonry is 6 inches and minimum thickness for grouted masonry is 8 inches. With poor soils—loosely compacted soils made of clays, silts, and organics—minimum wall thickness can be as large as 12 inches. See 2015 IRC Table R404.1.1(1).

R404.4

Retaining Walls

CHANGE TYPE: Modification

CHANGE SUMMARY: Retaining walls, freestanding walls not supported at the top, with more than 48 inches of unbalanced backfill must be designed by an engineer. Retaining walls resisting additional lateral loads and with more than 24 inches of unbalanced backfill must also be designed in accordance with accepted engineering practice.

2015 CODE: **R404.4 Retaining Walls.** Retaining walls that are not laterally supported at the top and that retain in excess of ~~24~~ 48 inches (~~610~~ 1219 mm) of unbalanced fill, or retaining walls exceeding 24 inches (610 mm) in height that resist lateral loads in addition to soil, shall be designed in accordance with accepted engineering practice to ensure stability against overturning, sliding, excessive foundation pressure, and water uplift. Retaining walls shall be designed for a safety factor of 1.5 against lateral sliding and overturning. This section shall not apply to foundation walls supporting buildings.

CHANGE SIGNIFICANCE: The type of wall addressed in Section R404.4 is a detached retaining wall of concrete or hollow, grouted or solid masonry, not supported at the top and laterally supported at the bottom against sliding and overturning by a footing covered by soil. The wall would typically be a site retaining wall primarily resisting lateral soil loads. When the wall must resist additional loads, such as vehicles parked above or fences built on top of the wall that are subject to wind loads, a wall with more than 24 inches of unbalanced backfill must be designed in accordance with accepted engineering practice.

Section R404.4 also has a new trigger height of 48 inches (previously 24 inches) for unbalanced backfill to be consistent with Section R404.1.3. This section specifically requires that concrete or masonry foundation walls supporting more than 48 inches of unbalanced fill and not laterally supported must have an engineered design.

The definition of a retaining wall within the provision is modified to clarify that this type of wall is not intended to support structural loads. A similar wall that does support structural loads is addressed by other sections.

Photo Courtesy of Peter Kulczyk

Retaining wall

Tables R502.3.1(1), R502.3.1(2)

Floor Joist Spans for Common Lumber Species

CHANGE TYPE: Modification

CHANGE SUMMARY: Changes to Southern Pine (SP), Douglas Fir-Larch (DFL), and Hemlock Fir (HF) lumber capacities have changed the floor joist span length in the prescriptive tables of the IRC. Span lengths for Southern Pine have decreased; lengths for DFL and HF joists have increased.

2015 CODE: Tables R502.3.1(1) and R502.3.1(2)

TABLE R502.3.1(1) Floor Joist Spans for Common Lumber Species (Residential sleeping areas, live load = 30 psf, L/Δ = 360)[a]

Joist Spacing (inches)	Species and Grade		Dead Load = 10 psf				Dead Load = 20 psf			
			2 × 6	2 × 8	2 × 10	2 × 12	2 × 6	2 × 8	2 × 10	2 × 12
			Maximum floor joist spans							
			(ft - in.)	(ft - in.)	(ft - in.)	(ft - in.)	(ft - in.)	(ft - in.)	(ft - in.)	(ft - in.)
12	Douglas fir-larch	SS	12-6	16-6	21-0	25-7	12-6	16-6	21-0	25-7
	Douglas fir-larch	#1	12-0	15-10	20-3	24-8	12-0	15-7	19-0	22-0
	Douglas fir-larch	#2	11-10	15-7	19-10	23-4	11-8	14-9	18-0	20-11
	Douglas fir-larch	#3	9-11	12-7	15-5	17-10	8-11	11-3	13-9	16-0
	Hem-fir	SS	11-10	15-7	19-10	24-2	11-10	15-7	19-10	24-2
	Hem-fir	#1	11-7	15-3	19-5	23-7	11-7	15-3	18-9	21-9
	Hem-fir	#2	11-0	14-6	18-6	22-6	11-0	14-4	17-6	20-4
	Hem-fir	#3	9-8	12-4	15-0	17-5	8-8	11-0	13-5	15-7
	Southern pine	SS	12-3	16-2	20-8	25-1	12-3	16-2	20-8	25-1
	Southern pine	#1	11-10	15-7	19-10	24-2	11-10	15-7	18-7	22-0
	Southern pine	#2	11-3	14-11	18-1	21-4	10-9	13-8	16-2	19-1
	Southern pine	#3	9-2	11-6	14-0	16-6	8-2	10-3	12-6	14-9
	Spruce-pine-fir	SS	11-7	15-3	19-5	23-7	11-7	15-3	19-5	23-7
	Spruce-pine-fir	#1	11-3	14-11	19-0	23-0	11-3	14-7	17-9	20-7
	Spruce-pine-fir	#2	11-3	14-11	19-0	23-0	11-3	14-7	17-9	20-7
	Spruce-pine-fir	#3	9-8	12-4	15-0	17-5	8-8	11-0	13-5	15-7

For SI: 1 inch = 25.4 mm, 1 foot = 304.8 mm, 1 pound per square foot = 0.0479 kPa.

Note: Check sources for availability of lumber in lengths greater than 20 feet.

a. Dead load limits for townhouses in Seismic Design Category C and all structures in Seismic Design Categories D_0, D_1, and D_2 shall be determined in accordance with Section R301.2.2.2.1.

(Portions of table not show for brevity and clarity.)

Lumber floor joists

Example—Floor Spans

#1 Bedroom

Dead load = 10 psf
2×10 joists
16″ o.c. spacing
Southern Pine (SP) #2

Maximum Span Allowed	2012	2015
	18′-0″	15′-8″

The SP #2 span length is significantly reduced from the 2012 IRC span length.

Note: An SP #1 joist will span about the same length in the 2015 IRC Table R502.3.1(1) or R502.3.1(2) as the SP #2 did in the tables in the 2012 IRC.

#2 Bathroom

Dead load = 20 psf
2×8 joists
16″ o.c. spacing
Douglas Fir-Larch (DFL) #2

Maximum Span Allowed	2012	2015
	11′-6″	11′-8″

The span has increased about 2 inches which is the typical increase in the table. Some cells for Douglas Fir and Hemlock have not changed. Others increased by 1-2 inches.

Floor joist span examples

R502.3.1(1), R502.3.1(2) continues

R502.3.1(1), R502.3.1(2) continued

CHANGE SIGNIFICANCE: New design values exist for Southern Pine lumber. These design values for all widths and grades of visually graded Southern Pine lumber became effective on June 1, 2013. The American Lumber Standards Committee (ALSC) approved the new design values as published in Southern Pine Inspection Bureau *Supplement No. 13* to the *2002 Standard Grading Rules for Southern Pine Lumber.* Values are a result of two years of testing current lumber available on the market to identify what changes had occurred in the strength of the Southern Pine lumber inventory.

Meanwhile, for Douglas Fir-Larch and Hemlock Fir, testing done in the 1990s slightly increased design values for bending. Revised design values for Select Structural, #2 and #3 grades of Douglas Fir-Larch, and #1 grade of Hemlock Fir increased by 25 psi. Testing to check current stock has validated the design values set in the 1990s. Although these values were updated in the wood standards, span tables incorporated into the 2000 *International Building Code* (IBC) and 2000 IRC were based on span tables predating the revised design values from the 1990s.

The 2015 IRC span tables will now be in agreement with the wood standards' span tables with the revisions for Southern Pine, Douglas Fir-Larch, and Hemlock Fir. The new design values apply only to new construction. The integrity of existing structures designed and built using design values meeting the applicable building codes in effect at the time of permitting is not a concern.

For Southern Pine, the changes reflect shorter spans. For Douglas Fir-Larch and Hemlock Fir, the changes result in slightly longer spans.

R502.10

Framing of Floor Openings

CHANGE TYPE: Modification

CHANGE SUMMARY: Requirements for header joist and trimmer connections in the framing of floor openings have been deleted. This section conflicted with Section R502.6, which contains minimum bearing lengths for all joists and headers.

2015 CODE: R502.10 Framing of Openings. Openings in floor framing shall be framed with a header and trimmer joists. ~~When~~ Where the header joist span does not exceed 4 feet (1219 mm), the header joist may be a single member the same size as the floor joist. Single trimmer joists may be used to carry a single header joist that is located within 3 feet (914 mm) of the trimmer joist bearing. When the header joist span exceeds 4 feet (1219 mm), the trimmer joists and the header joist shall be doubled and of sufficient cross section to support the floor joists framing into the header. ~~Approved hangers shall be used for the header joist to trimmer joist connections when the header joist span exceeds 6 feet (1829 mm). Tail joists over 12 feet (3658 mm) long shall be supported at the header by framing anchors or on ledger strips not less than 2 inches by 2 inches (51 mm by 51 mm).~~

CHANGE SIGNIFICANCE: There was conflicting language in the 2012 IRC regarding support of framing members at floor openings. Section R502.10 required that header joists be provided with approved

R502.10 continues

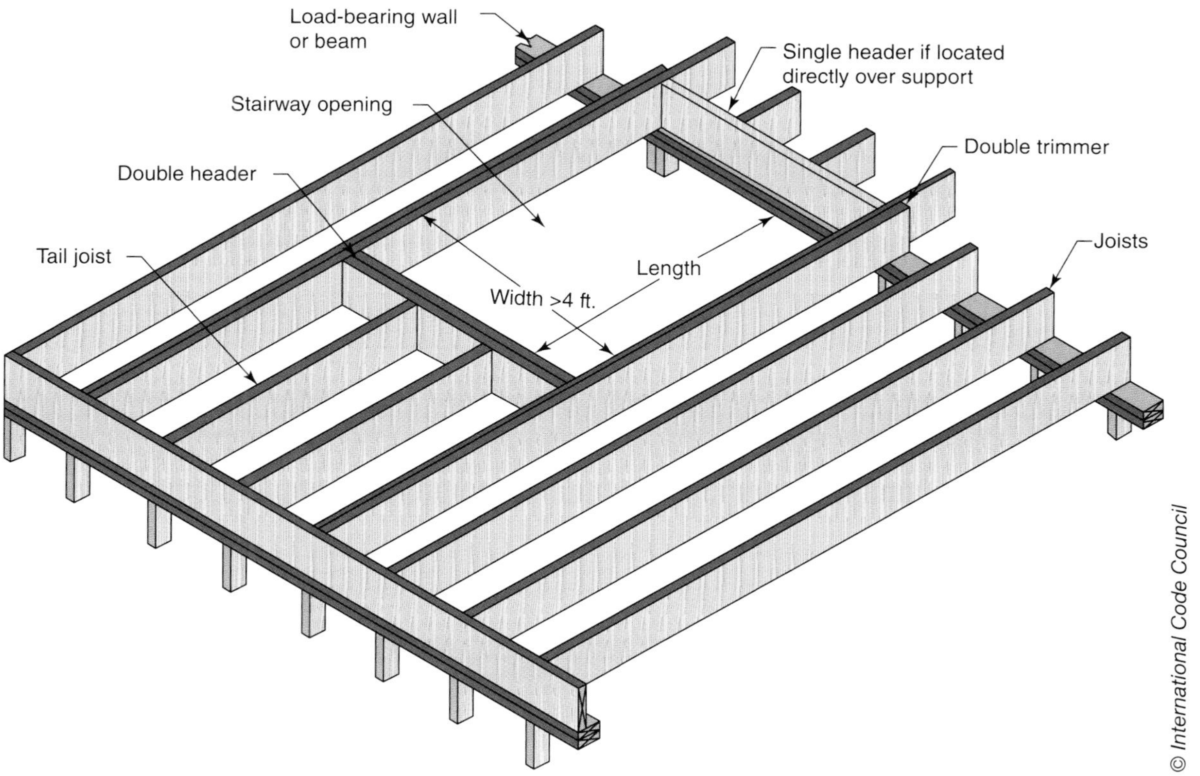

Opening in floor framing

R502.10 continued

hangers only when they exceeded 6 feet in length and that joists be supported on framing anchors or ledger strips only when they were over 12 feet long.

Section R502.6 requires all joists, beams, and girders to have not less than 1.5 inches of bearing regardless of length. Applying 2012 IRC language literally, a 10-foot-long joist framed into a stairway opening at one end and into the face of a beam at the other would require a joist hanger where it connects to the beam but not at the stairway header. The loads are assumed to be distributed evenly along the joist. After deleting the language in Section R502.10, framing at openings now has the same bearing requirements as other joists per R502.6.

R507.1, R507.4

Decking

CHANGE TYPE: Modification

CHANGE SUMMARY: The code sets the maximum allowable spacing for deck joists supporting the various types of common decking materials.

2015 CODE: R507.1 Decks. Wood-framed decks shall be in accordance with this section or Section R301 for materials and conditions not prescribed herein. Where supported by attachment to an exterior wall, decks shall be positively anchored to the primary structure and designed for both vertical and lateral loads.

Such attachment shall not be accomplished by the use of toenails or nails subject to withdrawal. Where positive connection to the primary building structure cannot be verified during inspection, decks shall be self-supporting. For decks with cantilevered framing members, connections to exterior walls or other framing members shall be designed and constructed to resist uplift resulting from the full live load specified in Table R301.5 acting on the cantilevered portion of the deck.

R507.4 Decking. Maximum allowable spacing for joists supporting decking shall be in accordance with Table R507.4. Wood decking shall be attached to each supporting member with not less than (2) 8d threaded nails or (2) No. 8 wood screws.

TABLE R507.4 Maximum joist spacing

Material type and nominal size	Maximum on-center joist spacing	
	Perpendicular to joist	Diagonal to joist[a]
$1^1/_4$-inch thick wood	16 inches	12 inches
2-inch thick wood	24 inches	16 inches
Plastic composite	In accordance with Section R507.3	In accordance with Section R507.3

For SI: 1 inch = 25.4 mm, 1 foot = 304.8 mm, 1 degree = 0.01745 rad.

a. Maximum angle of 45 degrees from perpendicular for wood deck boards

R507.1, R507.4 continues

Decking

R507.1, R507.4 continued

CHANGE SIGNIFICANCE: The new Table R507.4 sets the maximum joist spacing for support of decking materials. The spacing is based on the type and thickness of the decking material and its orientation to the joist. Decking placed diagonally to the direction of the joists must span a greater distance than decking installed perpendicular to the joists. Therefore, a diagonal installation requires reduced spacing of the supports. The joist spacing values reflect current construction conventions and recommended best practices. The new table mirrors the organization and format of Table R503.1, Minimum Thickness of Lumber Floor Sheathing Based on the Support Spacing. However, the new spacing values for support of decking are based on typical decking materials which perform satisfactorily in deck construction and match current construction practices.

Lumber decking with a 2-inch nominal thickness allows a joist spacing of 24 inches on center when applied perpendicular to the supports and 16 inches on center when applied diagonally. For nominal 1¼-inch wood decking, the spacing is reduced to 16 inches and 12 inches, respectively. Plastic composite decking must comply with the requirements of ASTM D7032 and be installed in accordance with the manufacturer's instructions, as prescribed in Section R507.3.

R507.2

Deck Ledger Connection to Band Joist

CHANGE TYPE: Clarification

CHANGE SUMMARY: The deck ledger section is reorganized to better describe the minimum requirements for connection of deck ledgers to band joists.

2015 CODE: R507.2 Deck Ledger Connection to Band Joist. ~~For decks supporting a total design load of 50 pounds per square foot (2394 Pa) [40 pounds per square foot (1915 Pa) live load plus 10 pounds per square foot (479 Pa) dead load], the connection between a deck ledger of pressure-preservative-treated Southern Pine, incised pressure-preservative-treated Hem-Fir, or *approved* decay-resistant species, and a 2-inch (51 mm) nominal lumber band joist bearing on a sill plate or wall plate shall be constructed with ½-inch (12.7 mm) lag screws or bolts with washers in accordance with Table R507.2. Lag screws, bolts and washers shall be hot-dipped galvanized or stainless steel.~~ Deck ledger connections to band joists shall be in accordance with this section, Tables R507.2 and R507.2.1, and Figures R507.2.1(1) and R507.2.1(2). For other grades, species, connection details and loading conditions, deck ledger connections shall be designed in accordance with Section R301.

~~**R507.2.1 Placement of lag screws or bolts in deck ledgers and band joists.** The lag screws or bolts in deck ledgers and band joists shall be placed in accordance with Table R507.2.1 and Figures R507.2.1(1) and R507.2.1(2).~~

R507.2.1 Ledger Details. Deck ledgers installed in accordance with Section R507.2 shall be a minimum 2-inch by 8-inch (51 mm by 203 mm) nominal, pressure-preservative-treated Southern Pine, incised pressure-preservative-treated Hem-fir, or approved, naturally durable, No. 2 grade or better lumber. Deck ledgers installed in accordance with Section R507.2 shall not support concentrated loads from beams or girders. Deck ledgers shall not be supported on stone or masonry veneer.

R507.2 continues

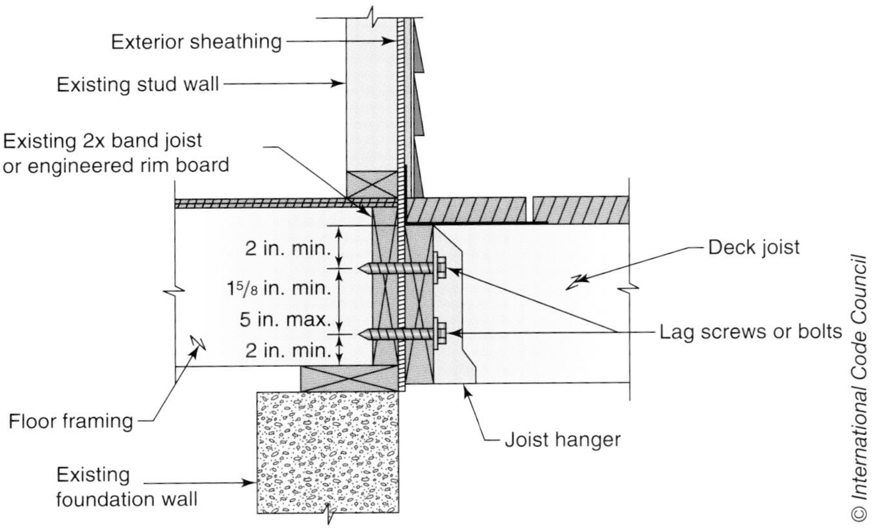

Deck ledger connection to band joist

R507.2 continued

~~**R507.2.2 Alternate Deck Ledger Connections.** Deck ledger connections not conforming to Table R507.2 shall be designed in accordance with accepted engineering practice. Girders supporting deck joists shall not be supported on deck ledgers or band joists. Deck ledgers shall not be supported on stone or masonry veneer.~~

R507.2.2 Band Joist Details. Band joists attached by a ledger in accordance with Section R507.2 shall be a minimum 2-inch-nominal (51 mm), solid-sawn, spruce-pine-fir lumber or a minimum 1-inch by 9½-inch (25 mm by 241 mm) dimensional, Douglas Fir, laminated veneer lumber. Band joists attached to a ledger in accordance with Section R507.2 shall be fully supported by a wall or sill plate below.

R507.2.3 Ledger to Band Joist Fastener Details. Fasteners used in deck ledger connections in accordance with Table R507.2 shall be hot-dipped galvanized or stainless steel and shall be installed in accordance with Table R507.2.1 and Figures R507.2.1(1) and R507.2.1(2).

TABLE R507.2 ~~**Fastener Spacing for a Southern Pine or Hem-fir Deck Ledger and A 2-Inch-Nominal Solid-Sawn Spruce-pine-fir Band Joist**~~~~[c, f, g]~~ **Deck Ledger Connection to Band Joist**[a, b] **(Deck live load = 40 psf, deck dead load = 10 psf, snow load ≤ 40 psf)**

	Joist Span						
	6′ and less	6′1″ to 8′	8′1″ to 10′	10′1″ to 12′	12′1″ to 14′	14′1″ to 16′	16′1″ to 18′
Connection Details	On-center spacing of fasteners[d,~~e~~]						
½ inch diameter lag screw with ~~15/32~~ ½ inch maximum sheathing [c,d]	30	23	18	15	13	11	10
½ inch diameter bolt with ~~15/32~~ ½ inch maximum sheathing[d]	36	36	34	29	24	21	19
½ inch diameter bolt with ~~15/32~~ 1 inch maximum sheathing ~~and ½ inch washers~~~~[b, h]~~ [e]	36	36	29	24	21	18	16

For SI: 1 inch = 25.4 mm, 1 foot = 304.8 mm. 1 pound per square foot = 0.0479 kPa.

a. Ledgers shall be flashed in accordance with Section R703.8 to prevent water from contacting the house band joist.
b. Snow load shall not be assumed to act concurrently with live load.
c. The tip of the lag screw shall fully extend beyond the inside face of the band joist.
d. Sheathing shall be wood structural panel or solid sawn lumber.
e. Sheathing shall be permitted to be wood structural panel, gypsum board, fiberboard, lumber, or foam sheathing. Up to ½-inch thickness of stacked washers shall be permitted to substitute for up to ½ inch of allowable sheathing thickness when combined with wood structural panel or lumber sheathing. ~~The maximum gap between the face of the ledger board and face of the wall sheathing shall be ½ inch.~~

~~b. Lag screws and bolts shall be staggered in accordance with Section R507.2.1~~
~~c. Deck ledger shall be minimum 2 × 8 pressure-preservative-treated No. 2 grade lumber, or other approved materials as established by standard engineering practice.~~
~~d. When solid-sawn pressure-preservative-treated deck ledgers are attached to a minimum 1-inch-thick engineered wood product (structural composite lumber, laminated veneer lumber or wood structural panel band joist), the ledger attachment shall be designed in accordance with accepted engineering practice.~~
~~f. A minimum 1 × 9 ½ Douglas Fir laminated veneer lumber rim board shall be permitted in lieu of the 2-inch nominal band joist.~~

CHANGE SIGNIFICANCE: Section R507.2 addressing the prescriptive method for connecting a deck ledger to the band joist (rim board) has been reorganized to clarify the requirements. Redundant language has been removed, prescriptive options have been moved from the table footnotes to the section text, and language describing the approved materials has been revised to provide consistency. The 2015 IRC Section R507.2 adds the IRC-defined term "naturally durable lumber" to the materials allowed for a deck ledger connection using the prescriptive provisions. In the 2012 IRC, the description of allowable species for ledger material was not consistent between the section text, the table title, and the table footnotes. Section R507.2 referred to decay-resistant properties of pressure-preservative-treated pine or hem-fir, and then continued with a reference to "approved decay-resistant species," leaving it to the building official to decide whether pine and hem-fir were approved. The heading of Table R507.2, however, referred only to pine and hem-fir and not the use of decay-resistant species. Lastly, table footnotes e and f referenced use of any pressure-preservative-treated, #2 grade lumber species or use of engineered lumber.

The 2012 IRC text required a nominal 2-inch band joist in Table R507.2. Although code users recognized that 2 inches was intended as a minimum dimension, thicker band joists were not specifically addressed. In the 2015 IRC, the term "minimum" is moved in front of the size description. The sheathing thickness of $^{15}/_{32}$ inch is updated to ½ inch to accommodate the thickness of common foam plastic sheathing.

Table R507.2 first appeared in the 2009 IRC to provide an easy-to-follow prescriptive means for attaching a deck to a dwelling. Other methods may still be used, and often are, to provide equivalent connection capacities, as long as the method is approved by the building official. For example, proprietary fasteners are commonly installed following the manufacturer's instructions and based on equivalent capacities. Testing to develop the prescriptive method in Table R507.2 was performed with three configurations:

1. ½ inch lag screw with $^{15}/_{32}$ inch OSB between the ledger and the band joist
2. ½ inch bolt with $^{15}/_{32}$ inch OSB between the ledger and the band joist
3. ½ inch bolt with ½ inch stack of washers and $^{15}/_{32}$ inch OSB between the ledger and the band joist

These three cases correspond to the three rows of the ledger table. Based on testing, for the first two configurations, the ledger, OSB, and band joist must be in direct contact with one another. For the third configuration, an additional gap filled by the washers is permitted between the ledger and the band joist. Minor changes have occurred to the table in the past two code cycles.

Note that the terms "band joist" and "rim board" are used synonymously and are interchangeable in this significant change and throughout the book.

R507.2.4

Alternative Deck Lateral Load Connection

CHANGE TYPE: Modification

CHANGE SUMMARY: When the prescriptive deck lateral load connection that has appeared in the previous editions of the code is chosen as a design option, the code now requires the two hold-down devices to be within 2 feet of the ends of the deck. A new lateral load connection option prescribes four hold-downs installed below the deck structure.

2015 CODE: **~~R507.2.3~~R507.2.4 Deck Lateral Load Connection.** The lateral load connection required by Section R507.1 shall be permitted to be in accordance with Figure R507.2.3(1) or R507.2.3(2). Where the lateral load connection is provided in accordance with Figure R507.2.3(1), hold-down tension devices shall be installed in not less than two locations per deck within 24 inches of each end of the deck. Each device shall have an allowable stress design capacity of not less than 1500 pounds (6672 N). Where the lateral load connections are provided in accordance with Figure R507.2.3(2), the hold-down tension devices shall be installed in not less than four locations per deck, and each device shall have an allowable stress design capacity of not less than 750 pounds (3336 N).

CHANGE SIGNIFICANCE: The prescriptive deck lateral load connections have frequently been misunderstood. Initially, a figure was added to the 2009 IRC to depict one possible connection of a deck structure to a dwelling to resist lateral loads. Since that edition, many designers and jurisdictions have mistakenly thought the connection was required. The connection comes from FEMA 232, the *Homebuilder's Guide to Earthquake Resistant Design and Construction*, and is a connection suitable to resist lateral loads during an earthquake in high seismic regions.

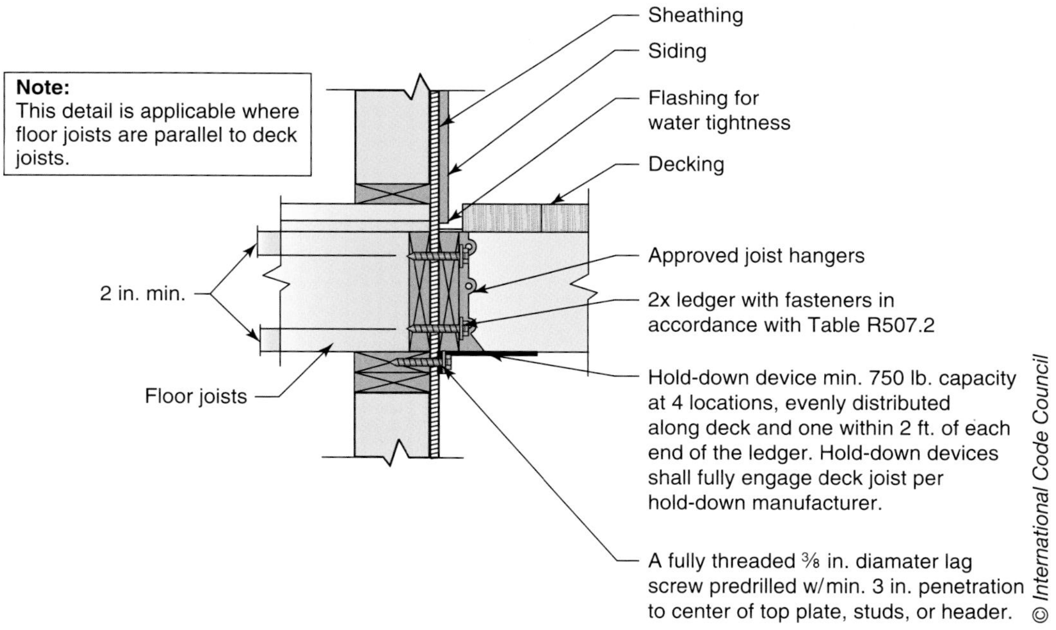

Alternative deck attachment for lateral loads

Testing had not been done to determine an appropriate connection for lateral loads in low-seismic regions or a connection suitable for easy construction in existing dwellings.

In the hearings for the 2015 IRC, a number of proposals were brought forward, including those related to lateral load connection options. This code change combines proposals accepted by the IRC committee and those modified during public comment hearings.

The connection using hold-downs to tie the deck joists at two locations to floor joists in a dwelling in Figure R507.2.3(1) remains one potential solution. New Figure R507.2.3(2) offers a second solution for lateral load connection of the deck to the dwelling. Four angles connect the base of the deck to the house structure. The angles are fastened to the bottom of the deck joist and a 3/8-inch lag screw is installed through the ledger and into blocking or the band joist (rim board) on the backside of the exterior wall sheathing. This solution is another prescriptive option that is not required but is a method to resist lateral loads by applying angles underneath the deck so that ceiling finishes and flooring materials in an existing house do not need removal as would be required by the first option.

The method with angles arises from the common occurrence of constructing decks for existing houses. Use of hold-downs applied to floor joists requires removal of finish ceiling materials when a deck is added after initial construction. In addition, the previous detail required additional nailing through the dwelling floor sheathing into the floor joist to improve the lateral load capacity, necessitating the removal of floor finishes as well as ceiling finishes.

R507.5, R507.6, R507.7

Deck Joists and Beams

CHANGE TYPE: Addition

CHANGE SUMMARY: New sections and tables provide prescriptive methods for joists and beams in deck construction. Section R507.5 describes requirements for deck joists, Section R507.6 lists requirements for deck beams, and Section R507.7 describes minimum bearing requirements for joists and beams.

2015 CODE: **R507.5 Deck Joists.** Maximum allowable spans for wood deck joists, as shown in Figure R507.5, shall be in accordance with Table R507.5. Deck joists shall be permitted to cantilever a maximum of one-fourth of the actual, adjacent joist span.

R507.5.1 Lateral Restraint at Supports. Joist ends and bearing locations shall be provided with lateral restraint to prevent rotation. Where lateral restraint is provided by joist hangers or blocking between joists, their depth shall equal not less than 60 percent of the joist depth. Where lateral restraint is provided by rim joists, they shall be secured to the end of each joist with not less than (3) 10d (3 inch by 0.128 inch) (76 mm by 3 mm) nails or (3) No. 10 by 3 inch (76 mm) long wood screws.

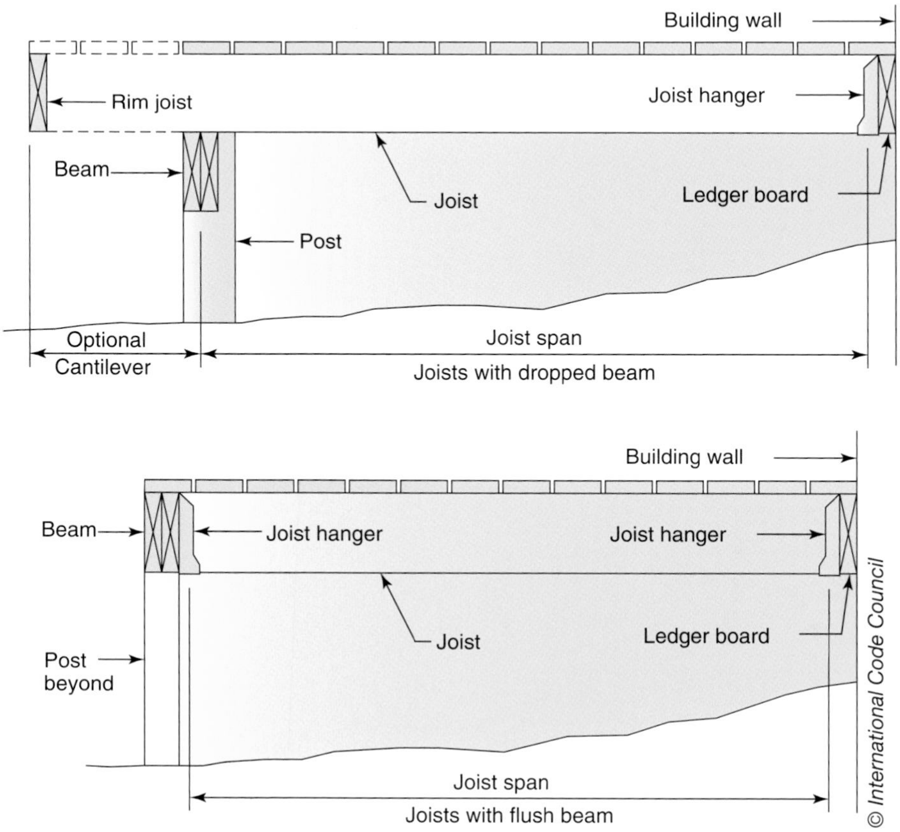

Typical deck joist spans

TABLE R507.5 Deck Joist Spans for Common Lumber Species[f] (ft.-in.)

Species[a]	Size	Spacing of Deck Joists With No Cantilever[b] (inches) 12	16	24	Spacing of Deck Joists With Cantilevers[c] (inches) 12	16	24
Southern pine	2 × 6	9-11	9-0	7-7	6-8	6-8	6-8
	2 × 8	13-1	11-10	9-8	10-1	10-1	9-8
	2 × 10	16-2	14-0	11-5	14-6	14-0	11-5
	2 × 12	18-0	16-6	13-6	18-0	16-6	13-6
Douglas fir-larch[d], hem-fir[d] spruce-pine-fir[d]	2 × 6	9-6	8-8	7-2	6-3	6-3	6-3
	2 × 8	12-6	11-1	9-1	9-5	9-5	9-1
	2 × 10	15-8	13-7	11-1	13-7	13-7	11-1
	2 × 12	18-0	15-9	12-10	18-0	15-9	12-10
Redwood, western cedars, ponderosa pine[e], red pine[e]	2 × 6	8-10	8-0	7-0	5-7	5-7	5-7
	2 × 8	11-8	10-7	8-8	8-6	8-6	8-6
	2 × 10	14-11	13-0	10-7	12-3	12-3	10-7
	2 × 12	17-5	15-1	12-4	16-5	15-1	12

For SI: 1 inch = 25.4 mm, 1 foot = 304.8 mm, 1 pound per square foot = 0.0479 kPa.

a. No. 2 grade with wet service factor.
b. Ground snow load, live load = 40 psf, dead load = 10 psf, L/Δ = 360.
c. Ground snow load, live load = 40 psf, dead load = 10 psf, L/Δ = 360 at main span, L/Δ = 180 at cantilever with a 220-pound point load applied to end.
d. Includes incising factor.
e. Northern species with no incising factor
f. Cantilevered spans not exceeding the nominal depth of the joist are permitted.

R507.5, R507.6, R507.7 continues

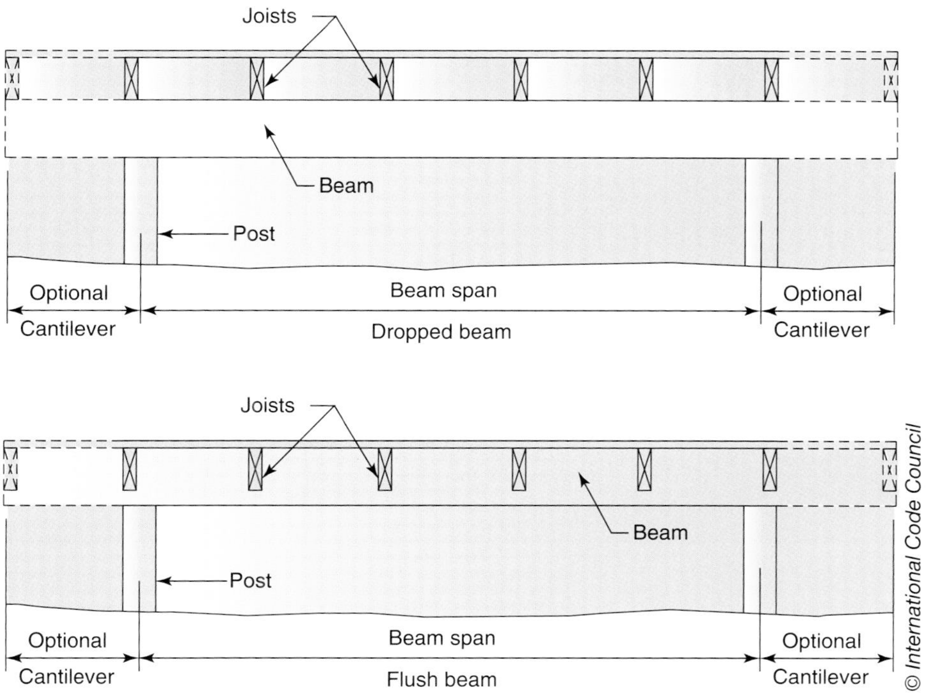

Typical deck beam spans

R507.5, R507.6, R507.7 continued

R507.6 Deck Beams. Maximum allowable spans for wood deck beams, as shown in Figure R507.6, shall be in accordance with Table R507.6. Beam plies shall be fastened with two rows of 10d (3 inch by 0.128 inch) (76 mm by 3 mm) nails minimum at 16 inches (406 mm) on center along each edge. Beams shall be permitted to cantilever at each end up to one-fourth of the actual beam span. Splices of multi-span beams shall be located at interior post locations.

R507.7 Deck Joist and Deck Beam Bearing. The ends of each joist and beam shall have not less than 1½ inches (38 mm) of bearing on wood or metal and not less than 3 inches (76 mm) on concrete or masonry for the entire width of the beam. Joist framing into the side of a ledger board or beam shall be supported by approved joist hangers. Joists bearing on a beam shall be connected to the beam to resist lateral displacement.

TABLE R507.6 Deck Beam Span Lengths[a, b] (ft.-in.)

		Deck Joist Span Less Than or Equal to: (feet)						
Species[c]	**Size[d]**	**6**	**8**	**10**	**12**	**14**	**16**	**18**
Southern pine	2 – 2 × 6	6-11	5-11	5-4	4-10	4-6	4-3	4-0
	2 – 2 × 8	8-9	7-7	6-9	6-2	5-9	5-4	5-0
	2 – 2 × 10	10-4	9-0	8-0	7-4	6-9	6-4	6-0
	2 – 2 × 12	12-2	10-7	9-5	8-7	8-0	7-6	7-0
	3 – 2 × 6	8-2	7-5	6-8	6-1	5-8	5-3	5-0
	3 – 2 × 8	10-10	9-6	8-6	7-9	7-2	6-8	6-4
	3 – 2 × 10	13-0	11-3	10-0	9-2	8-6	7-11	7-6
	3 – 2 × 12	15-3	13-3	11-10	10-9	10-0	9-4	8-10
Douglas fir-larch[e], hem-fir[e], spruce-pine-fir[e], redwood, western cedars, ponderosa pine[f], red pine[f]	3 × 6 or 2 – 2 × 6	5-5	4-8	4-2	3-10	3-6	3-1	2-9
	3 × 8 or 2 – 2 × 8	6-10	5-11	5-4	4-10	4-6	4-1	3-8
	3 × 10 or 2 – 2 × 10	8-4	7-3	6-6	5-11	5-6	5-1	4-8
	3 × 12 or 2 – 2 × 12	9-8	8-5	7-6	6-10	6-4	5-11	5-7
	4 × 6	6-5	5-6	4-11	4-6	4-2	3-11	3-8
	4 × 8	8-5	7-3	6-6	5-11	5-6	5-2	4-10
	4 × 10	9-11	8-7	7-8	7-0	6-6	6-1	5-8
	4 × 12	11-5	9-11	8-10	8-1	7-6	7-0	6-7
	3 – 2 × 6	7-4	6-8	6-0	5-6	5-1	4-9	4-6
	3 – 2 × 8	9-8	8-6	7-7	6-11	6-5	6-0	5-8
	3 – 2 × 10	12-0	10-5	9-4	8-6	7-10	7-4	6-11
	3 – 2 × 12	13- 11	12-1	10-9	9-10	9-1	8-6	8-1

For SI: 1 inch = 25.4 mm, 1 foot = 304.8 mm, 1 pound per square foot = 0.0479 kPa.

a. Ground snow load, live load = 40 psf, dead load = 10 psf, L/Δ = 360 at main span, L/Δ = 180 at cantilever with a 220-pound point load applied at the end.
b. Beams supporting deck joists from one side only.
c. No 2 grade, wet service factor.
d. Beam depth shall be greater than or equal to depth of joists with a flush beam condition.
e. Includes incising factor.
f. Northern species. Incising factor not included.

R507.7.1 Deck Post to Deck Beam. Deck beams shall be attached to deck posts in accordance with Figure R507.8.7.1 or by other equivalent means capable of resisting lateral displacement. Manufactured post-to-beam connectors shall be sized for the post and beam sizes. All bolts shall have washers under the head and nut.

> **Exception**: Where deck beams bear directly on footings in accordance with Section R507.8.1.

CHANGE SIGNIFICANCE: The 2015 IRC includes additional details for decks in an effort to provide prescriptive methods for conventional wood deck construction. There are a large number of construction methods that have long been in practice and are widely accepted. Designers and builders have used available information for determining joist and beam spans, as well as support and connection details. The new information and span tables in the code reflect a desire by many code users for more prescriptive guidance specific to decks.

Deck support provisions now describe maximum joist and beam spans, minimum connections between beams and posts, and minimum bearing length. New span tables specifically for decks are introduced. The span tables addressing joist and beam length are not based on the existing tables in IRC Chapters 5 and 6. Spans are shorter than listed in the current floor joist tables. The deck tables assume use of the joists in outdoor, potentially wet, conditions.

The new tables are based on wood capacity using the *National Design Specification for Wood Construction* (NDS). Additional wood species have also been included, such as Redwood, western cedar, ponderosa pine and red pine, that are not included in the existing joist and beam span tables. The deck joist and beam tables assume #2 grade wood, wet use, and incising, when applicable. Incising is done to assist chemical additives to soak deeper into preservative-treated lumber. Incising is only assumed in lumber species that are preservative treated, such as Douglas Fir and Hemlock Fir, and resistant to pressure treatment. Southern Pine more easily absorbs preservatives during the pressure-treating process and does not require incising.

R507.5, R507.6, R507.7 continues

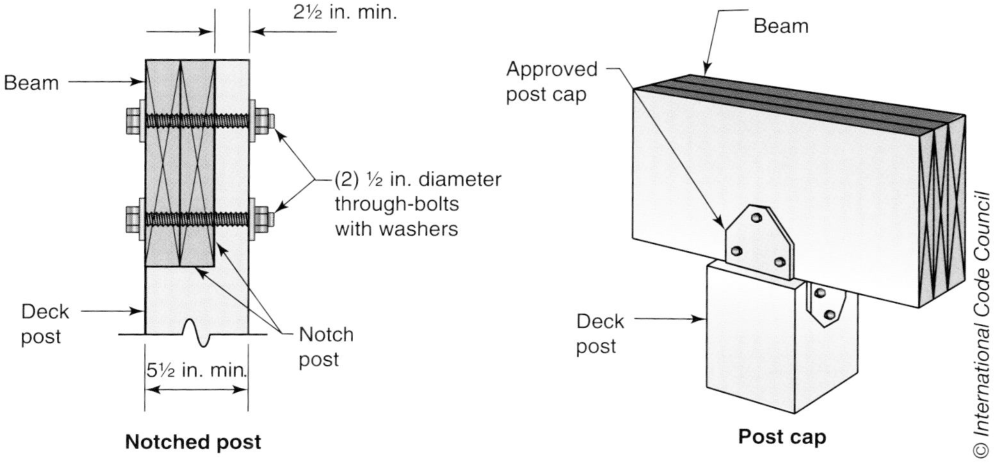

Connection of deck posts to deck beam

R507.5, R507.6, R507.7 continued

In addition to the new span tables, the code now addresses connection details between beams and posts specific to deck construction. The connection details illustrated in Figure R507.7.1 offer two options. The first requires support by a notched post with two ½-inch-diameter through-bolts with washers. The notch must provide 3 inches of bearing for the width of the beam with a minimum 2½ inches of the post remaining for the through-bolt connection. This type of connection will require a minimum nominal 4 × 6 post to provide the necessary 5½-inch cross section. The second option for beam-to-post connection is a manufactured connector commonly called a "post cap." The connector must be approved by the building official and must be sized for the post and beam sizes. Other equivalent connection details are also acceptable.

Bearing requirements for deck beams and joists in Section R507.7 are consistent with bearing requirements in Section R502.6 of the IRC.

R507.8

Deck Posts

CHANGE TYPE: Addition

CHANGE SUMMARY: New Section R507.8 establishes minimum sizes of wood posts supporting wood decks and describes the requirements for connection of deck posts to the footing.

2015 CODE: **R507.8 Deck Posts.** For single-level, wood-framed decks with beams sized in accordance with Table R507.6, deck post size shall be in accordance with Table R507.8.

TABLE R507.8 Deck Post Height[a]

Deck Post Size	Maximum Height
4 × 4	8′
4 × 6	8′
6 × 6	14′

For SI: 1 foot = 304.8 mm

a. Measured to the underside of the beam.

R507.8.1 Deck Post to Deck Footing. Posts shall bear on footings in accordance with Section R403 and Figure R507.8.1. Posts shall be restrained to prevent lateral displacement at the bottom support. Such lateral restraint shall be provided by manufactured connectors installed in accordance with Section R507 and the manufacturers' instructions or a minimum post embedment of 12 inches (305 mm) in surrounding soils or concrete piers.

CHANGE SIGNIFICANCE: As part of a more detailed prescriptive deck design option, the 2015 IRC adds provisions for sizing wood posts and connecting posts to the foundation for a deck. The post-sizing provisions are presented in tabular form. Depending on the height of the post, the code permits nominal 4 × 4, 4 × 6, and 6 × 6 wood posts. In practice, nominal 6 × 6 posts are most commonly used. The code does not prescribe the species or grade for deck posts. Section R317 addresses protection of wood against decay. A minimum post-to-footing connection is required to provide lateral restraint and prevent lateral displacement. The code requires manufactured connectors to be installed in accordance with the manufacturer's requirements when less than 12 inches of footing embedment exists.

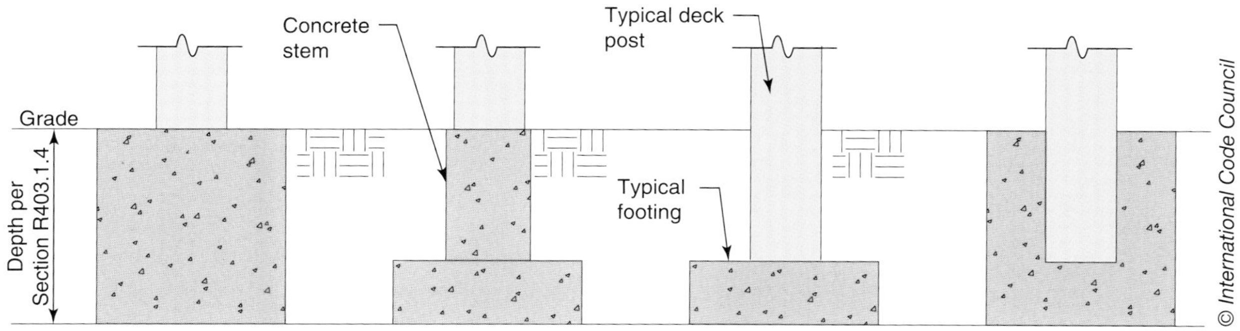

Typical deck posts to deck footings

Table R602.3(1)

Fastening Schedule—Roof Requirements

CHANGE TYPE: Modification

CHANGE SUMMARY: The Fastening Schedule now contains multiple nail size options. Clarification of roof rafter connections at ridge, valley, and hip is added.

2015 CODE:

TABLE R602.3(1) Fastening Schedule ~~for Structural Members~~

Item	Description of Building Elements	Number and Type of Fastener[a, b, c]	Spacing and Location ~~of Fasteners~~
		Roof	
1	Blocking between ceiling joists or rafters to top plate~~, toe nail~~	~~3-8d~~ 4-8d box (2½″ × 0.113″); or 3-8d common (2½″ × 0.131″); or 3-10d box (3″ × 0.128″); or 3-3″ × 0.131″ nails	Toe nail
2	Ceiling joists to top plate~~, toe nail~~	~~3-8d~~ 4-8d box (2½″ × 0.113″); or 3-8d common (2½″ × 0.131″); or 3-10d box (3″ × 0.128″); or 3-3″ × 0.131″ nails	Per joist, toe nail
3	Ceiling joist not attached to parallel rafter, laps over partitions~~, face nail~~ [See Sections R802.3.1, R802.3.2, Table R802.5.1(9)]	~~3-10d~~ 4-10d box (3″ × 0.128″); or 3-16d common (3½″ × 0.162″); or 4-3″ × 0.131″ nails	Face nail
4	Ceiling joist attached to parallel rafter (heel joint) [See Sections R802.3.1, R802.3.2, Table R802.5.1(9)]	Per Table R802.5.1(9)	Face nail
5	Collar tie to rafter, face nail or 1¼″ × 20 gage ridge strap to rafter	~~3-10d~~ 4-10d box (3″ × 0.128″); or 3-10d common (3″ × 0.148″); or 4-3″ × 0.131″ nails	Face nail each rafter
6	Rafter or roof truss to plate~~, toe nail~~	3-16d box nails (3½″ × 0.135″); or 3-10d common nails (3″ × 0.148″); or 4-10d box (3″ × 0.128″); or 4-3″ × 0.131 nails	2 toe nails on one side and 1 toe nail on opposite side of each rafter or truss[i]
7	Roof rafters to ridge, valley or hip rafters or roof rafter to minimum 2″ ridge beam~~, toe nail face nail~~	4-16d box (3½″ × 0.135″); or 3-10d common (3½″ × 0.148″); or 4-10d box (3″ × 0.128″); or 4-3″ × 0.131″ nails	Toe nail
		3-16d box (3½″ × 0.135″) 2-16d common (3½″ × 0.162″); or 3-10d box (3″ × 0.128″); or 3-3″ × 0.131″ nails	End nail

(Footnotes not shown for brevity and clarity.)

© Portlandia/Shutterstock.com.

Roof framing

CHANGE SIGNIFICANCE: IRC Table R602.3(1), wood frame nailing schedule, is reformatted to give typical nailing options to make the table consistent with 2015 IBC Table 2304.10.1. The change states minimum size and number of fasteners for each connection. Changes in both the IBC and IRC tables create increased consistency of minimum nailing requirements for wood frame construction. Nailing requirements are clarified using the exact dimensions of commonly used power-driven, box and common nail sizes in the table.

In many cases, the IRC minimum nailing for roofs remains unchanged except for addition of IBC nailing options. For instance, the base nailing of the following remain unchanged: Item 6, Rafter or roof truss to plate; Item 7, Roof rafters to ridge, valley, or hip rafters. In other cases, there is an increase in the number of smaller-diameter nails by 1 nail in order to maintain a minimum connection of approximately equal strength to that provided by IBC nailing requirements.

Nailing values are based on normal load duration and calculated assuming framing with a minimum Specific Gravity of 0.42, for example, using Spruce-Pine-Fir (SPF). However, some minimum nailing requirements are recommended as good practice and are not associated with a standard minimum load or calculation basis.

An item for ceiling joists attached to parallel rafters has been added to point the user to the minimum nailing requirements located in Chapter 8. Rafter-to-ridge-beam connections are also added to the fastener schedule.

Table R602.3(1)

Fastening Schedule—Wall Requirements

CHANGE TYPE: Modification

CHANGE SUMMARY: The Fastening Schedule for Structural Members now contains multiple nail size options. Clarification of double top plate splicing is added. Descriptions are updated in the IRC and the *International Building Code* (IBC) for attachments in walls as well.

2015 CODE:

TABLE R602.3(1) Fastening Schedule ~~for Structural Members~~

Item	Description of Building Elements	Number and Type of Fastener[a, b, c]	Spacing ~~of Fasteners~~ and Location
		Wall	
8	Stud to stud (not at braced wall panels) ~~Built-up studs—face nail~~	~~10d (3″ × 0.128″)~~ 16d common (3½″ × 0.162″)	24″ o.c. face nail
		10d box (3″ × 0.128″); or 3″ × 0.131″ nails	16″ o.c. face nail
9	Stud to stud and abutting studs at intersecting wall corners (at braced wall panels)~~, face nail~~	16d box (3½″ × 0.135″); or 3″ × 0.131″ nails	12″ o.c. face nail
		16d common (3½″ × 0.162″)	16″ o.c. face nail
10	Built-up header~~, two pieces with~~ (2″ to 2″ header with ½″ spacer)	~~16d (3½″ × 0.135″)~~ 16d common (3½″ × 0.162″)	16″ o.c. each edge face nail
		16d box (3½″ × 0.135″)	12″ o.c. each edge face nail
11	Continuous header to stud~~, toe nail~~	~~4-8d~~ 5-8d box (2½″ × 0.113″); or 4-8d common (2½″ × 0.131″); or 4-10d box (3″ × 0.128″)	Toe nail
12	Top plate to top plate ~~Double top plates, face nail~~	~~10d (3″ × 0.128″)~~ 16d common (3½″ × 0.162″)	~~24″ o.c.~~ 16″ o.c. face nail
		10d box (3″ × 0.128″); or 3″ × 0.131″ nails	12″ o.c. face nail
13	Double top plate splice for SDCs A-D$_2$ with seismic braced wall line spacing < 25′	~~8-16d (3-½″ × 0.135″)~~ 8-16d common (3½″ × 0.162″); or 12-16d box (3½″ × 0.135″); or 12-10d box (3″ × 0.128″); or 12-3″ × 0.131″ nails	Face nail on each side of end joint (minimum 24″ lap splice length each side of end joint)
	Double top plate splice SDCs D$_0$, D$_1$ or D$_2$; and braced wall line spacing ≥ 25′	12-16d (3-½″ × 0.135″)	
14	Bottom plate to joist, rim joist, band joist or blocking (not at braced wall panels) ~~Sole plate to joist or blocking, face nail~~	~~16d (3½″ × 0.135″)~~ 16d common (3½″ × 0.162″)	16″ o.c. face nail
		16d box (3½″ × 0.135″); or 3″ × 0.131″ nails	12″ o.c. face nail
15	~~Sole plate to~~ Bottom plate to joist, rim joist, band joist, or blocking (at braced wall panel)~~, face nail~~	3-16d box (3½″ × 0.135″);	3 each 16″ o.c. face nail
		or 2-16d common (3½″ × 0.162″);	2 each 16″ o.c. face nail
		or 4-3″ × 0.131″ nails	4 each 16″ o.c. face nail
16	Top or bottom plate to stud ~~Stud to sole plate, toe nail~~	~~3-8d~~ 4-8d box (2½″ × 0.113″);or ~~2-16d~~ 3-16d box (3½″ × 0.135″); or 4-8d common (2½″ × 0.131″); or 4-10d box (3″ × 0.128″); or 4-3″ × 0.131″ nails	Toe nail
		~~2-16d~~ 3-16d box (3½″ × 0.135″); or 2-16d common (3½″ × 0.162″); or 3-10d box (3″ × 0.128″); or 3-3″ × 0.131″ nails	End nail

Item	Description of Building Elements	Number and Type of Fastener[a, b, c]	Spacing ~~of Fasteners~~ and Location
		Wall	
17	Top plates, laps at corners and intersections~~, face nail~~	~~2-10d~~ 3-10d box (3″ × 0.128″); or 2-16d common (3½″ × 0.162″); or 3-3″ × 0.131″ nails	Face nail
18	1″ brace to each stud and plate~~, face nail~~	~~2-8d~~ 3-8d box (2½″ × 0.113″); or 2-8d common (2½″ × 0.131″); or 2-10d box (3″ × 0.128″); or 2 staples 1¾″	Face nail
19	1″ × 6″ sheathing to each bearing~~, face nail~~	~~2-8d~~ 3-8d box (2½″ × 0.113″); or 2-8d common (2½″ × 0.131″); or 2-10d box (3″ × 0.128″); or 2 staples ~~1¾″~~, 1″ crown, 16 ga., 1¾″ long	Face nail
20	1″ × 8″ and wider sheathing to each bearing~~, face nail~~	~~2-8d~~ 3-8d box (2½″ × 0.113″); or 3-8d common (2½″ × 0.131″); or 3-10d box (3″ × 0.128″); or 3 staples ~~1¾″~~, 1″ crown, 16 ga., 1¾″ long	Face nail
	Wider than 1″ × 8″	~~3-8d~~ 4-8d box (2½″ × 0.113″); or 3-8d common (2½″ × 0.131″); or 3-10d box (3″ × 0.128″); or 4 staples ~~1¾″~~, 1″ crown, 16 ga., 1¾″ long	Face nail

(Footnotes not shown for brevity and clarity.)

CHANGE SIGNIFICANCE: IRC Table R602.3(1), wood frame nailing schedule, is reformatted to give typical nailing options and to make the table consistent with 2015 IBC Table 2304.10.1. The change states the minimum size and number of fasteners for each connection. Changes in both the IBC and IRC tables create increased consistency of minimum nailing requirements for wood frame construction. Nailing requirements are clarified using the exact dimensions of commonly used power-driven, box and common nail sizes in the table.

In many cases, the IRC minimum nailing requirements remain unchanged except for addition of IBC nailing options. For instance, base nailing of the following remain unchanged: Item 9, Stud to stud and abutting studs at intersecting wall corners (at braced wall panels); and Item 15, Bottom plate to joist, rim joist, band joist, or blocking (at braced wall panel). In other cases, there is an increase in the number of smaller nails by 1 in order to maintain a minimum connection of approximately equal strength to that provided by IBC nailing. Nailing values are based on normal load duration and calculated assuming framing with Specific Gravity equal to 0.42, for example, using Spruce-Pine-Fir (SPF). However, some minimum nailing requirements are recommended as good practice and are not associated with a standard minimum load or calculation basis.

Low resistance of IRC minimum nailing relative to applied loads occurs with connection details such as bottom plate to joist and top plate to top plate, particularly where loads are based on upper IRC limits (for example, wind pressures associated with 140 mph Exposure B and 10-foot stud heights). In many cases, the increased strength of IBC minimum fastening provides a better match to loads than the 2012 IRC fastening schedule.

In previous editions, the IRC had two requirements for double top plate splices. In 2012 IRC Table R602.3(1), Item 14, the requirement for a double

Table R602.3(1) continues

Table R602.3(1) continued

Wall framing

top plate splice was a minimum 24 inches offset at the splice between the top and bottom plates, attached with (8) 16d nails. This conflicted with the requirement in Table R602.10.3(4), footnote c. The footnote required use of (12) 16d nails on each side of the splice. To correct the conflict, former Item 14 of R602.3(1) is divided into two separate line items, to differentiate the appropriate number of nails. In addition, language now indicates that fasteners are required on each side of the splice location. A corresponding change for footnote c of Table R602.10.3(4) refers the user back to Table R602.3(1) for splice-plate attachment guidance.

CHANGE TYPE: Modification

CHANGE SUMMARY: The Fastening Schedule for Structural Members now contains multiple nail size options. Clarification of the joist-to-band-joist (rim board) connection is added.

2015 CODE:

Table R602.3(1)

Fastening Schedule—Floor Requirements

TABLE R602.3(1) Fastening Schedule for Structural Members

Item	Description of Building Elements	Number and Type of Fastener[a, b, c]	Spacing ~~of Fasteners~~ and Location
		Floor	
21	Joist to sill, top plate or girder	4-8d box (2½″ × 0.113″); or 3-8d common (2½″ × 0.131″); or 3-10d box (3″ × 0.128″); or 3-3″ × 0.131″ nails	Toe nail
22	Rim joist, band joist, or blocking to sill or top plate (roof applications also)	8d box (2½″ × 0.113″)	4″ o.c. toe nail
		8d common (2½″ × 0.131″); or 10d box (3″ × 0.128″); or 3″ × 0.131″ nails	6″ o.c. toe nail
23	1″ × 6″ subfloor or less to each joist	3-8d box (2½″ × 0.113″); or 2-8d common (2½″ × 0.131″); or 3-10d box (3″ × 0.128″); or 2 staples, 1″ crown, 16 ga. 1 ¾″ long	Face nail
24	2″ subfloor to joist or girder~~, blind and face nail~~	~~2-16d~~ 3-16d box (3½″ × 0.135″); or 2-16d common (3½″ × 0.162″)	Blind and face nail
25	2″ planks (plank & beam - floor & roof)	~~2-16d~~ 3-16d box (3½″ × 0.135″); or 2-16d common (3½″ × 0.162″)	At each bearing, face nail
26	Band or rim Joist to joist	3 – 16d common (3½″ × 0.162″); or 4 -10 box (3″ × 0.128); or 4 – 3″ × 0.131″ nails; or 4 – 3″ × 14 gage staples, 7/16″ crown	End nail
27	Built-up girders and beams, 2-inch lumber layers	20d common (4″ × 0.192″); or	Nail each layer as follows: 32″ o.c. at top and bottom and staggered. ~~Two nails at ends and at each splice.~~
		10d box (3″ × 0.128″); or 3-3″ × 0.131″ nails	24″ o.c. face nail at top and bottom staggered on opposite sides
		And: 2-20d common (4″ × 0.192″); or 3-10d box (3″ × 0.128″); or 3-3″ × 0.131″ nails	Face nail at ends and at each splice
28	Ledger strip supporting joists or rafters	~~3-16d~~ 4-16d box (3½″ × 0.135″); or 3-16d common (3½″ × 0.162″); or 4-10d box (3″ × 0.128″); or 4-3″ × 0.131″ nails	At each joist or rafter, face nail
29	Bridging to joist	2-10d (3″ × 0.128″)	Each end, toenail

(Footnotes not shown for brevity and clarity.)

Table R602.3(1) continues

Table R602.3(1) continued

Floor framing

CHANGE SIGNIFICANCE: IRC Table R602.3(1), wood frame nailing schedule, is reformatted to give typical nailing options for floors and to make the table consistent with 2015 IBC Table 2304.10.1. The change states the minimum size and number of fasteners for each connection. Changes in both the IBC and IRC tables create increased consistency of minimum nailing requirements for wood frame construction. Nailing requirements are clarified using the exact dimensions of commonly used power-driven, box and common nail sizes in the table.

In many cases, the IRC minimum nailing for floors remains unchanged except for addition of IBC nailing options. In other cases, there is an increase in the number of smaller-diameter nails by 1 nail or a reduced spacing in order to maintain a minimum connection of approximately equal strength to that provided by IBC nailing.

Nailing values are based on normal load duration and calculated assuming framing with a Specific Gravity of 0.42, for example, using Spruce-Pine-Fir (SPF). However, some minimum nailing requirements are recommended as good practice and are not associated with a standard minimum load or calculation basis.

The subsection on floors has also been updated with nailing of rim joist applications combined into Item 22 including rim or band joist or blocking to sill or top plate. The new category includes all the connections toe nailed at 4 or 6 inches on center at the edges of floors.

CHANGE TYPE: Modification

R602.3.1

Stud Size, Height, and Spacing

CHANGE SUMMARY: Table R602.3.1 is deleted and the exception for walls greater than 10 feet tall is added to the text of Section R602.3.1. If studs in a tall wall meet Exception 2, they meet the requirements of the IRC and do not need engineering or use of an alternate standard.

2015 CODE: R602.3.1 Stud Size, Height, and Spacing. The size, height, and spacing of studs shall be in accordance with Table R602.3(5).

Exceptions:

1. Utility grade studs shall not be spaced more than 16 inches (406 mm) on center, shall not support more than a roof and ceiling, and shall not exceed 8 feet (2438 mm) in height for exterior walls and load-bearing walls or 10 feet (3048 mm) for interior nonload-bearing walls.
2. ~~Studs more than 10 feet in height which are in accordance with Table R602.3.1.~~ Where snow loads are less than or equal to 25 pounds per square foot (1.2 kPa), and the ultimate design wind speed is less than or equal to 130 mph (58 m/s), 2-inch by 6-inch (38 mm by 140 mm) studs supporting a roof load with not more than 6 feet (1829 mm) of tributary length shall have a maximum height of 18 feet (5486 mm) where spaced at 16 inches (406 mm) on center, or 20 feet (6096 mm) where spaced at 12 inches (305 mm) on center. Studs shall be minimum No. 2 grade lumber.

R602.3.1 continues

~~**TABLE R602.3.1 Maximum Allowable Length of Wood Studs Expose to Wind Speeds of 100 MPH or Less in Seismic Design Categories A, B, C, D_0, D_1, and D_2**~~[b,c]

	On-Center Spacing (inches)			
Height (feet)	24	16	12	8
	Supporting a Roof Only			
>10	2 × 4	2 × 4	2 × 4	2 × 4
12	2 × 6	2 × 4	2 × 4	2 × 4
14	2 × 6	2 × 6	2 × 6	2 × 4
16	2 × 6	2 × 6	2 × 6	2 × 4
18	NA[a]	2 × 6	2 × 6	2 × 6
20	NA[a]	NA[a]	2 × 6	2 × 6
24	NA[a]	NA[a]	NA[a]	2 × 6
	Supporting one Floor and a Roof			
>10	2 × 6	2 × 4	2 × 4	2 × 4
12	2 × 6	2 × 6	2 × 6	2 × 4
14	2 × 6	2 × 6	2 × 6	2 × 6
16	NA[a]	2 × 6	2 × 6	2 × 6
18	NA[a]	2 × 6	2 × 6	2 × 6
20	NA[a]	NA[a]	2 × 6	2 × 6
24	NA[a]	NA[a]	NA[a]	2 × 6

continues

R602.3.1 continued

~~TABLE R602.3.1~~ *(Continued)*

~~Height (feet)~~	~~On-Center Spacing (inches)~~			
	~~24~~	~~16~~	~~12~~	~~8~~
	~~Supporting Two Floors and a Roof~~			
~~> 10~~	~~2 × 6~~	~~2 × 6~~	~~2 × 4~~	~~2 × 4~~
~~12~~	~~2 × 6~~	~~2 × 6~~	~~2 × 6~~	~~2 × 6~~
~~14~~	~~2 × 6~~	~~2 × 6~~	~~2 × 6~~	~~2 × 6~~
~~16~~	~~NA[a]~~	~~NA[a]~~	~~2 × 6~~	~~2 × 6~~
~~18~~	~~NA[a]~~	~~NA[a]~~	~~2 × 6~~	~~2 × 6~~
~~20~~	~~NA[a]~~	~~NA[a]~~	~~NA[a]~~	~~2 × 6~~
~~22~~	~~NA[a]~~	~~NA[a]~~	~~NA[a]~~	~~NA[a]~~
~~24~~	~~NA[a]~~	~~NA[a]~~	~~NA[a]~~	~~NA[a]~~

~~For SI: 1 inch = 25.4 mm, 1 foot = 304.8 mm, 1 pound per square foot = 0.0479 kPa, 1 pound per square inch = 6.895 kPa, 1 mile per hour = 0.447 m/s.~~

~~a. Design required.~~

~~b. Applicability of this table assumes the following: Snow load not exceeding 25 psf, f_b not less than 1310 psi determined by multiplying the AF&PA NDS tabular base design value by the repetitive use factor, and by the size factor for all species except southern pine, E not less than 1.6 × 106 psi, tributary dimensions for floors and roofs not exceeding 6 feet, maximum span for floors and roof not exceeding 12 feet, eaves not over 2 feet in dimension and exterior sheathing. Where the conditions are not within these parameters, design is required.~~

~~c. Utility, standard, stud and No. 3 grade lumber of any species are not permitted.~~

CHANGE SIGNIFICANCE: Stud height is limited by multiple provisions within the IRC. In the 2012 IRC, Table R602.3(5) allowed load-bearing studs up to 10 feet tall. Non-load-bearing studs were as tall as 20 feet if certain limits were met. An exception in Section R602.10, wall bracing, allowed load-bearing studs up to 12 feet tall. And, in Section R602.3.1, load-bearing studs could be 20 feet in height when the limits of Table R602.3.1 were met. If wall studs did not meet the limits of these three sections, the wall or story was engineered.

Table R602.3.1 in the 2012 IRC has been a major source of confusion. Footnote b, referring to engineering loads and properties, was confusing and often misunderstood. The text described engineering limits that aren't considered in conventional construction or in the prescriptive design process of the IRC.

In the 2015 IRC, a code change eliminates Table R602.3.1 and places an exception in Section R602.3.1 for tall walls. Tall walls meeting the limits in the exception do not require engineering. Reference to engineering properties is deleted. Note that this exception does not apply in locations with more than 25 psf snow load, 2 × 4 construction, or roof tributary widths greater than 6 feet.

The process for determining whether walls studs, a wall, or a story must be engineered based on stud height now checks:

1. Table R602.3(5)
2. Section R602.10, and, if necessary
3. Section R602.3.1, Exception 2

If wall studs fail to meet the requirements of the three sections, the wall or story is engineered.

Example—Prescriptive Tall Walls

In the following three cases, tall walls meeting the IRC's limits are illustrated.

Case 1: 2 × 6 Continuous Studs Used in an 18-Foot Gable End Wall

The gable end wall studs do not support a roof load. They form a non-load-bearing wall. From Table R602.3(5), non-bearing walls may have studs up to 20 feet tall when using 2 × 6 lumber.

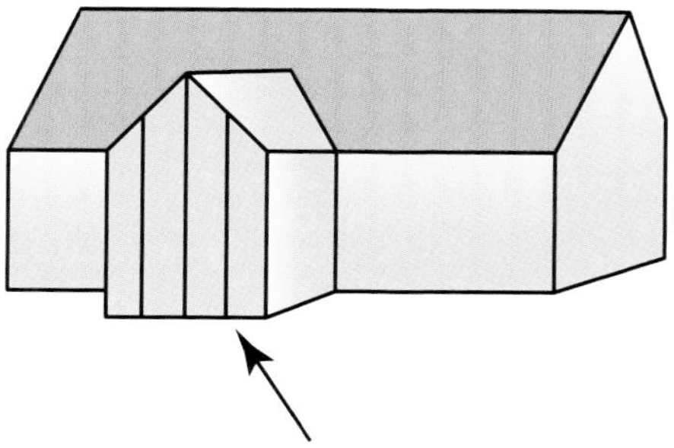

Tall wall studs in gable, no roof load

Case 2: 2 × 6 Continuous Studs Used in a 20-Foot Tall Wall Supporting a Projection (Roof Framing Parallel to Wall)

The studs used for a two-story projection where the roof framing runs parallel to the wall will carry a roof load. If the studs can meet all the limits of Section R602.3.1, Exception 2, then no engineering of the wall is required. The following four limits must be met:

1. Snow load ≤25 psf
2. Wind speed ≤130 mph
3. 2 × 6 construction
4. Roof load tributary width ≤6 feet

Assuming the first three conditions are met, the roof load tributary width limit must be met.

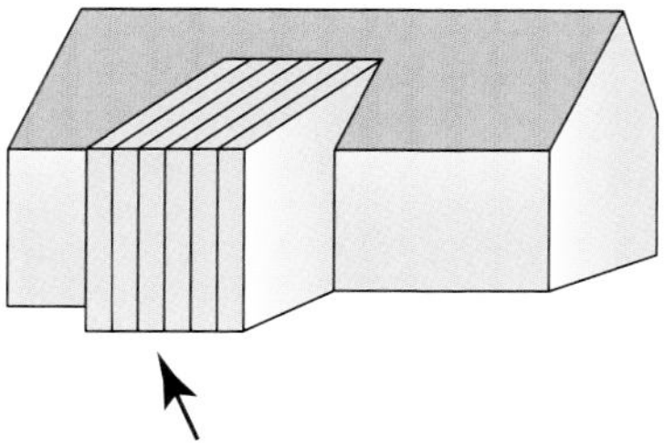

Tall wall studs parallel to roof framing – studs carry roof load

R602.3.1 continues

R602.3.1 continued

Example—Prescriptive Tall Walls *(Continued)*

What is a roof load tributary width? Tributary width describes the width of the roof surface that will transfer live or snow loads to nearby walls. To determine tributary width in a roof, measure the total distance between roof supports, typically walls in residential construction, and divide the distance in half. If walls supporting the roof are 12 feet apart or less (a roof tributary width of 6 feet), Exception 2 limits are met.

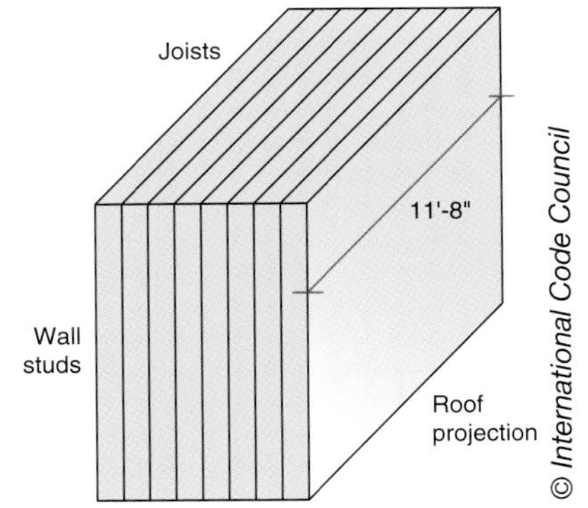

Framing: Wall studs and roof projection joists at 16″ o.c.
Total distance = 11′-8″, Tributary width = 5′-10″

Case 3: 2 × 6 Continuous Studs Used in a Variable-Height Wall Supporting a Projection (Roof Framing Perpendicular to Wall)

The studs where the roof framing runs perpendicular to the wall do not carry a roof load. They form a non-load-bearing wall. From Table R602.3(5), non-bearing walls may have studs up to 20 feet tall when using 2 × 6 lumber.

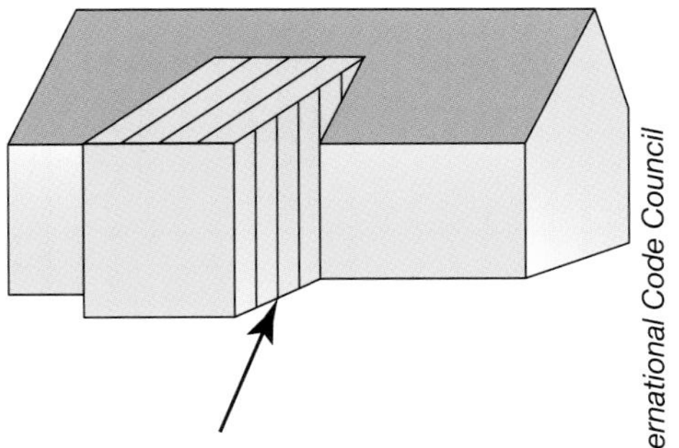

Tall wall studs perpendicular to roof framing, no roof load

CHANGE TYPE: Modification

CHANGE SUMMARY: The girder and header span tables of Chapter 5 have been moved into Chapter 6, to the header section. Multi-ply and single header tables are combined. A new section describing rim board headers is added.

R602.7

Headers

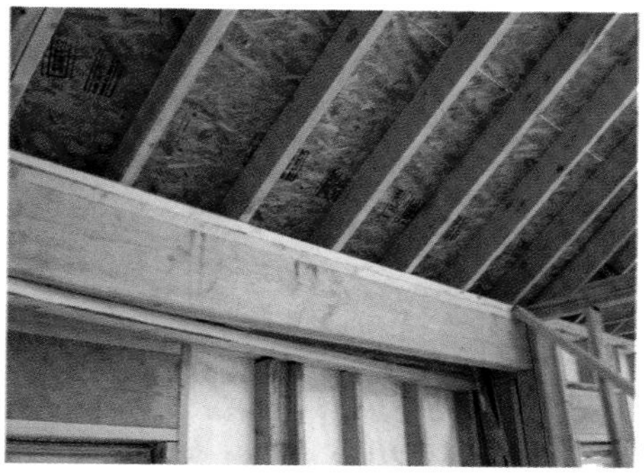

Solid-sawn lumber header

2015 CODE: R602.7 Headers. For header spans see Tables R602.7(1), R602.7(2) and R602.7(3) ~~R502.5(1), R502.5(2),and R602.7.1~~.

R602.7.1 Single Member Headers. Single headers shall be framed with a single flat 2-inch-nominal (51 mm) member or wall plate not less in width than the wall studs on the top and bottom of the header in accordance with Figures R602.7.1(1) and R602.7.1(2) and face nailed to the top and bottom of the header with 10d box nails (3 inches × 0.128 inches) (76 mm by 3 mm) spaced 12 inches (305 mm) on center.

R602.7.2 Rim Board Headers. Rim board header size, material, and span shall be in accordance with Table R602.7(1) for rim board headers. Rim board headers shall be constructed in accordance with Figure R602.7.2 and shall be supported at each end by full height studs. The number of full height studs at each end shall be not less than the number of studs displaced by half of the header span based on the maximum stud spacing in accordance with Table R602.3(5). Rim board headers supporting concentrated loads shall be designed in accordance with accepted engineering practice.

R602.7 continues

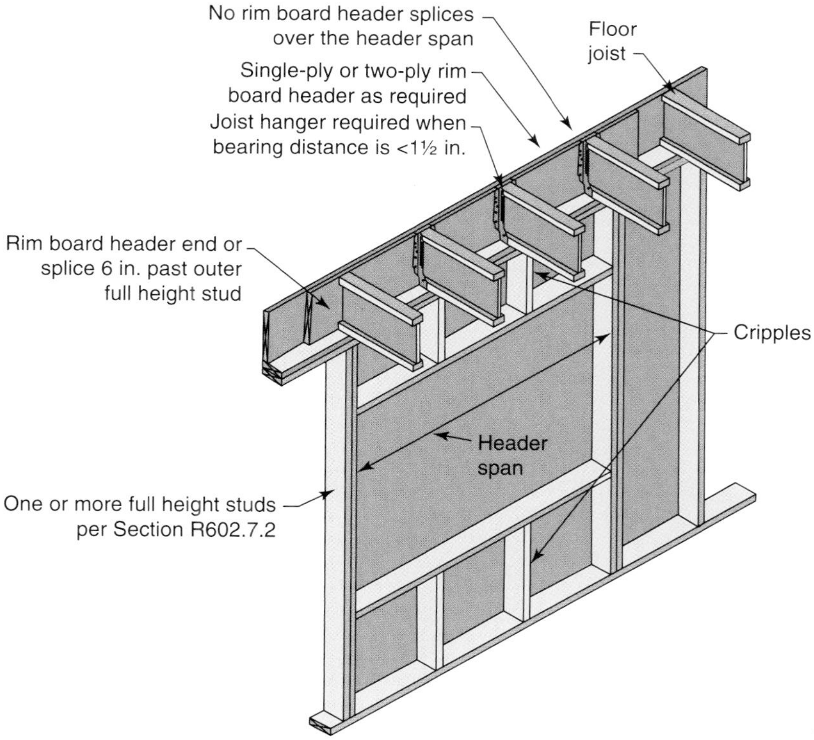

Rim board header construction

R602.7 continued

TABLE ~~R502.5(1)~~ R602.7(1) Girder Spans[a] and Header Spans[a] for Exterior Bearing Walls

(Maximum Spans for Douglas Fir-Larch, Hem-Fir, Southern Pine and Spruce-Pine-Fir[b] and Required Number of Jack Studs)

		Ground Snow Load (psf)[e]					
		30					
		Building width[c] (feet)					
		20		28		36	
Girders and Headers Supporting	Size	Span	NJ[d]	Span	NJ[d]	Span	NJ[d]
Roof and ceiling	1-2 × 8	4-6	1	3-10	1	3-5	1
	1-2 × 10	5-8	1	4-11	1	4-4	1
	1-2 × 12	6-11	1	5-11	2	5-3	2
	2-2 × 4	3-6	1	3-2	1	2-10	1
	2-2 × 6	5-5	1	4-8	1	4-2	1
	2-2 × 8	6-10	1	5-11	2	5-4	2
	2-2 × 10	8-5	2	7-3	2	6-6	2
	2-2 × 12	9-9	2	8-5	2	7-6	2
	3-2 × 8	8-4	1	7-5	1	6-8	1
	3-2 × 10	10-6	1	9-1	2	8-2	2
	3-2 × 12	12-2	2	10-7	2	9-5	2
	4-2 × 8	9-2	1	8-4	1	7-8	1
	4-2 × 10	11-8	1	10-6	1	9-5	2
	4-2 × 12	14-1	1	12-2	2	10-11	2

(Portions of table not shown for brevity and clarity. No changes to footnotes.)

TABLE R602.7(3) Girder and Header Spans[a] for Open Porches (Maximum Span for Douglas Fir-Larch, Hem-Fir, Southern Pine, and Spruce-Pine-Fir[b])

	Supporting Roof						Supporting Floor	
	Ground Snow Load[c] (psf)							
	30		50		70			
	Depth of Porch[d] (feet)							
Size	8	14	8	14	8	14	8	14
2-2 × 6	7-6	5-8	6-2	4-8	5-4	4-0	6-4	4-9
2-2 × 8	10-1	7-7	8-3	6-2	7-1	5-4	8-5	6-4
2-2 × 10	12-4	9-4	10-1	7-7	8-9	6-7	10-4	7-9
2-2 × 12	14-4	10-10	11-8	8-10	10-1	7-8	11-11	9-0

For SI: 1 inch = 25.4 mm, 1 foot = 304.8 mm, 1 pound per square foot = 0.0479 kPa

a. Spans are given in feet and inches.
b. Tabulated values assume #2 grade lumber, wet service and incising for refractory species. Use 30 psf ground snow load for cases in which ground snow load is less than 30 psf and the roof live load is equal to or less than 20 psf.
c. Porch width is measured horizontally from building face to the centerline of the header. For widths between those shown, spans are permitted to be interpolated.

R602.7.5 Supports for Headers. Headers shall be supported on each end with one or more jack studs or with approved framing anchors in accordance with Table R602.7(1) or R602.7(2). The full height stud adjacent to each end of the header shall be end nailed to each end of the header with 4-16d nails (3.5 inches × 0.135 inches). The minimum number of full height studs at each end of a header shall be in accordance with Table R602.7.5.

TABLE R602.7.5 Minimum Number of Full Height Studs at Each End of Headers in Exterior Walls

Header Span (feet)	Maximum Stud Spacing (in.) per Table R602.3(5)	
	16	24
≤3	1	1
4	2	1
8	3	2
12	5	3
16	6	4

CHANGE SIGNIFICANCE: The header section of the IRC has been reorganized to combine single- and multi-ply headers into one table. Tables R502.5(1) and R502.5(2) of the 2012 IRC were moved from Chapter 5 to Section R602.7. Because the tables are frequently referenced in the wall provisions and the tables are used for headers more frequently than for girders, a proposal was brought forward and approved to move the tables. The change makes the tables conveniently located for header sizing.

Table R602.7.1 of the 2012 IRC is combined with Table R502.5(1) to give a single table containing 1-, 2-, and 3-ply headers in 2015 IRC Table R602.7(1). The table also contains minimum jack stud requirements for all header options. Within the table, span lengths for Southern Pine, Douglas Fir-Larch, and Hemlock Fir are updated.

New Southern Pine lumber design values for all widths and grades of visually graded Southern Pine lumber became effective on June 1, 2013. The American Lumber Standards Committee (ALSC) approved the new design values as published in Southern Pine Inspection Bureau *Supplement No. 13* to the *2002 Standard Grading Rules for Southern Pine Lumber*. These values are a result of two years of testing current lumber available on the market to see what changes, if any, had occurred in the strength of the Southern Pine lumber inventory.

Meanwhile, testing for Douglas Fir-Larch and Hemlock Fir, done in the 1990s, slightly increased design values for bending. Revised design values for Select Structural, #2, and #3 grades of Douglas Fir-Larch and #1 grade of Hem-Fir all increased by 25 psi. Testing to check current stock has validated the design values set in the 1990s. Although these values were updated in the wood standards, span tables incorporated into the

R602.7 continues

R602.7 continued

2000 *International Building Code* (IBC) and 2000 IRC were based on span tables predating revised design values from the 1990s.

The 2015 IRC span tables are now in agreement with the wood standard span tables with the revisions for Southern Pine, Douglas Fir-Larch, and Hemlock Fir. For Southern Pine, the changes reflect shorter spans. For Douglas Fir and Hemlock, the changes result in slightly longer spans. New design values apply only to new construction.

Headers made of rim board are added as a new subsection. Section R602.7.2 contains the requirements for rim board headers. Table R602.7(1) is used to determine the maximum span of the header, and Figure R602.7.2 illustrates construction of the header. A rim board header is a solid-sawn band or rim joist, or is an approved alternate material, that surrounds or contains the floor joists. Rather than placing the header in a wall cavity, it is acceptable to use the rim board in the floor system as a header above a window or door. This is not a new application and has been approved locally for many years.

Due to the interest in creating more energy-efficient dwellings and the use of the *International Energy Conservation Code* (IECC) across the country, header options that encourage increased use of insulation have become more common. Single-ply header spans based on #2 Hem-Fir design values were introduced in the 2012 IRC. The rim board header spans introduced in the 2015 IRC also assume #2 Hem-Fir lumber and may be single-ply, allowing space for insulation behind the lumber, in addition to allowing more cavity insulation in the wall space below because the header has been moved up into the floor space.

A new table, Table R602.7(3) for girders and headers in open porches, is added. The table is based on post construction supporting girders for porches with an 8-foot or 14-foot width. The span lengths in the table are based on the 2005 AF&PA *National Design Specification for Wood (NDS)* using wood species in the IRC span tables.

An additional new table, Table R602.7.5, lists minimum full-height stud (king stud) requirements for wall openings. The intent of the minimum full-height stud requirements is to ensure structural integrity in resisting loads by compensating for removal of full-height layout studs over the span of the wall opening. Full-height studs are used to stabilize a header by nailing the first stud into the end of the header, preventing header rotation. The number of full-height studs required is based on out-of-plane wind and gravity loads. An out-of-plane wind load is the load a wall resists due to wind pressing directly on it. Gravity loads passing down from the roof or wall above are transferred by both king studs and headers. Gravity loads passing through the header are then transferred by trimmers or jack studs to the foundation or story below.

Lastly, box headers remain in Section R602.7.3 as a useful option for energy-efficient designs.

CHANGE TYPE: Modification

CHANGE SUMMARY: Table values in Table R602.10.3(1) have changed slightly due to use of ultimate design wind speed values to calculate required bracing length.

2015 CODE:

Table R602.10.3(1)

Bracing Requirements Based on Wind Speed

TABLE R602.10.3(1) Bracing Requirements Based on Wind Speed

- Exposure Category B
- 30-Foot Mean Roof Height
- 10-Foot Eave-to-Ridge Height
- 10-Foot Wall Height
- 2 Braced Wall Lines

			Minimum Total Length (Feet) of Braced Wall Panels Required Along Each Braced Wall Line[a]			
Ultimate Design Wind Speed (mph)	Story Location	Braced Wall Line Spacing (feet)	Method LIB[b]	Method GB	Methods DWB, WSP, SFB, PBS, PCP, HPS, BV-WSP, ABW, PFH, PFC, CS-SFB[c]	Methods CS-WSP, CS-G, CS-PF
≤115	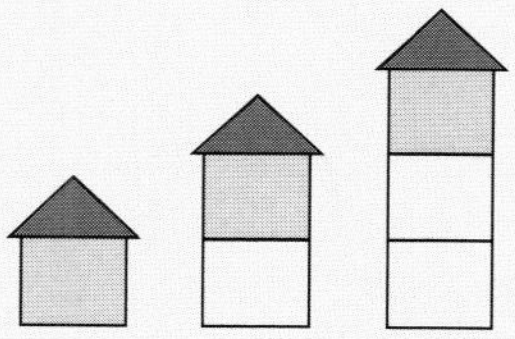	10	3.5	3.5	2.0	2.0
		20	6.5	6.5	3.5	3.5
		30	9.5	9.5	5.5	4.5
		40	12.5	12.5	7.0	6.0
		50	15.0	15.0	9.0	7.5
		60	18.0	18.0	10.5	9.0
	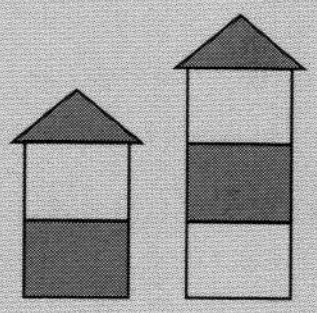	10	7.0	7.0	4.0	3.5
		20	12.5	12.5	7.5	6.5
		30	18.0	18.0	10.5	9.0
		40	23.5	23.5	13.5	11.5
		50	29.0	29.0	16.5	14.0
		60	34.5	34.5	20.0	17.0
	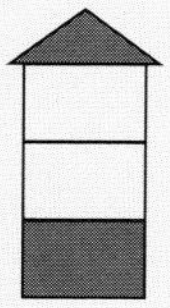	10	NP	10.0	6.0	5.0
		20	NP	18.5	11.0	9.0
		30	NP	27.0	15.5	13.0
		40	NP	35.0	20.0	17.0
		50	NP	43.0	24.5	21.0
		60	NP	51.0	29.0	25.0

(Portions of table not shown for brevity and clarity.)

Table R602.10.3(1) continues

Table R602.10.3(1) continued

Wall bracing to resist wind forces

CHANGE SIGNIFICANCE: Values in Table R602.10.3(1) for required minimum bracing length changed slightly as the new ultimate design wind speeds were used to calculate bracing. Previously, there were four wind speed categories—85, 90, 100, and 110 mph. In the 2015 IRC there are five categories—110, 115, 120, 130, and 140 mph. With an additional category, the bracing lengths are not identical and, due to rounding, lengths may vary by 0.5 feet from the 2012 IRC table value. In the table included in this section, bracing lengths for 115 mph are used to illustrate the similarity to bracing lengths found in 2012 IRC Table R602.10.3(1) for 90-mph basic wind speeds. Compare the values to your copy of the 2012 IRC Table R602.10.3(1).

An additional change to this section adds BV-WSP, ABW, PFG, and PFH to the intermittent methods column of Table R602.10.3(1). All bracing methods are now specifically listed in the table.

Table R602.10.5

Contributing Length of Method CS-PF Braced Wall Panels

CHANGE TYPE: Modification

CHANGE SUMMARY: The contributing length of continuously sheathed portal frames (Method CS-PF) in low-seismic regions has increased by 50 percent.

2015 CODE:

TABLE R602.10.5 Minimum Length of Braced Wall Panels

Method (See Table R602.10.4)		Minimum Length[a] (in.) Wall Height 8 ft.	9 ft.	10 ft.	11 ft.	12 ft.	Contributing Length (in.)
CS-PF	SDC A, B, and C	16	18	20	22[e]	24[e]	1.5 × Actual[b]
	SDC D_0, D_1 and D_2	16	18	20	22[e]	24[e]	Actual[b]

(Portions of table and footnotes not shown for brevity and clarity.)

CHANGE SIGNIFICANCE: In the 2012 IRC, Method CS-PF, continuously sheathed portal frame, has a contributing bracing length that matches the actual length of braced panel in the portal frame. This bracing length is based in part on a 6:1 aspect ratio of the panels and two rows of nails into braced wall panel studs.

Method PFG (Portal Frame at Garage) is permitted, in the 2012 IRC Table R602.10.5, a 1.5 multiplier for contributing bracing length. This panel has a 4:1 aspect ratio and is required to have a braced wall panel at

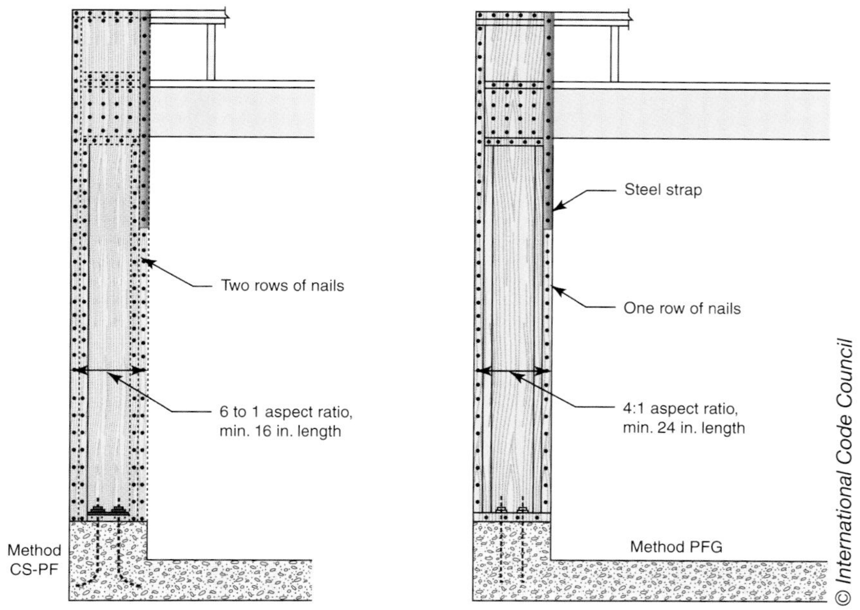

Comparison of nailing of CS-PF and PFG

Table R602.10.5 continues

Table R602.10.5 continued

the end of a single portal frame. Method PFG has only a single row of nailing along studs in the braced wall panel and is permitted with intermittent bracing. The 1.5 multiplier has been permitted because Method PFG is restricted for use in areas of low seismicity (Seismic Design Categories [SDCs] A, B, and C).

Cyclic testing of Method CS-PF (Continuous Sheathed—Portal Frame) has shown that the CS-PF has a design strength as high as Method PFG. Based on the results of this testing, the same multiplier may be applied to Method CS-PF when similarly restricted to areas of low seismicity.

CS-PF can have a leg length as small as 16 inches, while PFG has a minimum leg length of 24 inches. What makes CS-PF perform as well or better than PFG, even with a shorter leg length, is the fact that CS-PF has nearly twice as many fasteners as PFG. It is the fastener interaction between the framing and sheathing that determines the ultimate capacity of this bracing system.

R602.10.6.2

Method PFH: Portal Frame with Hold-Downs

CHANGE TYPE: Modification

CHANGE SUMMARY: Due to recent testing of Method PFH (Portal Frame with Hold-downs), the minimum required capacity of the hold-downs is lowered to 3500 lbs in the 2015 IRC. Additionally, the new testing confirms that two sill plates are sufficient under each braced wall panel of the portal rather than the three plates used in Method PFH for the 2012 IRC.

2015 CODE: R602.10.6.2 Method PFH: Portal Frame with Hold-Downs. Method PFH braced wall panels shall be constructed in accordance with Figure R602.10.6.2.

CHANGE SIGNIFICANCE: Two technical changes occur in Figure R602.10.6.2, Method PFH: Portal Frame with Hold-Downs. First, the minimum required capacity of the hold-downs is reduced from 4200 lbs to 3500 lbs. Second, the third bottom plate has been removed from the base of the braced wall panel.

Initial testing conducted on the portal frames utilized the 4200-lb. hold-down because the straps were available and in common use by the

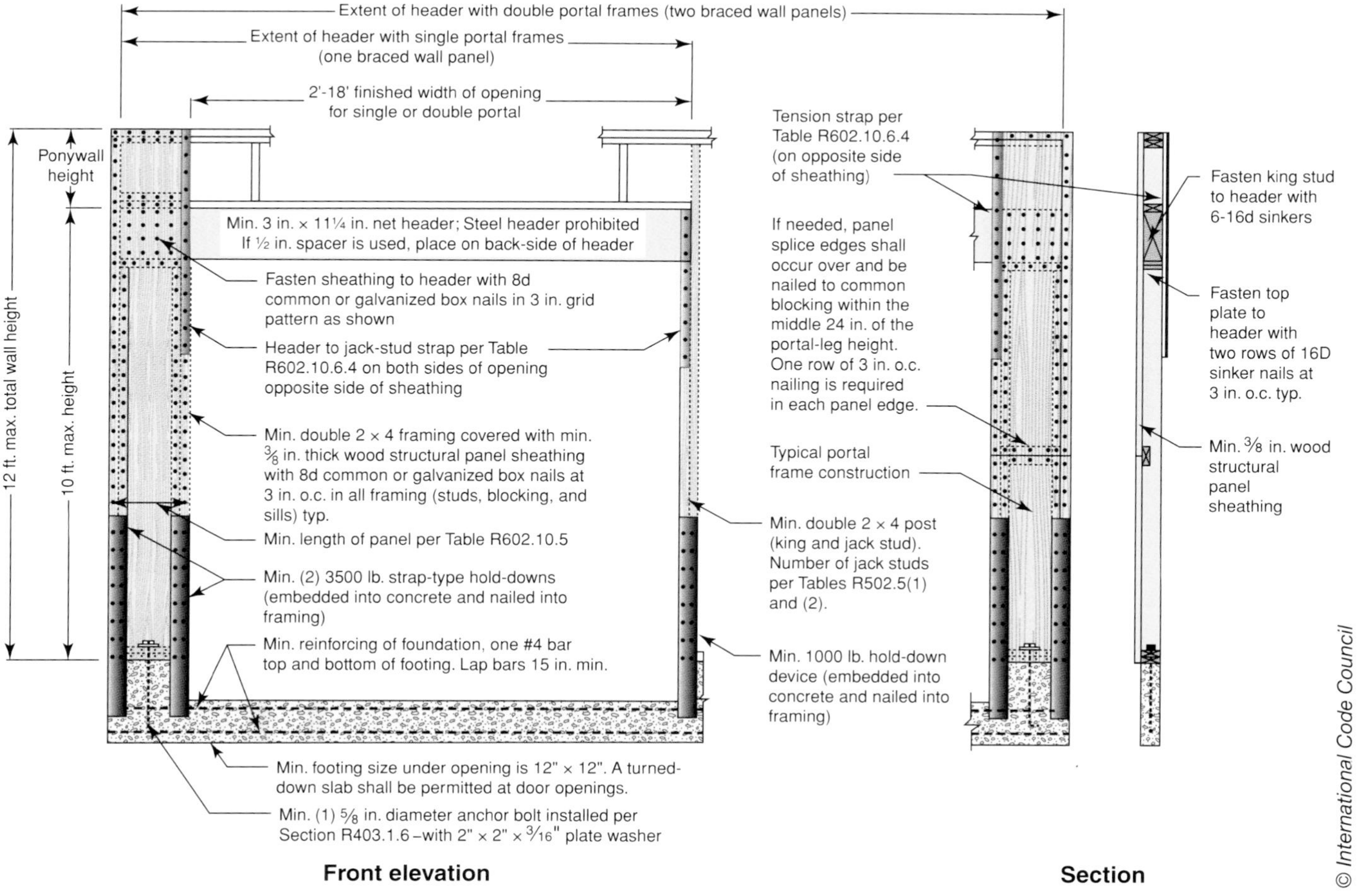

Method PFH: Portal frame with hold-downs

R602.10.6.2 continues

R602.10.6.2 continued

construction industry. During initial testing, no attempt was made to determine minimum hold-down capacity. When design for anchorage in cracked concrete became required in accordance with the American Concrete Institute standard ACI 318, *Building Code Requirements for Structural Concrete*, retesting of the portal frames with 4200-lb. hold-downs became necessary to determine the impact on the performance of the system, if any. As 4200-lb. hold-downs were no longer easily available, other common strap sizes were considered. The 3500-lb. hold-down was selected for testing.

Eight-foot-tall portals with braced wall panels 24 inches long and 10-foot-tall portals with braced wall panels 16 inches long were tested. Double portals were tested with 4200-lb. hold-downs and then retested with 3500-lb. hold-downs. The results indicate that the portal frame is not sensitive to reduction in hold-down capacity in the 4200 lbs to 3500 lbs range. No testing was done to determine a minimum hold-down capacity.

The second technical change removed the third bottom plate. Original testing was conducted with a third plate in place. The third plate causes difficulties in the field—typical anchors are too short to accommodate the third plate and provide the required depth of penetration into the foundation. This results in inadequate anchor depth-of-embedment or use of threaded sleeves and threaded rods to extend the bolt length to accommodate a third plate. When investigating the change in 3500-lb. hold-down capacity, tests were run with double bottom plates. The results of testing indicate that a third bottom plate has little impact on performance of portal frames.

R602.10.11

Cripple Wall Bracing

CHANGE TYPE: Modification

CHANGE SUMMARY: A reduction is no longer required in determining the maximum distance between braced wall panels in a cripple wall. References to the bracing length adjustment tables clarify that increased bracing is required if gypsum wall finish is not applied to the cripple walls.

2015 CODE: R602.10.11 Cripple Wall Bracing. Cripple walls shall be constructed in accordance with Section R602.9 and braced in accordance with this section. Cripple walls shall be braced with the length and method of bracing used for the wall above in accordance with Tables R602.10.3(1) and R602.10.3(3), and the applicable adjustment factors in Tables R602.10.3(2) or R602.10.3(4), respectively, except the length of the cripple wall bracing shall be multiplied by a factor of 1.15. ~~The maximum distance between adjacent edges of braced wall panels shall be reduced from 20 feet (6069 mm) to 14 feet (4267 mm).~~ Where gypsum wall board is not used on the inside of the cripple wall bracing, the length adjustments for the elimination of the gypsum wallboard, or equivalent, shall be applied as directed in Tables R602.10.3(2) and R602.10.3(4) to the length of cripple wall bracing required. This adjustment shall be taken in addition to the 1.15 increase.

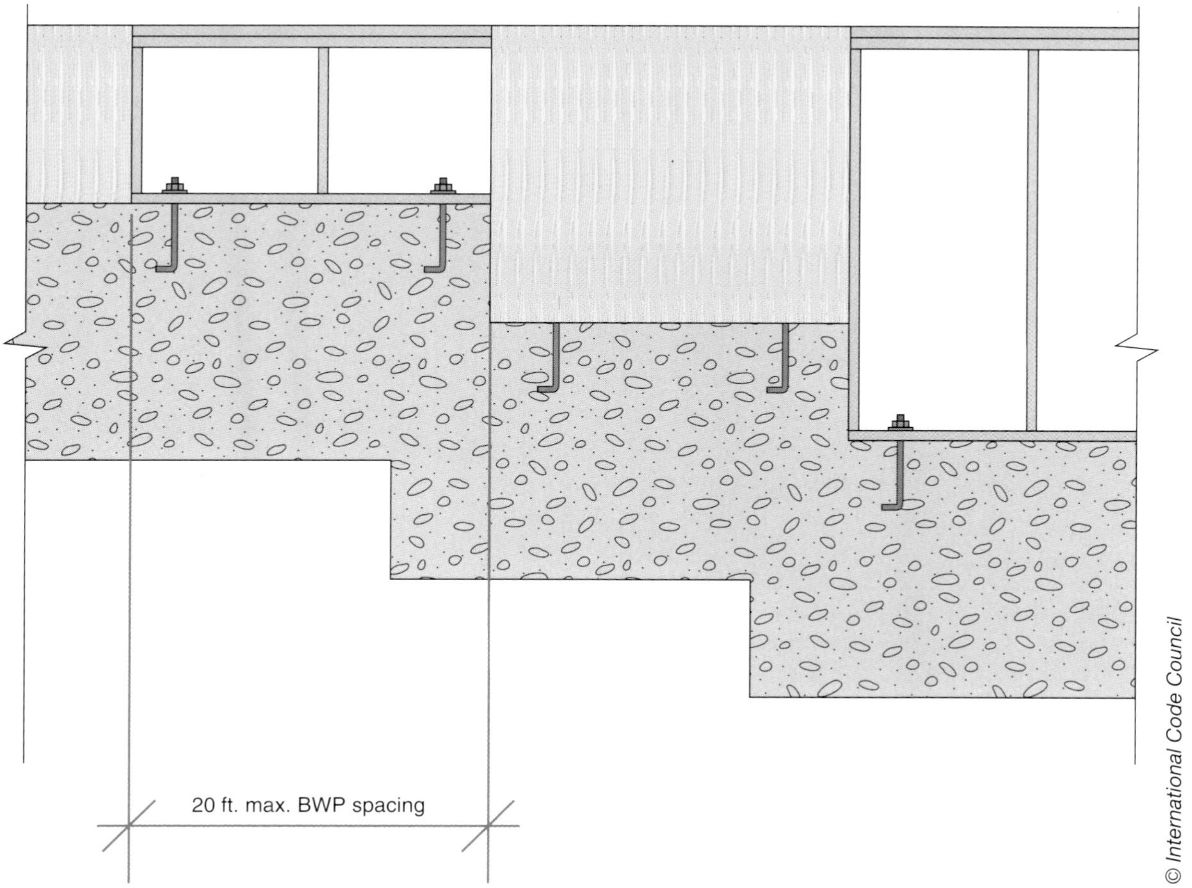

Braced wall panel spacing on cripple walls

R602.10.11 continues

R602.10.11 continued

CHANGE SIGNIFICANCE: Reduction in spacing between braced wall panels in a cripple wall originated from cripple wall failures observed in seismic events such as the 1994 Northridge Earthquake. For low-seismic regions (SDC A and B) and moderate-seismic regions (SDC C), the reduction of cripple wall braced panel spacing is no longer required.

The 2012 IRC cripple wall provisions, Section R602.10.11, require the bracing length of a cripple wall to be increased by a factor of 1.15, based on the length of bracing used in the wall above a cripple wall. Cripple walls are often not fabricated with a gypsum wall board finish on the inside. In order to make resistance to wind or seismic forces equal to the walls above, required bracing lengths in cripple walls must be increased by the applicable adjustment factors (1.4 for wind and 1.5 for seismic) to account for the lack of gypsum board finish. In the 2015 IRC, the section now references Tables R602.10.3(2) and (4) to clarify that the increase applies. The alternative to the bracing increase is to sheath the inside of cripple wall framing with gypsum wall board or an equivalent finish.

Note that the 1.15 increase for cripple walls was part of the code before wall bracing lengths were based on gypsum board, or an equivalent, being required on the inside of braced wall panels. The intent of the IRC is that cripple walls, being below the first story, have an increased bracing length and resistance to lateral loads similar to the increase of bracing from the second story to the first story. The increase for the absence of gypsum board finish is in addition to the 1.15 increase.

CHANGE TYPE: Modification

R602.12

Simplified Wall Bracing

CHANGE SUMMARY: Simplified wall bracing is now allowed for one- to three-story dwellings and townhouses in Wind Exposure Category B or C with ultimate design wind speeds (V_{ult}) of 130 mph or less.

2015 CODE: R602.12 Simplified Wall Bracing. Buildings meeting all of the conditions listed below shall be permitted to be braced in accordance with this section as an alternate to the requirements of

TABLE R602.12.4 Minimum Number of Bracing Units on Each Side of the Circumscribed Rectangle

			Minimum Number of Bracing Units on Each Long Side [a,b,d]					
			Length of short side (ft)[c]					
Ultimate Design Wind Speed (mph)	Story Level	Eave-To Ridge Height (feet)	10	20	30	40	50	60
	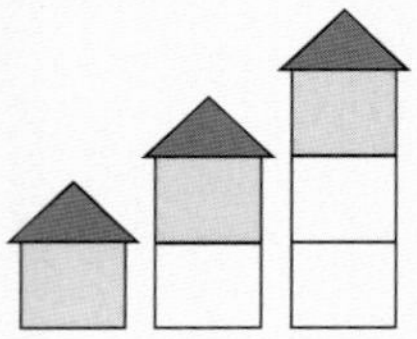		1	2	2	2	3	3
115	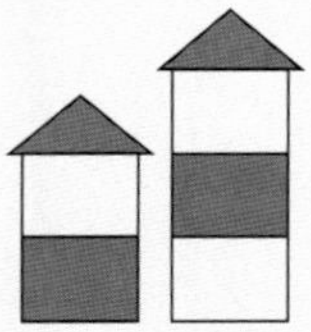	10	2	3	3	4	5	6
	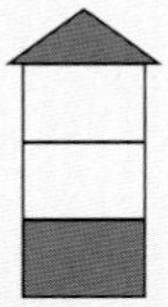		2	3	4	6	7	8
			1	2	3	3	4	4
115	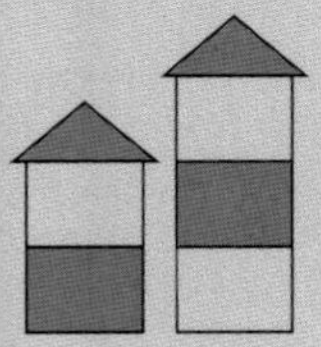	15	2	3	4	5	6	7
	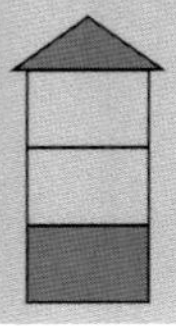		2	4	5	6	7	9

R602.12 continues

R602.12 continued

TABLE R602.12.4 *(Continued)*

Ultimate Design Wind Speed (mph)	Story Level	Eave-To Ridge Height (feet)	Minimum Number of Bracing Units on Each Long Side [a,b,d] Length of short side (ft)[c] 10	20	30	40	50	60
	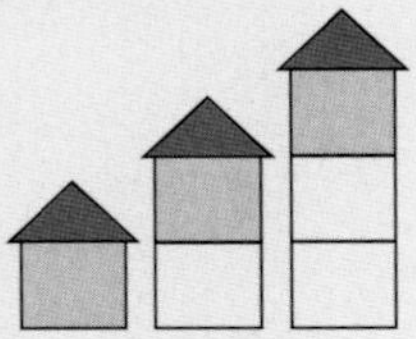		1	2	2	3	3	4
130	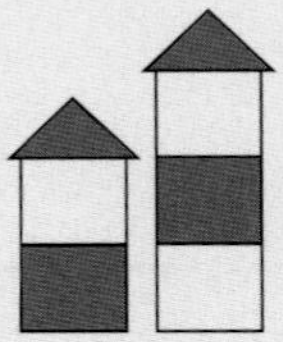	10	2	3	4	5	6	7
	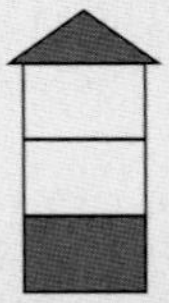		2	4	5	7	8	10
			2	3	3	4	4	6
130	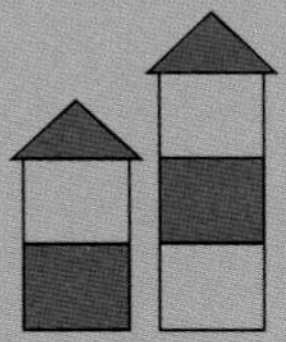	15	3	4	6	7	8	10
			3	6	7	10	11	13

For SI: 1 ft. = 304.8 mm, 1 in. = 25.4 mm

a. Interpolation shall not be permitted.
b. Cripple walls or wood-framed basement walls in a walk-out condition ~~of a one-story structure~~ shall be ~~designed~~ designated as the first ~~floor~~ story ~~of a two-story house~~ and the stories above shall be redesignated as the second and third stories, respectively, and shall be prohibited in a three-story structure.
c. Actual lengths of the sides of the circumscribed rectangle shall be rounded to the next highest unit of 10 when using this table.
d. For exposure category C, multiply bracing units by a factor of 1.20 for a one-story building, 1.30 for a two-story building, and 1.40 for a three-story building.

(Portions of table not shown for brevity and clarity.)

Section R602.10. The entire building shall be braced in accordance with this section; the use of other bracing provisions of R602.10, except as specified herein, shall not be permitted.

1. There shall be not more than ~~two~~ three stories above the top of a concrete or masonry foundation or basement wall. Permanent wood foundations shall not be permitted.
2. Floors shall not cantilever more than 24 inches (607 mm) beyond the foundation or bearing wall below.
3. Wall height shall not be greater than 10 feet (2743 mm).
4. The building shall have a roof eave-to-ridge height of 15 feet (4572 mm) or less.
5. Exterior walls shall have gypsum board with a minimum thickness of ½ inches (12.7 mm) installed on the interior side fastened in accordance with Table R702.3.5.
6. The structure shall be located where the ~~basic~~ ultimate design wind speed is less than or equal to ~~90~~ 130 mph (~~40~~ 58 m/s), and the Exposure Category is ~~A,~~ B or C.
7. The structure shall be located in Seismic Design Category of A, B or C for detached one- and two-family dwellings or Seismic Design Category A or B for townhouses.
8. Cripple walls shall not be permitted in ~~two~~ three-story buildings.

CHANGE SIGNIFICANCE: Using the wall bracing table values for basic wind speeds of 90 mph and 100 mph (ASD values) from 2012 IRC Table R602.10.3(1), use of Simplified Wall Bracing is expanded to a wider range of geographic areas and building types. The 90-mph values in Table R602.12.4 in the 2012 IRC were calculated from Table R602.10.3(1). These values form the basis of the 115-mph V_{ult} values in 2015 IRC Table R602.12.4. The required bracing values for regions with 130-mph wind speed in Table R602.12.4 of the 2015 IRC are also based on Table R602.10.3(1). The wind speed limits have been converted to ultimate design wind speed with a limit of 130 mph (V_{ult}) that is equivalent to the former 100-mph basic wind speed.

Using the adjustments for Wind Exposure Category C from Table R602.10.3(2), use of Simplified Wall Bracing is expanded. Because the values in Table R602.12.4 are calculated from Table R602.10.3(1), adjustment factors create an accurate required bracing length for Wind Exposure Category C similar to the increase in bracing length using Table R602.10.3(2).

R603.9.5

Structural Sheathing over Steel Framing for Stone and Masonry Veneer

CHANGE TYPE: Modification

CHANGE SUMMARY: Section R603.9.5 addressing the bracing requirements for cold-formed steel framing with stone or masonry veneer has been expanded to include the higher seismic design categories. This section directs the user to increase bracing length when a structure is located in SDC C, D_0, D_1 or D_2 and has stone or masonry veneer.

2015 CODE: **R603.9.5 Structural Sheathing for Stone and Masonry Veneer.** ~~In Seismic Design Category C, w~~Where stone and masonry veneer ~~is~~ are installed in accordance with Section R703.8, the length of full-height sheathing for exterior and interior wall lines backing or perpendicular to and laterally supporting walls with veneer shall comply with this section.

R603.9.5.1 Seismic Design Category C. In Seismic Design Category C, the length of structural sheathing for walls supporting one story, roof and ceiling shall be the greater of the amounts required by Section R603.9.2, except Section R603.9.2.2 shall be permitted.

R603.9.5.2 Seismic Design Categories D_0, D_1 and D_2. In Seismic Design Categories D_0, D_1 and D_2, the required length of structural sheathing and overturning anchorage shall be determined in accordance with Tables R603.9.5(1), R603.9.5(2), R603.9.5(3) and R603.9.5(4). Overturning anchorage shall be installed on the doubled studs at the end of each full height wall segment.

TABLE R603.9.5(1) Required Length of Full Height Sheathing and Associated Overturning Anchorage for Walls Supporting Walls with Stone or Masonry Veneer and Using 33-mil Cold-Formed Steel Framing and 6-inch Screw Spacing on the Perimeter of Each Panel of Structural Sheathing

Seismic Design Category	Story	Braced Wall Line Length (feet)						Single-Story Hold-Down Force (pounds)	Cumulative Hold-Down Force (pounds)
		10	20	30	40	50	60		
		Minimum Total Length of Braced Wall Panels Required Along Each Braced Wall Line (feet)							
	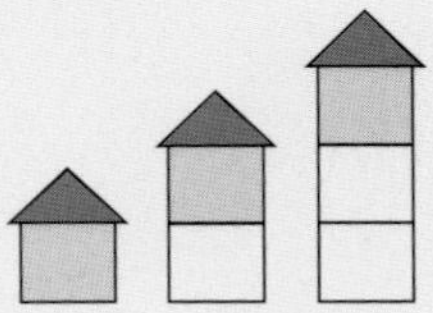	3.3	4.7	6.1	7.4	8.8	10.2	3,360	—
D_0	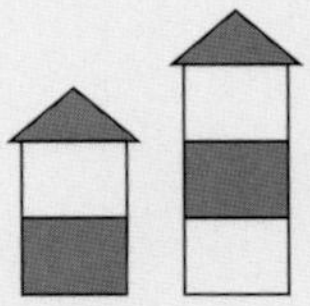	5.3	8.7	12.1	15.4	18.8	22.2	3,360	6,720
	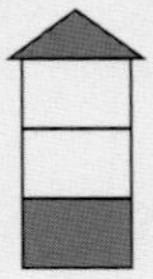	7.3	12.7	18.0	23.4	28.8	34.2	3,360	10,080

TABLE R603.9.5(1) *(Continued)*

Seismic Design Category	Story	Braced Wall Line Length (feet): 10	20	30	40	50	60	Single-Story Hold-Down Force (pounds)	Cumulative Hold-Down Force (pounds)
		Minimum Total Length of Braced Wall Panels Required Along Each Braced Wall Line (feet)							
D1	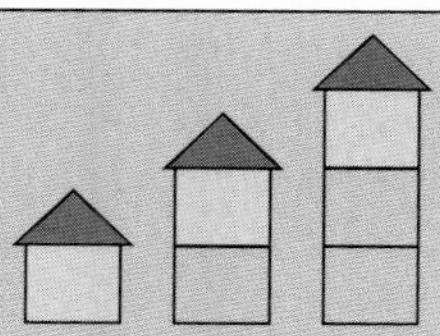	4.1	5.8	7.5	9.2	10.9	12.7	3,360	—
	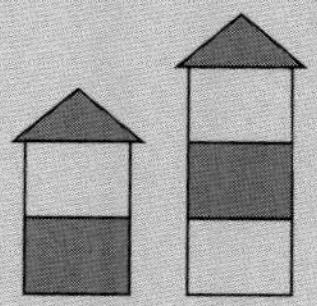	6.6	10.7	14.9	19.1	23.3	27.5	3,360	6,720
	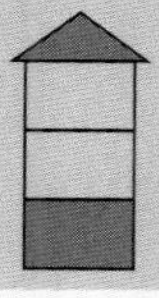	9.0	15.7	22.4	29.0	35.7	42.2	3,360	10,080
D2	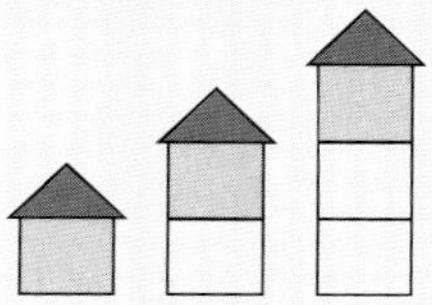	5.7	8.2	10.6	13.0	15.4	17.8	3,360	—
	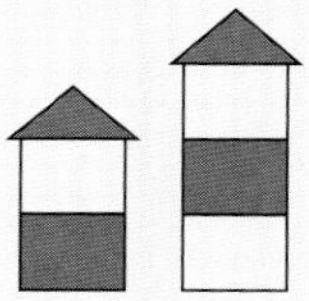	9.2	15.1	21.1	27.0	32.9	38.8	3,360	6,720
	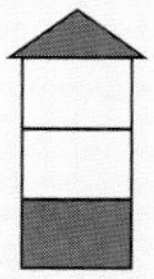	12.7	22.1	31.5	40.9	50.3	59.7	3,360	10,080

(Tables R603.9.5(2)-(4) not shown for brevity.)

CHANGE SIGNIFICANCE: This code change ensures that cold-formed steel framing can resist the higher lateral loads associated with use of masonry veneer in high-seismic regions. The original provisions for anchoring masonry veneer to residential buildings were developed with the concept of anchoring to wood framing. Cold-formed steel framing can also be used to support veneer. Connections between cold-formed steel members must be appropriate for the fasteners used to tie veneer to walls.

In Section R603.9.2 an exception has been added to the method of determining the minimum length of sheathing (bracing) required to account for the increased lateral loads associated with stone or masonry veneer. This increase in the weight of the wall is only considered in high-seismic regions. An exception points to Tables R603.9.5(1) through

R603.9.5 continues

R603.9.5 continued

R603.9.5(4) providing required minimum sheathing length for each seismic design category.

Values in the four tables are based on allowable design values provided in AISI S213-07, *North American Standard for Cold-Formed Steel Framing—Lateral Design*, including Supplement No. 1. The values are located in Table C2.1-3 for 33-mil and 43-mil stud thicknesses and 6-inch and 4-inch screw spacing around the perimeter. It is assumed that the in-line framing concept of cold-formed steel light-frame construction provides the continuous load path required to transfer overturning loads to the foundation. Walls from one story to another must be aligned.

Assumptions for building configuration for the tables include:

1. Twenty percent of the wall area has door and window openings.
2. Wall veneer seismic weight perpendicular to the direction of analysis contributes to lateral forces. The masonry veneer in-plane, on walls resisting the lateral load, supports its own seismic weight.
3. Maximum story height for each floor is 10 feet.
4. Masonry is 4-inch-thick clay masonry with 40-psf dead load.

CHANGE TYPE: Reorganization

CHANGE SUMMARY: Sections R606, R607, R608, and R609 have been organized into one section providing requirements for masonry construction of single- and two-family dwellings and townhouses. Table 6-1 summarizes the relocation of sections in the 2015 IRC and lists sections that are new.

R606

Masonry Walls

2015 CODE: **Section R606**

TABLE 6-1 Summary of Changes to the Masonry Wall Construction Provisions of Section R606

Section Title	2015 IRC	2012 IRC
MASONRY CONSTRUCTION MATERIALS.	R606.2	NEP
Concrete masonry units.	R606.2.1	NEP
Clay or shale masonry units.	R606.2.2	NEP
AAC masonry.	R606.2.3	NEP
Stone masonry units.	R606.2.4	NEP
Architectural cast stone.	R606.2.5	NEP
Second-hand units.	R606.2.6	NEP
Mortar.	R606.2.7	R607.1
Foundation walls.	R606.2.7.1	R607.1.1
Masonry in SDC A, B and C.	R606.2.7.2	R607.1.2
Masonry in SDC D_0, D_1 and D_2.	R606.2.7.3	R607.1.3
Mortar Proportions.	Table R606.2.7	Table R607.1
Surface-bonding mortar.	R606.2.8	NEP
Mortar for AAC masonry.	R606.2.9	NEP
Mortar for adhered masonry veneer.	R606.2.10	NEP
Grout.	R606.2.11	R609.1.1
Grout Proportions by Volume for Masonry Construction.	Table R606.2.11	Table R609.1.1
Metal Reinforcement and Accessories.	R606.2.12	R606.15
CONSTRUCTION REQUIREMENTS.	R606.3	NEP
Bed and head joints.	R606.3.1	R607.2.1
Mortar joint thickness tolerance.	—	R607.2.1.1
Masonry unit placement.	R606.3.2	R607.2.2
Solid masonry.	R606.3.2.1	R607.2.2.1
Hollow masonry.	R606.3.2.2	R607.2.2.2
Installation of wall ties.	R606.3.3	R607.3
Protection for reinforcement.	R606.3.4	R606.13
Corrosion protection.	R606.3.4.1	R606.15.1
Minimum Corrosion Protection.	Table R606.3.4.1	Table R606.15.1
Grouting requirements.	R606.3.5	R609.1.2
Grout placement.	R606.3.5.1	R609.1.4
Grout Space Dimensions and Pour Heights.	Table R606.3.5.1	Table R609.1.2
Cleanouts.	R606.3.5.2	R609.1.5
Construction.	R606.3.5.3	R609.3.1, R609.4.1

R606 continues

R606 continued

TABLE 6-1 (*Continued*)

Section Title	2015 IRC	2012 IRC
Grouted multiple-wythe masonry.	R606.3.6	R609.2
Bonding of backup wythe.	R606.3.6.1	R609.2.1
Grout barriers.	R606.3.6.2	R609.2.3
Masonry bonding pattern.	R606.3.7	R608.2
Masonry laid in running bond.	R606.3.7.1	R608.2.1
Masonry laid in stack bond.	R606.3.7.2	R608.2.2
THICKNESS OF MASONRY.	R606.4	R606.2
Minimum thickness.	R606.4.1	R606.2.1
Rubble stone masonry wall.	R606.4.2	R606.4.2
Change in thickness.	R606.4.3	R606.2.3
Parapet walls.	R606.4.4	R606.2.4
CORBELED MASONRY.	R606.5	R606.3
Units.	R606.5.1	R606.3.1
Corbel projection.	R606.5.2	R606.3.2
Corbeled masonry supporting floor or roof-framing members.	R606.5.3	R606.3.3
SUPPORT CONDITIONS.	R606.6	R606.4
Bearing on support.	R606.6.1	R606.4.1
Support at foundation.	R606.6.2	R606.4.2
Beam supports.	R606.6.3	R606.14
Joist bearing.	R606.6.3.1	R606.14.1
Lateral support.	R606.6.4	R606.9
Spacing of Lateral Support for Masonry Walls.	Table R606.6.4	Table R606.9
Horizontal lateral support.	R606.6.4.1	R606.9.1
Bonding pattern.	R606.6.4.1.1	R606.9.1.1
Metal reinforcement.	R606.6.4.1.2	R606.9.1.2
Vertical lateral support.	R606.6.4.2	R606.9.2
Roof structures.	R606.6.4.2.1	R606.9.2.1
Floor diaphragms.	R606.6.4.2.2	R606.9.2.2
PIERS.	R606.7	R606.6
Pier cap.	R606.7.1	R606.6.1
CHASES.	R606.8	R606.7
ALLOWABLE STRESSES.	R606.9	R606.5
Combined units.	R606.9.1	R606.5.1
Allowable Compressive Stresses for Empirical Design of Masonry.	Table R606.9	Table R606.5
LINTELS.	R606.10	R606.10
ANCHORAGE.	R606.11	R606.11
Anchorage requirements for masonry walls located in Seismic Design Category A, B, or C and where wind loads are less than 30 psf.	Figure R606.11(1)	Figure R606.11(1)
Requirements for reinforced grouted masonry construction in Seismic Design Category C.	Figure R606.11(2)	Figure R606.11(2)
Requirements for reinforced masonry construction in Seismic Design Category D_0, D_1 or D_2.	Figure R606.11(3)	Figure R606.11(3)

Section Title	2015 IRC	2012 IRC
SEISMIC REQUIREMENTS.	R606.12	R606.12
General.	R606.12.1	R606.12.1
Floor and roof diaphragm construction.	R606.12.1.1	R606.12.1.1
Seismic design category C.	R606.12.2	R606.12.2
Minimum length of wall without openings.	R606.12.2.1	R606.12.2.1
Minimum solid wall length along exterior wall lines.	Table R606.12.2.1	Table R606.12.2.1
Design of elements not part of the lateral force-resisting system.	R606.12.2.2	R606.12.2.2
Load-bearing frames or columns.	R606.12.2.2.1	R606.12.2.2.1
Masonry partition walls.	R606.12.2.2.2	R606.12.2.2.2
Reinforcement requirements for masonry elements.	R606.12.2.2.3	R606.12.2.2.3
Design of elements part of the lateral force-resisting system.	R606.12.2.3	R606.12.2.3
Connections to masonry shear walls.	R606.12.2.3.1	R606.12.2.3.1
Connections to masonry columns.	R606.12.2.3.2	R606.12.2.3.2
Minimum reinforcement requirements for masonry shear walls.	R606.12.2.3.3	R606.12.2.3.3
Seismic Design Category D_0 or D_1.	R606.12.3	R606.12.3
Design requirements.	606.12.3.1	606.12.3.1
Minimum reinforcement requirements for masonry walls.	R606.12.3.2	R606.12.3.2
Minimum Distributed Wall Reinforcement for Buildings Assigned to Seismic Design Category D_0 or D_1.	Table R606.12.3.2	Table R606.12.3.2
Shear wall reinforcement requirements.	R606.12.3.2.1	R606.12.3.2.1
Minimum reinforcement for masonry columns.	R606.12.3.3	R606.12.3.3
Material restrictions.	R606.12.3.4	R606.12.3.4
Lateral tie anchorage.	R606.12.3.5	R606.12.3.5
Seismic Design Category D_2.	R606.12.4	R606.12.4
Design of elements not part of the lateral force-resisting system.	R606.12.4.1	R606.12.4.1
Minimum Reinforcing for Stacked Bonded Masonry Walls in Seismic Design Category D_2.	Table R606.12.4.1	Table R606.12.4.1
Design of elements part of the lateral force-resisting system.	R606.12.4.2	R606.12.4.2
Minimum Reinforcing for Stacked Bonded Masonry Walls in Seismic Design Category D_2.	Table R606.12.4.2	Table R606.12.4.2
MULTIPLE-WYTHE MASONRY.	R606.13	R608.1
Bonding with masonry headers.	R606.13.1	R608.1.1
Solid units.	R606.13.1.1	R608.1.1.1
Hollow units.	R606.13.1.2	R608.1.1.2
Bonding with wall ties or joint reinforcement.	R606.13.2	R608.1.2
Bonding with wall ties.	R606.13.2.1	R608.1.2.1
Bonding with adjustable wall ties.	R606.13.2.2	R608.1.2.2
Bonding with prefabricated joint reinforcement.	R606.13.2.3	R608.1.2.3
Bonding with natural or cast stone.	R606.13.3	R608.1.3
Ashlar masonry.	R606.13.3.1	R608.1.3.1
Rubble stone masonry.	R606.13.3.2	R608.1.3.2
ANCHORED AND ADHERED MASONRY VENEER.	R606.14	NEP
Anchored veneer.	R606.14.1	NEP
Adhered veneer.	R606.14.2	NEP

NEP = No equivalent provision, new to the 2015 IRC.

R606 continues

R606 continued

CHANGE SIGNIFICANCE: Section R606 is a cleanup and consolidation of the masonry design and construction requirements scattered throughout Sections R606, R607, R608, and R609 in the 2012 IRC. Provisions in Section R606, covering above-ground masonry wall construction, evolved independently resulting in conflicting and disconnected code provisions. For example, mortar requirements for masonry construction were covered in Section R607; however, these requirements were not referenced by Sections R606, R608, or R609.

Given the substantial reorganization, there are some technical differences compared to the 2012 IRC requirements in Sections R606, R607, R608, and R609:

- Section R606.2 has been added to define minimum requirements for masonry materials. Whereas the IRC had covered material requirements for mortar and grout, masonry unit requirements were not explicitly defined. The new provisions mirror material requirements in the IBC.
- Grout lift requirements triggering cleanouts are combined. A new trigger of 64 inches for all masonry construction is used. This is consistent with current IBC requirements. Similarly, grout lift requirements triggering special inspection are increased from 60 to 64 inches for consistency.
- Section R606.12.3 introduces a limit on the use of Autoclaved Aerated Concrete (AAC) masonry in shear walls assigned to SDC D_0–D_2.
- Section R606.14 is added to provide a pointer to the anchored and adhered veneer provisions of Chapter 7.

Note: Section R606 does not address masonry veneer, which is located in Chapter 7, or masonry foundations, which are described in Chapter 4.

R606.3.5

Grouting Requirements for Masonry Construction

CHANGE TYPE: Modification

CHANGE SUMMARY: With reorganization of the masonry wall provisions in the 2015 IRC, the section covering provisions for grouting above-ground masonry walls now combines all the requirements for single, multi-wythe, and reinforced masonry construction in one section. Clarified provisions address grout placement, cleanouts, and construction for all three types of masonry construction.

2015 CODE: **R606.3.5 Grouting Requirements.**

~~R609.1.4~~ R606.3.5.1 Grout Placement. Grout shall be a plastic mix suitable for pumping without segregation of the constituents and shall be mixed thoroughly. Grout shall be placed by pumping or by an *approved* alternate method and shall be placed before any initial set occurs and ~~in no case~~ not more than 1½ hours after water has been added. ~~Grouting shall be done in a continuous pour, in lifts not exceeding 5 feet (1524 mm). It~~ Grout shall be consolidated by puddling or mechanical vibrating during placing and reconsolidated after excess moisture has been absorbed but before plasticity is lost. Grout shall not be pumped through aluminum pipes.

~~R609.1.2 Grout Requirements.~~ Maximum pour heights and the minimum dimensions of spaces provided for grout placement shall conform to Table R606.3.5.1. Grout shall be poured in lifts of 8-foot (2438 mm) maximum height. When a total grout pour exceeds 8 feet (2438 mm) in height, the grout shall be placed in lifts not exceeding 64 inches (1626 mm) and special inspection during grouting shall be required. If the work is stopped for one hour or longer, the horizontal construction joints shall be formed by stopping all tiers at the same elevation and with the grout 1 inch (25 mm) below the top.

Photo courtesy of David Guinther of San Jose First United Methodist Church

Grouted cells

R606.3.5 continues

R606.3.5 continued

~~R609.1.3 Grout Space (Cleaning)~~ R606.3.5.2 Cleanouts. Provision shall be made for cleaning the space to be grouted. Mortar that projects more than ½ inch (12.7 mm) into the grout space and any other foreign matter shall be removed from the grout space prior to inspection and grouting.

~~R609.1.5 Cleanouts.~~ Where required by the *building official*, cleanouts shall be provided in the bottom course of masonry for each grout pour where the grout pour height exceeds 64 inches (1626 mm). In solid grouted masonry, cleanouts shall be spaced horizontally not more than 32 inches (813 mm) on center. The cleanouts shall be sealed before grouting and after inspection.

~~R609.4.1~~ R606.3.5.3 Construction. Requirements for grouted masonry construction shall be as follows:

1. ~~Reinforced hollow-unit masonry shall be built to preserve the unobstructed vertical continuity of the cells to be filled.~~ Masonry shall be built to preserve the unobstructed vertical continuity of the cells or spaces to be filled. In partially grouted construction, cross webs forming cells to be filled shall be full-bedded in mortar to prevent leakage of grout. Head and end joints shall be solidly filled with mortar for a distance in from the face of the wall or unit not less than the thickness of the longitudinal face shells.

~~2. Cells to be filled shall have vertical alignment sufficient to maintain a clear, unobstructed continuous vertical cell of dimensions prescribed in Table R609.1.2.~~

~~3.~~2. Vertical reinforcement shall be held in position at top and bottom and at intervals not exceeding 200 diameters of the reinforcement.

~~4.~~3. Cells containing reinforcement shall be filled solidly with grout. ~~Grout shall be poured in lifts of 8-foot (2438 mm) maximum height. When a total grout pour exceeds 8 feet (2438 mm) in height, the grout shall be placed in lifts not exceeding 5 feet (1524 mm) and special inspection during grouting shall be required.~~

~~5. Horizontal steel shall be fully embedded by grout in an uninterrupted pour.~~

4. **~~R609.3.1 Construction.~~** The thickness of grout or mortar between masonry units and reinforcement shall not be less than ¼ inch (6.4 mm), except that ¼-inch (6.4-mm) bars shall be permitted to be laid in horizontal mortar joints ~~at least~~ not less than ½ inch (12.7 mm) thick, and steel wire reinforcement shall be permitted to be laid in horizontal mortar joints ~~at least~~ not less than twice the thickness of the wire diameter.

CHANGE SIGNIFICANCE: In the 2012 IRC there were several conflicts in the grouting requirements. Grout pour height triggering cleanouts varied depending upon whether the masonry construction was multi-wythe, single-wythe, or reinforced. Now the grout lift height triggering cleanouts is 64 inches for all masonry construction. This height is also consistent with current IBC requirements. Similarly, grout lift requirements triggering special inspection are increased from 60 to 64 inches for consistency.

In Section R606.3.5.1, grout placement, maximum grouting height, and grout lift height are all addressed. This section also contains one of the few locations in the IRC requiring special inspection. Special inspection is required when multiple grout lifts are needed, in other words, when the wall height is greater than 8 feet. Special inspection is not defined in the IRC, but the intent is that this installation must be checked by an independent third-party inspector who has expertise in masonry wall construction.

Construction requirements for different masonry construction types are combined into Section R606.3.5.3 with four total requirements. Alignment of units, minimum reinforcement, and minimum grout thickness are all addressed in this section.

R610.7

Drilling and Notching in Structural Insulated Panels

Holes made onsite in structural insulated panels

CHANGE TYPE: Modification

CHANGE SUMMARY: Drilling and notching provisions for structural insulated panels (SIP) are clarified.

2015 CODE: **R610.7 Drilling and Notching**. The maximum vertical chase penetration in SIPs shall have a maximum side dimension of 2 inches (51 mm) centered in the panel. Vertical chases shall have a minimum spacing of 24 inches (610 mm) on center. Maximum of two horizontal chases shall be permitted in each wall panel—one at 14 inches (360 mm) plus or minus 2 inches (51 mm) from the bottom of the panel and one ~~at mid-height of the wall panel core~~ at 48 inches (1220 mm) plus or minus 2 inches (51 mm) from the bottom edge of the SIPs panel. ~~The maximum allowable penetration size in a wall panel shall be as shown on the manufacturer's shop drawings circular or rectangular with a maximum dimension of 12 inches (300 mm). Overcutting of holes in facing panels shall not be permitted.~~ Additional penetrations are permitted where justified by analysis.

CHANGE SIGNIFICANCE: The wording of Section R613.7 in the 2012 IRC was based on an 8-foot-tall structural insulated panel (SIP). As Section R613 permitted up to 10-foot-tall walls, the horizontal chases, which are used for switch-box wiring, need to be placed 48 inches above the bottom edge of the SIP, which is standard switch-box height. A chase in SIP construction is a drilled hole or slot through which cables are run, typically for electrical connections. Chases may be created by the manufacturer or drilled onsite.

In 2015 IRC Section R610.7, a tolerance is added to the dimension for ease of use in the field. Requirements for manufacturers' shop drawings showing holes have been deleted. Prescriptive maximum size and location are now the only requirements for vertical and horizontal chases. Any additional holes or chases require engineered design.

CHANGE TYPE: Modification

R703.3

Siding Material Thickness and Attachment

CHANGE SUMMARY: Table R703.4, Weather Resistant Siding Attachment and Minimum Thickness, is simplified. New code language is added to Section R703 to clarify limitations of use of the table and to describe fastener type, length, and penetration.

2015 CODE: ~~**R703.4**~~ **R703.3 Nominal Thickness and Attachments.** ~~Unless specified otherwise, all~~ The nominal thickness and attachment of exterior wall coverings shall be ~~securely fastened~~ in accordance with Table ~~R703.4~~ R703.3(1), the wall covering material requirements of this section, and the wall covering manufacturer's installation instructions ~~or with other *approved* aluminum, stainless steel, zinc-coated or other *approved* corrosion-resistive fasteners~~. Cladding attachment over foam sheathing shall comply with the additional requirements and limitations of Sections R703.15 through R703.17. Nominal material thicknesses in Table R703.3(1) are based on a maximum stud spacing of 16 inches (406 mm) on center. Where specified by the siding manufacturer's instructions and supported by a test report or other documentation, attachment to studs with greater spacing is permitted. Fasteners for exterior wall coverings attached to wood framing shall be in accordance with Section R703.3.2 and Table R703.3(1). Exterior wall coverings shall be attached to cold-formed steel light frame construction in accordance with the cladding manufacturer's installation instructions, the requirements of Table R703.3(1) using screw fasteners substituted for the nails specified in accordance with Table R703.3(2), or an approved design.

R703.3.1 Wind Limitations. Where the ~~basic wind speed in accordance with Figure R301.2(4)A is 110 miles per hour (49 m/s) or higher~~ design wind pressure exceeds 30 psf (1436 Pa), or where the limits of Table R703.3.1 are exceeded, the attachment of wall coverings shall be designed to resist the component and cladding loads specified in Table R301.2(2), adjusted for height and exposure in accordance with Table R301.2(3). For the determination of wall covering attachment, component and cladding loads shall be determined using an effective wind area of 10 ft^2 (0.93 m^2).

TABLE R703.3.1 Limits for Attachment per Table R703.3(1)

Maximum Mean Roof Height			
Ultimate Wind Speed (mph, 3-second gust)	Exposure		
	B	C	D
115	NL	50′	20′
120	NL	30′	DR
130	60′	15′	DR
140	35′	DR	DR

For SI: 1 foot = 304.8 mm, 1 mile per hour = 0.447 m/s
NL = not limited by Table R703.3.1, DR = Design Required

R703.3 continues

R703.3 continued

R703.3.2 Fasteners. Exterior wall coverings shall be securely fastened with aluminum, galvanized, stainless steel, or rust-preventative coated nails or staples in accordance with Table R703.3(1) or with other approved corrosion-resistant fasteners in accordance with the wall covering manufacturer's installation instructions. Nails and staples shall comply with ASTM F1667. Nails shall be T-head, modified round head, or round head with smooth or deformed shanks. Staples shall have a minimum crown width of 7/16-inch (11.1 mm) outside diameter and be manufactured of minimum 16 gage wire. Where fiberboard, gypsum, or foam plastic sheathing backing is used, nails or staples shall be driven into the studs. Where wood or wood structural panel sheathing is used, fasteners shall be driven into studs unless otherwise permitted to be driven into sheathing in accordance with either the siding manufacturer's installation instructions or Table R703.3.2.

TABLE R703.3.2 Optional Siding Attachment Schedule for Fasteners Where No Stud Penetration Necessary

Application	Number and type of Fastener	Spacing of Fasteners[b]
Exterior wall covering (weighing 3 psf or less) attachment to wood structural panel sheathing, either direct or over foam sheathing a maximum of 2 inches thick.[a]	Ring shank roofing nail (0.120″ min. dia.)	12″ o.c.
	Ring shank nail (0.148″ min. dia.)	15″ o.c.
Note: Does not apply to vertical siding.	No. 6 screw (0.138″ min. dia.)	12″ o.c.
	No. 8 screw (0.164″ min. dia.)	16″ o.c.

a. Fastener length shall be sufficient to penetrate back side of the wood structural panel sheathing by at least ¼ inch. The wood structural panel sheathing shall be not less than 7/16 inch in thickness.

b. Spacing of fasteners is per 12 inches of siding width. For other siding widths, multiply spacing of fasteners above by a factor of 12/s, where s is the siding width in inches. Fastener spacing shall never be greater than the manufacturer's minimum recommendations.

R703.3.3 Minimum Fastener Length and Penetration. Fasteners shall have the greater of the minimum length specified in Table R703.3(1) or as required to provide a minimum penetration into framing as follows:

1. Fasteners for horizontal aluminum siding, steel siding, particleboard panel siding, wood structural panel siding in accordance with ANSI/APA-PRP 210, fiber-cement panel siding, and fiber-cement lap siding installed over foam plastic sheathing shall penetrate not less than 1½ inches (38 mm) into framing or shall be in accordance with the manufacturer's installation instructions.
2. Fasteners for hardboard panel and lap siding shall penetrate not less than 1½ inches (38 mm) into framing.
3. Fasteners for vinyl siding and insulated vinyl siding installed over wood or wood structural panel sheathing shall penetrate not less than 1¼ inches (32 mm) into sheathing and framing combined. Vinyl siding and insulated vinyl siding shall be permitted to be installed with fasteners penetrating into or through wood or wood structural sheathing of minimum thickness as specified by

TABLE R703.3(1) Siding Minimum Attachment and Minimum Thickness

Siding Material		Nominal Thickness (inches)	Joint Treatment	Type of Supports for the Siding Material and Fasteners: Wood or wood structural panel sheathing into stud	Type of Supports for the Siding Material and Fasteners: Fiberboard sheathing into stud
Anchored veneer: brick, concrete, masonry or stone (See Section R703.8)		2	Section R703.8	Section R703.8	
Adhered veneer: concrete, stone or masonry (See Section R703.12)		—	Section R703.12	Section R703.12	
Fiber-cement siding	Panel siding (See Section R703.10.1)	$^{5}/_{16}$	Section R703.10.1	6d common (2″ × 0.113″)	6d common (2″ × 0.113″)
	Lap siding (See Section R703.10.2)	$^{5}/_{16}$	Section R703.10.2	6d common (2″ × 0.113″)	6d common (2″ × 0.113″)
Hardboard panel siding (See Section R703.3)		$^{7}/_{16}$	—	0.120″ nail (shank) with 0.225″ head	0.120″ nail (shank) with 0.225″ head
Hardboard lap siding (See Section R703.3)		$^{7}/_{16}$	Note e	0.099″ nail (shank) with 0.240″ head	0.099″ nail (shank) with 0.240″ head
Horizontal aluminum[a]	Without insulation	0.019[b]	Lap	Siding nail 1½″ × 0.120″	Siding nail 2″ × 0.120″
	Without insulation	0.024	Lap	Siding nail 1½″ × 0.120″	Siding nail 2″ × 0.120″
	With insulation	0.019	Lap	Siding nail 1½″ × 0.120″	Siding nail 2½″ × 0.120″
Insulated Vinyl Siding[j] (See Section R703.13)		0.035 (vinyl siding layer only)	Lap	0.120 nail (shank) with a 0.313 head or 16 gauge crown[h,i]	0.120 nail (shank) with a 0.313 head or 16 gauge crown[h]
Particleboard panels		$^{3}/_{8}$	—	6d box nail (2″ × 0.099″)	6d box nail (2″ × 0.099″)
		$^{1}/_{2}$	—	6d box nail (2″ × 0.099″)	6d box nail (2″ × 0.099″)
		$^{5}/_{8}$	—	6d box nail (2″ × 0.099″)	8d box nail (2½″ × 0.113″)
Polypropylene siding[k]		Not applicable	Lap	Section R703.14.1	Section R703.14.1
Steel[c]		29 ga.	Lap	Siding nail (1¾″ × 0.113″) Staple–1¾″	Siding nail (2¾″ × 0.113″) Staple–2½″
Vinyl siding (see Section R703.11)		0.035	Lap	0.120″ nail (shank) with a 0.313″ head or 16 gauge staple with $^{3}/_{8}$ to ½-inch crown[h,i]	0.120″ nail (shank) with a 0.313″ head or 16 gauge staple with $^{3}/_{8}$ to ½-inch crown[h]
Wood siding (See Section R703.3)	Wood rustic, drop	$^{3}/_{8}$ Min	Lap	6d box or siding nail (2″ × 0.099″)	6d box or siding nail (2″ × 0.099″)
	Shiplap	$^{19}/_{32}$ Average	Lap		
	Bevel	$^{7}/_{16}$	Lap		
	Butt tip	$^{3}/_{16}$	Lap		
Wood structural panel ANSI/APA PRP-210 siding (exterior grade) (See Section R703.3)		$^{3}/_{8}$ – ½	Note e	2″ × 0.099″ siding nail	2½″ × 0.113″ siding nail
Wood structural panel lap siding (See Section R703.3)		$^{3}/_{8}$ – ½	Note e Note g	2″ × 0.099″ siding nail	2½″ × 0.113″ siding nail

For SI: 1 inch = 25.4 mm.

h. Minimum fastener length must be sufficient to penetrate sheathing, other nailable substrate and framing a total of a minimum of 1¼ inches or in accordance with the manufacturer's installation instructions.

i. Where specified by the manufacturer's instructions and supported by a test report, fasteners are permitted to penetrate into or fully through nailable sheathing or other nailable substrate of minimum thickness specified by the instructions or test report, without penetrating into framing.

j. Insulated vinyl siding shall comply with ASTM D7793.

k. Polypropylene siding shall comply with ASTM D7254.

(Portions of table not shown for brevity and clarity.)

R703.3 continues

R703.3 continued

the manufacturer's instructions or test report, with or without penetration into the framing. Where the fastener penetrates fully through the sheathing, the end of the fastener shall extend not less than ¼ inch (6.4 mm) beyond the opposite face of the sheathing. Fasteners for vinyl siding and insulated vinyl siding installed over foam plastic sheathing shall be in accordance with Section R703.11.2. Fasteners for vinyl siding and insulated vinyl siding installed over fiberboard or gypsum sheathing shall penetrate not less than 1¼ inches (32 mm) into framing.

4. Fasteners for vertical or horizontal wood siding shall penetrate not less than 1½ inches (38 mm) into studs, studs and wood sheathing combined, or blocking.
5. Fasteners for siding material installed over foam plastic sheathing shall have sufficient length to accommodate foam plastic sheathing thickness and to penetrate framing or sheathing and framing combined as specified in Items 1 through 4.

TABLE R703.3(2) Screw Fastener Substitution for Siding Attachment to Cold-Formed Steel Light Frame Construction[a,b,c,d,e]

Nail Diameter per Table R703.3(1)	Minimum Screw Fastener Size
0.099″	No. 6
0.113″	No. 7
0.120″	No. 8

For SI: 1 inch = 25.4 mm

a. Screws shall comply with ASTM C1513 and shall penetrate a minimum of three threads through minimum 33 mil (20 gauge) cold-formed steel frame construction.
b. Screw head diameter shall not be less than the nail head diameter required by Table R703.3(1).
c. Number and spacing of screw fasteners shall comply with Table R703.3(1).
d. Pan head, hex washer head, modified truss head, or other screw head types with a flat attachment surface under the head shall be used for vinyl siding attachment.
e. Aluminum siding shall not be fastened directly to cold-formed steel light frame construction.

CHANGE SIGNIFICANCE: The 2012 IRC Table R703.4 is replaced with a simplified table, and former footnotes and column text are added to the provisions relating to siding attachment.

Section R703.3 covers nominal thickness and attachment of siding. Attachment is based on 16 inch on center stud spacing. Table R703.3(1) is referenced for nail specifications, which have been reformatted to match Table R602.3(1). Minimum fastener size and penetration requirements, along with other installation details, are in Table R703.3(1) (former Table R703.4) and mirror current installation guides. The *water-resistive barrier required* column is deleted. All products in Table R703.3(1) require a water-resistive barrier. Exceptions remain in Section R703.2 for detached accessory buildings and certain paper-backed stucco lath products.

Section R703.3.1 references Table R301.2(2) for required component and cladding loads given the dwelling's location. New language requires design wind pressures to be determined using an effective wind area of 10 square feet. For wall cladding, the effective wind area will be governed by the effective wind area of an individual fastener, which is less than 10 square feet.

Table R703.3.1 has been added to simplify the determination of whether the prescriptive fastening provisions of Table R703.3(1) apply to a specific building. Fasteners are limited to a maximum design pressure of 30 psf. The limits in the table indicate where component and cladding pressures exceed 30 psf as a function of wind speed exposure and mean roof height. In most cases, especially in areas with lower wind speeds, the prescriptive fastening requirements in Table R703.3(1) will be applicable.

According to Table R301.2(2), for Zone 5 and an effective wind area of 10 square feet, the maximum negative pressure for an ultimate design wind speed of 140 mph is 28.0 psf (this is approximately equal to the old 110-mph basic wind speed). This value—less than 30 psf—correlates directly with the 140-mph limitation in Table R703.3.1. However, the tabulated pressures in Table R301.2(2) are for an assumed Wind Exposure B site condition and a mean roof height of 30 feet. For residential buildings with an ultimate design wind speed of 140 mph and Exposure C or D, or a mean roof height greater than 30 feet, the maximum negative pressure is higher than 30 psf.

Although mean roof heights of 50 feet and 60 feet are listed in the table as upper limits for the wind pressure, the IRC limit of three stories above-grade plane (R102.1) still applies. Chapter 7 of ICC 600 includes prescriptive attachment schedules for exterior wall coverings that may be applied when mean roof height limits per Table R703.5 are exceeded.

FEMA P-499, *Home Builder's Guide to Coastal Construction* (FEMA, 2009), includes Technical Fact Sheet 5.3, which addresses the attachment of siding in areas where wind loads for wall cladding exceed 30 psf as a result of wind speed, exposure category, or roof mean height by recommending selection of a siding product rated for those conditions. The manufacturer's product literature or installation instructions should specify fastener type, size and spacing, and any other installation details such as requirements for sheathing materials behind vinyl siding that are needed to achieve the product rating.

Section R703.3.2, a new subsection, describes minimum fastener requirements by combining 2012 IRC Table R703.4 footnotes b, c, d, g, and r. The section requires all nails and staples to comply with ASTM F1667. The 2012 IRC Table R703.4 footnotes i and j move to former Section R703.3, now Section R703.5, describing wood, hardboard, and wood structural panel siding. New subsections have been added to describe the specific requirements relevant to horizontal wood siding, vertical wood siding, and panel siding products.

Section R703.3.3, a new subsection, details fastener length and penetration. The penetration requirements from 2012 IRC Table R703.4 footnotes m and o for hardboard siding and footnotes v, y, and z are in this new section. The shank and head diameters in footnotes m and o for hardboard siding are moved into the main table, Table R703.3(1).

The 2012 IRC Table R703.4 footnotes q and s on fiber-cement are now located in their respective material listings. The 2012 IRC Table R703.4 footnote w reference to TMS 402 is now in the adhered veneer section, Section R703.12.

Insulated vinyl siding installation practices are updated, including several requirements placed in Table R703.3(1). Installation specifications are very similar to those for vinyl siding, including a minimum thickness requirement from ASTM D7793, a water-resistive barrier, nail penetration depth and size, fastener spacing, and provisions for installation over foam sheathing.

R703.5

Wood, Hardboard, and Wood Structural Panel Siding

CHANGE TYPE: Modification

CHANGE SUMMARY: Minimum spacing based on siding thickness has been moved from 2012 IRC Table R703.4 footnote i, siding attachment and minimum thickness, to 2015 IRC Section R703.5.2, panel siding. Requirements for vertical wood siding have moved from 2012 IRC footnote j to 2015 IRC Section R703.5.1, vertical wood siding.

2015 CODE: **~~R703.3~~ R703.5 Wood, Hardboard, and Wood Structural Panel Siding.** Wood, hardboard, and wood structural panel siding shall be installed in accordance with this section and Table R703.3(1). Hardboard siding shall comply with CPA/ANSI A135.6. Hardboard siding used as architectural trim shall comply with CPA/ANSI A 135.7.

R703.5.1 Vertical Wood Siding. Wood siding applied vertically shall be nailed to horizontal nailing strips or blocking set not more than 24 inches (610 mm) on center.

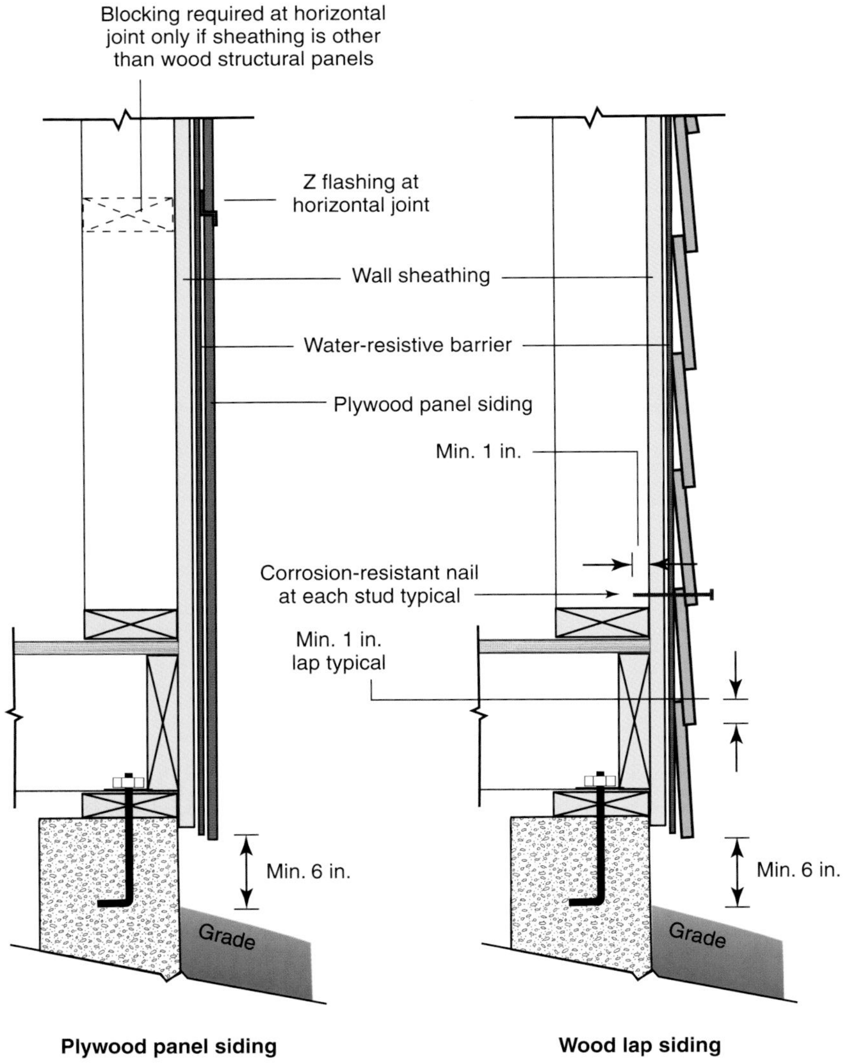

© International Code Council

Wood siding options

~~R703.3.1~~ R703.5.2 Panel Siding. 3/8-inch (9.5 mm) wood structural panel siding shall not be applied directly to studs spaced more than 16 inches (406 mm) on center where the long dimension is parallel to studs. Wood structural panel siding 7/16-inch (11.1 mm) or thinner shall not be applied directly to studs spaced more than 24 inches (610 mm) on center. The stud spacing shall not exceed the panel span rating provided by the manufacturer unless the panels are installed with the face grain perpendicular to the studs or over sheathing approved for that stud spacing.

Joints in wood, hardboard, or wood structural panel siding shall be made as follows unless otherwise approved. Vertical joints in panel siding shall occur over framing members, unless wood or wood structural panel sheathing is used, and shall be shiplapped or covered with a batten. Horizontal joints in panel siding shall be lapped ~~a minimum of~~ not less than 1 inch (25 mm) or shall be shiplapped or flashed with Z-flashing and occur over solid blocking, wood, or wood structural panel sheathing.

~~R703.3.2~~ R703.5.3 Horizontal Wood Siding. Horizontal lap siding shall be installed in accordance with the manufacturer's recommendations. Where there are no recommendations the siding shall be lapped ~~a minimum of~~ not less than 1 inch (25 mm), or ½ inch (13 mm) if rabbeted, and shall have the ends caulked, covered with a batten or sealed and installed over a strip of flashing.

CHANGE SIGNIFICANCE: The 2012 IRC Table 703.4 footnotes i and j moved to Section 703.5 describing wood, hardboard, and wood structural panel siding. New subsections describe the specific requirements for stud spacing and minimum siding lap relevant to horizontal wood siding, vertical wood siding, and panel siding products.

Minimum fastener size and penetration requirements, along with other installation details, are in Table R703.3(1), former Table R703.4, and mirror current Western Wood Products Association (WWPA) and Western Red Cedar Lumber Association (WRCLA) installation guides.

R703.6

Wood Shakes and Shingles on Exterior Walls

CHANGE TYPE: Modification

CHANGE SUMMARY: The provisions for the application of wood shakes and shingles on exterior walls have been reorganized to give more information within tables for ease of use.

2015 CODE: **R703.6 Wood Shakes and Shingles.** Wood shakes and shingles shall conform to CSSB *Grading Rules for Wood Shakes and Shingles.*

R703.6.1 Application. Wood shakes or shingles shall be applied either single-course or double-course over nominal ½-inch (12.7-mm) wood-based sheathing or to furring strips over ½-inch (12.7-mm) nominal nonwood sheathing. A ~~permeable~~ water-resistive barrier shall be provided over all sheathing, with horizontal overlaps in the membrane of not less than 2 inches (51 mm) and vertical overlaps of not less than 6 inches (152 mm). Where horizontal furring strips are used, they shall be 1 inch by 3 inches or 1 inch by 4 inches (25 mm by 76 mm or 25 mm by 102 mm) and shall be fastened ~~horizontally~~ to the studs with minimum 7d or 8d box nails and shall be spaced a distance on center equal to the actual weather exposure of the shakes or shingles, not to exceed the maximum exposure specified in Table R703.6.1. When installing shakes or shingles over a non-permeable water resistive barrier, furring strips shall be placed first

Wood shingles as exterior siding

TABLE R703.6.1 Maximum Weather Exposure for Wood Shakes and Shingles on Exterior Walls[a,b,c] (Dimensions are in inches)

Length	Exposure for Single Course	Exposure for Double Course
Shingles[a]		
16	~~7½~~ 7	12[b]
18	~~8½~~ 8	14[c]
24	~~11½~~ 10½	16[d]
Shakes[a]		
18	~~8½~~ 8	14
24	~~11½~~ 10½	18

For SI: 1 inch = 25.4 mm.

a. Dimensions given are for No. 1 grade.
b. A maximum ~~10-inch~~ 9-inch exposure is permitted for No. 2 grade.
c. A maximum ~~11-inch~~ 10-inch exposure is permitted for No. 2 grade.
d. A maximum 14-inch exposure is permitted for No. 2 grade.

vertically over the barrier and, in addition, horizontal furring strips shall be fastened to the vertical furring strips prior to attaching the shakes or shingles to the horizontal furring strips. The spacing between adjacent shingles to allow for expansion shall be 1/8 inch (3.2 mm) to ¼ inch (6.4 mm) apart and between adjacent shakes, shall be 3/8 inch (9.5 mm) to ½ inch (12.7 mm) apart. The offset spacing between joints in adjacent courses shall be ~~a minimum of~~ not less than 1½ inches (38 mm).

R703.6.2 Weather Exposure. The maximum weather exposure for shakes and shingles shall not exceed that specified in Table R703.6.1.

~~**R703.5.3 Attachment.** Each shake or shingle shall be held in place by two hot-dipped zinc-coated, stainless steel, or aluminum nails or staples. The fasteners shall be long enough to penetrate the sheathing or furring strips by a minimum of ½ inch (13 mm) and shall not be overdriven.~~

R703.6.3 Attachment. Wood shakes or shingles shall be installed according to this chapter and the manufacturer's installation instructions. Each shake or shingle shall be held in place by two stainless steel Type 304, Type 316, or hot-dipped zinc-coated galvanized corrosion-resistant box nails in accordance with Table R703.6.3(1) or R703.6.3(2). The hot-dipped zinc-coated galvanizing shall conform to minimum standard ASTM A153D, 1.0 ounce per square foot. Alternatively, 16 gauge stainless steel Type 304 or Type 316 staples with crown widths 7/16 inch (11 mm) minimum, ¾ inch (19 mm) maximum shall be used and the crown of the staple shall be placed parallel with the butt of the shake or the shingle. In single-course application, the fasteners shall be concealed by the course above and shall be driven approximately 1 inch (25 mm) above the butt line of the succeeding course and ¾ inch (19 mm) from the edge. In double-course applications, the exposed shake or shingle shall be face-nailed with two fasteners, driven approximately 2 inches (51 mm) above the butt line and ¾ inch (19 mm) from each edge. Fasteners installed within 15 miles (24 km) of salt water coastal areas shall be stainless steel Type 316. Fasteners for fire-retardant-treated shakes or shingles in accordance with Section R902 or pressure-impregnated-preservative-treated

R703.6 continues

R703.6 continued

shakes or shingles in accordance with AWPA U1 shall be, stainless steel Type 316. The fasteners shall penetrate the sheathing or furring strips by not less than ½ inch (13 mm) and shall not be overdriven. Fasteners for untreated (natural) and treated products shall comply with ASTM F1667.

R703.5.3.1 Staple Attachment. Wood shakes or shingles shall be installed according to this chapter and the manufacturer's installation instructions. Staples for untreated (natural) wood shakes or wood shingles shall be 16 gauge Stainless Steel Type 304,Type 316 (Fasteners installed within 15 miles of salt water coastal areas shall be stainless steel Type 316.) Staples shall not be less than 16 gauge and shall have a crown width of not less than minimum 7/16 inch (11 mm), maximum of ¾ inch and the crown of the staples shall be parallel with the butt of the shake or shingle.

In single-course application, the fasteners shall be concealed by the course above and shall be driven approximately 1 inch (25 mm) above the butt line of the succeeding course and ¾ inch (19 mm) from the edge. In double-course applications, the exposed shake or shingle shall be face-nailed with two casing nails staples, driven approximately 2 inches (51 mm) above the butt line and ¾ inch (19 mm) from each edge. In all application, staples shall be concealed by the course above. With shingles wider than 810 inches (203, 254 mm) two additional nails shall be required and shall be nailed driven approximately 1 inch (25 mm) apart near the center of the shingle. Fasteners for fire-retardant-treated (as defined in section R902.2) shingles, shakes or pressure-impregnated-preservative-treated shingles or shakes in accordance with AWPA U1 shall be Stainless Steel Type 316, applied as above. Fasteners for untreated (natural) and treated products shall comply with ASTM F1667.

TABLE R703.6.3(1) Single Course Sidewall Fasteners

Product Type	Nail Type and Minimum Length (inches)	Minimum Head Diameter (inches)	Minimum Shank Thickness (inches)
R & R and Sanded Shingles	**Type**		
16″ and 18″ shingles	3d Box 1¼	0.19	0.08
24″ shingles	4d Box 1½	0.19	0.08
Grooved Shingles	**Type**		
16″ and 18″ shingles	3d Box 1¼	0.19	0.08
24″ shingles	4d Box 1½	0.19	0.08
Split and Sawn Shakes	**Type**		
18″ Straight-Split Shakes	5d Box 1¾	0.19	0.08
18″ and 24″ Handsplit Shakes	6d Box 2	0.19	0.0915
24″ Tapersplit Shakes	5d Box 1¾	0.19	0.08
18″ and 24″ Tapersawn Shakes	6d Box 2	0.19	0.0915

TABLE R703.6.3(2) Double Course Sidewall Fasteners

Product Type	Nail Type and Minimum Length (inches)	Minimum Head Diameter (inches)	Minimum Shank Thickness (inches)
R & R and Sanded Shingles	**Type**		
16", 18" and 24" shingles	5d Box 1¾ or same size casing nails	0.19	0.08
Grooved Shingles	**Type**		
16", 18" and 24" shingles	5d Box 1¾	0.19	0.08
Split and Sawn Shakes	**Type**		
18" Straight-Split Shakes	7d Box 2¼ or 8d 2½	0.19	0.099
18" and 24" Handsplit Shakes	7d Box 2¼ or 8d 2½	0.19	0.099
24" Tapersplit Shakes	7d Box 2¼ or 8d 2½	0.19	0.099
18" and 24" Tapersawn Shakes	7d Box 2¼ or 8d 2½	0.19	0.099

CHANGE SIGNIFICANCE: This code change clarifies the fasteners required for attachment of wood shakes and shingles for wall cladding. The type of fastener to be used is determined by environmental factors and product types. For naturally durable wood, two types of stainless steel nail and hot-dipped galvanized box nails are approved. Alternatively, 16-gage type 304 or 316 stainless steel staples may also be used. For fasteners installed on dwellings within 15 miles of a saltwater coastal area, only type 316 stainless steel nails are allowed. Fasteners of fire-retardant treated wood must be type 316 stainless steel nails.

Shingles may not be applied with the vertical edges tight together. Tightly applied shingles do not leave room for expansion, causing fishmouthing, cupping, and curling. The code prescribes minimum and maximum spacing for shingles and shakes. Wood shingles must be attached in accordance with the manufacturer's directions. Vertical and horizontal furring over water-resistive barriers is now required. New Tables R703.6.3(1) and R703.6.3(2) list minimum nail type, length, and head and shank diameter for single and double courses, respectively.

Table R703.6.1 reduces the exposure length of the shingles and shakes. The change is required as the longer exposure lengths allowed in the code are no longer considered practical. Exposure lengths have been decreased in accordance with manufacturers' installation requirements.

R703.9

Exterior Insulation and Finish Systems (EIFS)

CHANGE TYPE: Modification

CHANGE SUMMARY: Limitations for exterior insulation and finish systems with and without drainage have been added to the 2015 IRC.

2015 CODE: R703.9 Exterior Insulation and Finish System (EIFS)/EIFS with Drainage. Exterior Insulation and Finish Systems (EIFS) shall comply with this chapter and Section~~s~~ R703.9.1. ~~and R703.9.3~~. EIFS with drainage shall comply with this chapter and Section~~s~~ R703.9.2~~, R703.9.3 and R703.9.4~~.

R703.9.1 Exterior Insulation and Finish Systems (EIFS). EIFS shall comply with ~~ASTM E 2568.~~ the following:

1. ASTM E2568.
2. EIFS shall be limited to applications over concrete or masonry wall assemblies (substrates).
3. Flashing of EIFS shall be provided in accordance with the requirements of Section R703.8.
4. EIFS shall be installed in accordance with the manufacturer's installation instructions.
5. EIFS shall terminate not less than 6 inches (152 mm) above the finished ground level.
6. Decorative trim shall not be face-nailed through the EIFS.

R703.9.2 Exterior Insulation and Finish System (EIFS) with Drainage. EIFS with drainage shall comply with the following: ~~ASTM E 2568 and shall have an average minimum drainage efficiency of 90 percent when tested in accordance with ASTM E 2273.~~

1. ASTM E2568.
2. EIFS with drainage shall be required over all wall assemblies with the exception of concrete and masonry wall assemblies.

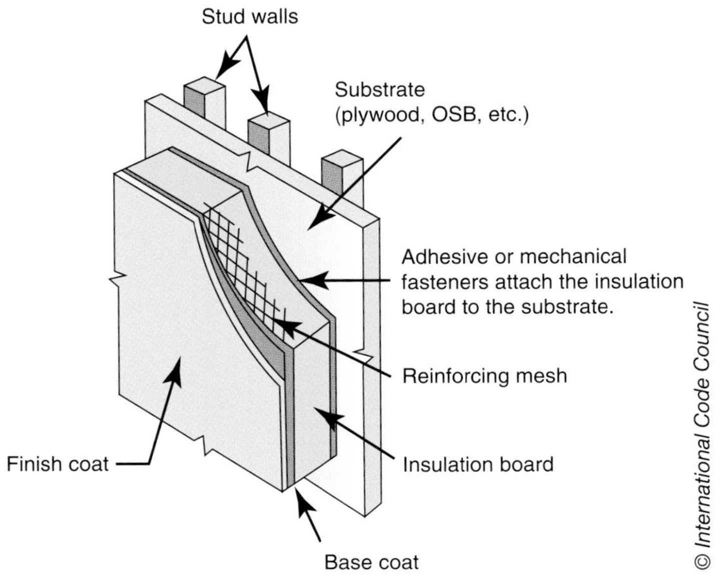

EIFS composition

3. EIFS with drainage shall have an average minimum drainage efficiency of 90 percent when tested in accordance with ASTM E 2273.
4. The water-resistive barrier shall comply with Section R703.2 or ASTM E 2570.
5. The water-resistive barrier shall be applied between the EIFS and the wall sheathing.
6. Flashing of EIFS with drainage shall be provided in accordance with the requirements of Section R703.8.
7. EIFS with drainage shall be installed in accordance with the manufacturer's installation instructions.
8. EIFS with drainage shall terminate not less than 6 inches (152 mm) above the finished ground level.
9. Decorative trim shall not be face nailed through the EIFS with drainage.

CHANGE SIGNIFICANCE: When the EIFS section was added to the IRC in the 2009 edition, it was industry's position that EIFS (also known as "barrier" EIFS or EIFS without drainage) would be limited to applications over concrete or masonry substrates. It was also the intent that EIFS with drainage would be required on light-framed walls constructed under the IRC. These applications are consistent with the ICC Evaluation Service Reports for the products.

In the 2012 IRC this intent was unclear. For example, in Section 703.1.1, Exception 2 allows an "opt-out" for the need for a means of drainage in the exterior wall envelope if it can meet the requirements of ASTM E331. Thus, although an EIFS "barrier" system could meet this requirement, the industry did not recommend this application on residential light-framed construction.

The 2015 IRC section on EIFS clarifies the use of EIFS. EIFS without drainage may be used in concrete and masonry wall construction. For light-frame construction, EIFS with drainage must be used. Reference to specific standards is added or clarified to assist the designer and builder in selecting the correct type of EIFS.

R703.11.1

Vinyl Siding Attachment

CHANGE TYPE: Addition

CHANGE SUMMARY: This code change clarifies nailing penetration and spacing requirements for horizontal and vertical vinyl siding.

2015 CODE: R703.11.1 Installation. Vinyl siding, soffit, and accessories shall be installed in accordance with the manufacturer's installation instructions.

R703.11.1.1 Fasteners. Unless specified otherwise by the manufacturer's instructions, fasteners for vinyl siding shall be 0.120-inch (3 mm) shank diameter nail with a 0.313-inch (8 mm) head or 16 gauge staple with a 3/8-inch (9.5 mm) to 1/2-inch (12.7 mm) crown.

R703.11.1.2 Penetration Depth. Unless specified otherwise by the manufacturer's instructions, fasteners shall penetrate into building framing. The total penetration into sheathing, furring framing or other nailable substrate shall be a minimum 1 1/4 inches (32 mm). Where specified by the manufacturer's instructions and supported by a test report, fasteners are permitted to penetrate into or fully through nailable sheathing or other nailable substrate of minimum thickness specified by the instructions or test report, without penetrating into framing. Where the fastener penetrates fully through the sheathing, the end of the fastener shall extend a minimum of 1/4 inch (6.4 mm) beyond the opposite face of the sheathing or nailable substrate.

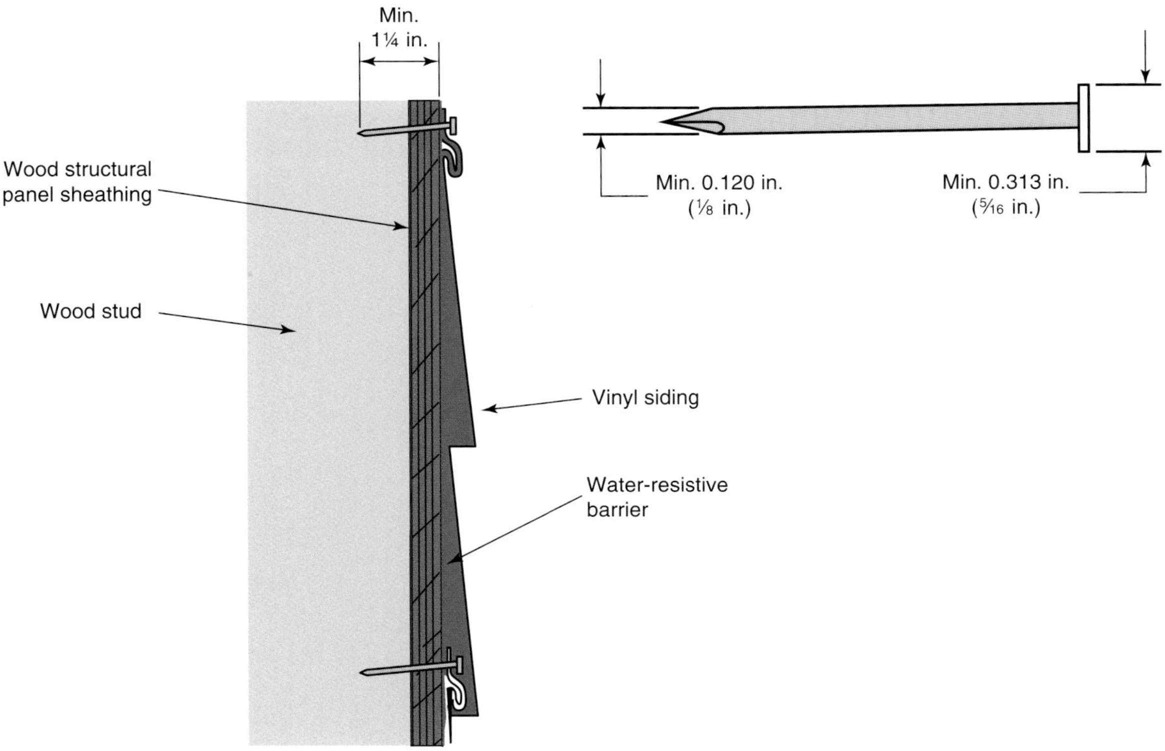

Vinyl siding attachment

© Bonita R. Cheshier/Shutterstock.com.

Vinyl siding installation

R703.11.1.3 Spacing. Unless specified otherwise by the manufacturer's instructions, the maximum spacing between fasteners for horizontal siding shall be 16 inches (406 mm), and for vertical siding 12 inches (305 mm) both horizontally and vertically. Where specified by the manufacturer's instructions and supported by a test report, greater fastener spacing is permitted.

~~R703.11.1.1~~ R703.11.1.4 Vinyl Soffit Panels. Soffit panels shall be individually fastened to a supporting component such as a nailing strip, fascia or subfascia component or as specified by the manufacturer's instructions.

SECTION R202
DEFINITIONS

NAILABLE SUBSTRATE. A product or material such as framing, sheathing, or furring, composed of wood or wood-based materials, or other materials and fasteners providing equivalent fastener withdrawal resistance.

CHANGE SIGNIFICANCE: In the 2012 IRC, information on vinyl siding fastener specifications, penetration, and spacing was found only in Table R703.4 and its footnotes. This code change places the requirements into the text of the code provision where they are easily located and more clearly stated.

R703.11.1 continues

R703.11.1 continued

Vinyl siding can be used in conjunction with a variety of sheathing types, some of which contribute to resisting fastener withdrawal. It is necessary to ensure that, regardless of the sheathing type, the total penetration into a material capable of holding fasteners is equivalent to what was used during testing of the siding. For typical siding installations, this is ¾-inch penetration into framing plus approximately ½ inch through wood sheathing, for a total of 1¼ inches of penetration into "nailable" material. This minimum penetration is required unless a different penetration is specified in the manufacturer's instructions. A definition of "nailable substrate" is added to define what is considered to be "nailable."

Where the siding is used over a non-nailable material the total penetration must still be achieved, in this case by using a fastener long enough to accommodate the thickness of non-nailable material and penetrate the full 1¼ inches into framing or a combination of framing and other nailable material. By stating the requirement in terms of the total required penetration, it is clear what penetration is needed for all installations.

In addition, the maximum fastener spacing for both horizontal and vertical siding has been added to the code text. The IRC previously had no provision for fastener spacing for vertical siding; the new requirement mirrors provisions in the IBC.

R703.13, R703.14

Insulated Vinyl Siding and Polypropylene Siding

CHANGE TYPE: Addition

CHANGE SUMMARY: New sections set minimum requirements for insulated vinyl siding and polypropylene siding.

2015 CODE: **703.13 Insulated Vinyl Siding.** Insulated vinyl siding shall be certified and labeled as conforming to the requirements of ASTM D7793 by an approved quality control agency.

703.13.1 Insulated Vinyl Siding and Accessories. Insulated vinyl siding and accessories shall be installed in accordance with manufacturer's installation instructions.

R703.14 Polypropylene Siding. Polypropylene siding shall be certified and labeled as conforming to the requirements of ASTM D7254 by an approved quality control agency.

R703.14.1 Polypropylene Siding and Accessories. Polypropylene siding and accessories shall be installed in accordance with manufacturer's installation instructions.

R703.14.1.1 Installation. Polypropylene siding shall be installed over and attached to wood structural panel sheathing with minimum thickness of 7/16 inch (11.1 mm), or other substrate, composed of wood or wood-based material and fasteners having equivalent withdrawal resistance.

R703.14.1.2 Fastener Requirements. Unless otherwise specified in the approved manufacturer's instructions, nails shall be corrosion resistant, with a minimum 0.120-inch (3 mm) shank and minimum 0.313-inch (8 mm) head diameter. Nails shall be a minimum of 1¼ inches (32 mm) long or as necessary to fully penetrate sheathing or substrate not less than ¾ inch (19.1 mm). Where the nail fully penetrates the sheathing or nailable substrate, the end of the fastener shall extend not less than ¼ inch (6.4 mm) beyond the opposite face of the sheathing or substrate. Staples are not permitted.

R703.14.2 Fire Separation. Polypropylene siding shall not be installed on walls with a fire separation distance of less than 5 feet (1524 mm) and walls closer than 10 feet (3048 mm) to a building on another lot.

> **Exception:** Walls perpendicular to the line used to determine the fire separation distance.

SECTION R202
DEFINITIONS

INSULATED VINYL SIDING. A vinyl cladding product with manufacturer-installed foam plastic insulating material as an integral part of the cladding product, having a minimum thermal resistance of not less than R-2.

POLYPROPYLENE SIDING. A shaped material, made principally from polypropylene homopolymer, or copolymer, that in some cases contains fillers or reinforcements, that is used to clad exterior walls or buildings.

R703.13, R703.14 continues

R703.13, R703.14 continued

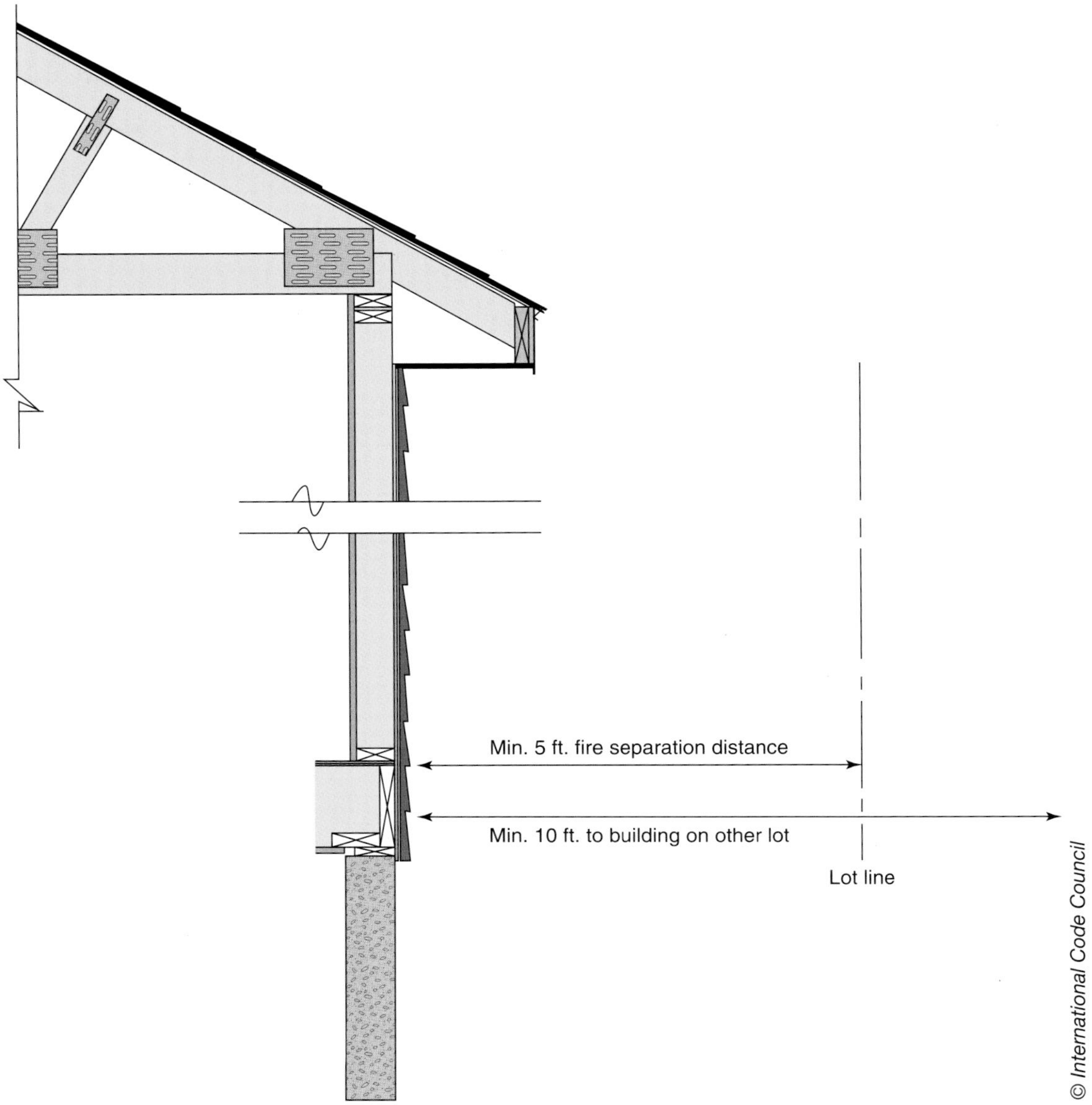

Minimum fire separation distance for polypropylene siding

CHANGE SIGNIFICANCE: New Section R703.13, Insulated Vinyl Siding, sets requirements based on the current ASTM standard for insulated vinyl siding, ASTM D7793. Insulated vinyl siding is certified to an ASTM standard by an approved third-party inspection agency. Performance requirements are specified by the ASTM standard, ensuring that insulated vinyl siding can meet minimum requirements as a cladding insulation. Insulated vinyl siding is vinyl siding with rigid foam insulation laminated or permanently attached to the panel.

This change provides a method for building officials to verify that insulated vinyl siding is code compliant. In energy codes and energy efficiency programs, insulated siding is recognized as a form of "continuous insulation," or insulation installed on the exterior of the building that helps reduce energy loss through framing or other building material.

The insulated siding provides a supplemental rain screen that reduces the amount of water reaching the underlying water-resistive barrier. With a properly applied water-resistive barrier, insulated siding minimizes moisture penetration from the exterior into a wall assembly and provides a way for moisture to readily drain and dry. The presence of a layer of thermal insulation filling the space between insulated siding and wall sheathing also aids in moisture management.

Section R703.14, Polypropylene Siding, sets minimum performance requirements for polypropylene siding and requires a third-party inspection agency to verify compliance with an internationally accepted ASTM standard. Additionally, appropriate installation and use of polypropylene siding are detailed. Use of polypropylene siding is limited to walls with a fire separation distance of 5 feet or more and walls 10 feet or more from a building on another lot.

R703.15, R703.16, R703.17

Cladding Attachment over Foam Sheathing

CHANGE TYPE: Addition

CHANGE SUMMARY: Three new sections set minimum requirements for cladding attachment over foam sheathing to wood framing (R703.15), cold-formed steel framing (R703.16), and masonry or concrete walls (R703.17). For light-frame construction, prescriptive requirements are given. Connection to concrete and masonry construction continues to require engineered design in most cases when placing foam over the concrete or masonry wall.

2015 CODE: **R703.15 Cladding Attachment over Foam Sheathing to Wood Framing.** Cladding shall be specified and installed in accordance with Section R703, the cladding manufacturer's approved installation instructions, including any limitations for use over foam plastic sheathing, or an approved design. In addition, the cladding or furring attachments through foam sheathing to framing shall meet or exceed the minimum fastening requirements of Section R703.15.1, Section R703.15.2, or an approved design for support of cladding weight.

Exceptions:

1. Where the cladding manufacturer has provided approved installation instructions for application over foam sheathing, those requirements shall apply.
2. For exterior insulation and finish systems, refer to Section R703.9.
3. For anchored masonry or stone veneer installed over foam sheathing; refer to Section R703.7.

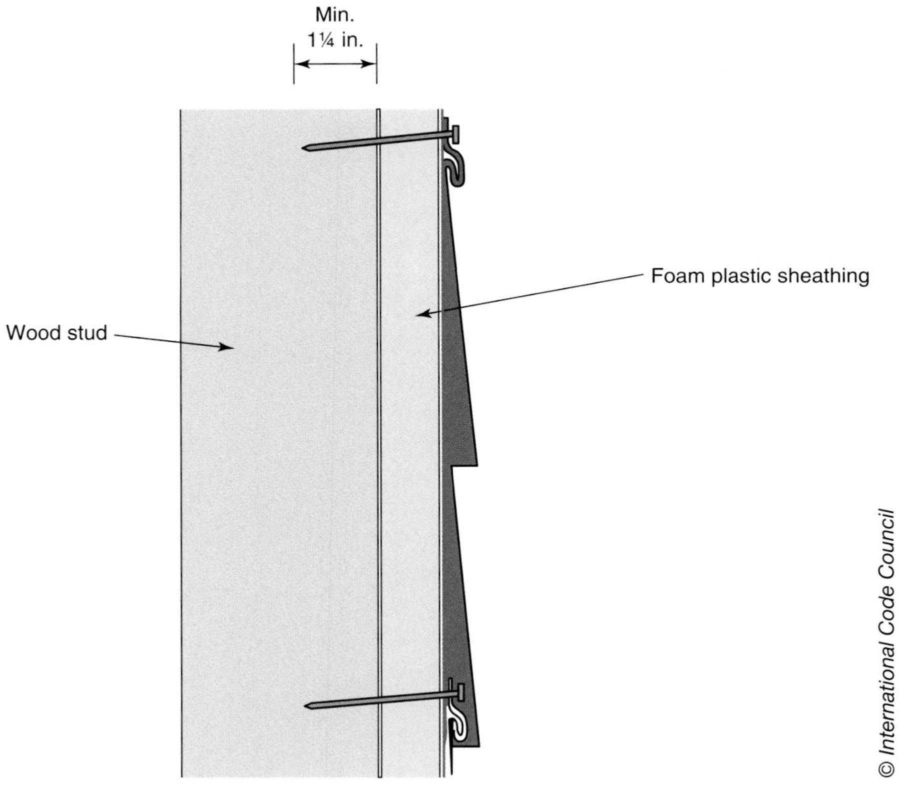

Siding attachment over foam sheathing to wood framing

R703.15.1 Direct Attachment. Where cladding is installed directly over foam sheathing without the use of furring, cladding minimum fastening requirements to support the cladding weight shall be as specified in Table R703.15.1.

R703.15.2 Furred Cladding Attachment. Where wood furring is used to attach cladding over foam sheathing, furring minimum fastening requirements to support the cladding weight shall be as specified in Table R703.15.2. Where placed horizontally, wood furring shall be preservative treated wood in accordance with Section R317.1 or naturally durable wood and fasteners shall be corrosion resistant in accordance with Section R317.3.

R703.17 Cladding Attachment over Foam Sheathing to Masonry or Concrete Wall Construction. Cladding shall be specified and installed in accordance with Section 703.3 and the cladding manufacturer's installation instructions or an approved design. Foam sheathing shall be attached to masonry or concrete construction in accordance with the insulation manufacturer's installation instructions or an approved design. Furring and furring attachments through foam sheathing into concrete or masonry substrate shall be designed to resist design loads determined in accordance with Section R301, including support of cladding weight as applicable. Fasteners used to attach cladding or furring through foam sheathing to masonry or concrete substrates shall be approved for application into masonry or concrete material and shall be installed in accordance with the fastener manufacturer's installation instructions.

TABLE R703.15.1 Cladding Minimum Fastening Requirements for Direct Attachment over Foam Plastic Sheathing to Support Cladding Weight[a]

Cladding Fastener Through Foam Sheathing	Cladding Fastener Type and Minimum Size[b]	Cladding Fastener Vertical Spacing (inches)	Maximum Thickness of Foam Sheathing[c] (inches)					
			16″oc Fastener Horizontal Spacing			24″oc Fastener Horizontal Spacing		
			Cladding Weight:			Cladding Weight:		
			3 psf	11 psf	25 psf	3 psf	11 psf	25 psf
Wood Framing (minimum 1¼ inch penetration)	0.113″ diameter nail	6	2	1	DR	2	0.75	DR
		8	2	1	DR	2	0.5	DR
		12	2	0.5	DR	2	DR	DR
	0.120″ diameter nail	6	3	1.5	0.5	3	0.75	DR
		8	3	1	DR	3	0.5	DR
		12	3	0.5	DR	2	DR	DR
	0.131″ diameter nail	6	4	2	0.75	4	1	DR
		8	4	1.5	0.5	4	0.75	DR
		12	4	0.75	DR	2	0.5	DR
	0.162″ diameter nail	6	4	4	1.5	4	2	1
		8	4	3	1	4	1.5	0.75
		12	4	2	0.75	4	1	DR

For SI: 1 inch = 25.4 mm; 1 pound per square foot (psf) = 0.0479 kPa
DR = design required
oc = on center

a. Wood framing shall be Spruce-Pine-Fir or any wood species with a specific gravity of 0.42 or greater in accordance with AWC NDS.
b. Nail fasteners shall comply with ASTM F1667, except nail length shall be permitted to exceed ASTM F1667 standard lengths.
c. Foam sheathing shall have a minimum compressive strength of 15 psi in accordance with ASTM C578 or ASTM C1289.

R703.15, R703.16, R703.17 continues

R703.15, R703.16, R703.17 continued

TABLE R703.15.2 Furring Minimum Fastening Requirements for Application over Foam Plastic Sheathing to Support Cladding Weight[a,b]

Furring Material	Framing Member	Fastener Type and Minimum Size[b]	Minimum Penetration into Wall Framing (inches)	Fastener Spacing in Furring (inches)	Maximum Thickness of Foam Sheathing[d] (inches)					
					16″oc Furring[e]			24″oc Furring[e]		
					Siding Weight:			Siding Weight:		
					3 psf	11 psf	25 psf	3 psf	11 psf	25 psf
Minimum 1× Wood Furring[c]	Minimum 2× Wood Stud	0.131″ diameter nail	1¼	8	4	2	1	4	1.5	DR
				12	4	1.5	DR	3	1	DR
				16	4	1	DR	3	0.5	DR
		0.162″ diameter nail	1¼	8	4	4	1.5	4	2	0.75
				12	4	2	0.75	4	1.5	DR
				16	4	1.5	DR	4	1	DR
		#10 wood screw	1	12	4	2	0.75	4	1.5	DR
				16	4	1.5	DR	4	1	DR
				24	4	1	DR	3	DR	DR
		¼″ lag screw	1½	12	4	3	1	4	2	0.5
				16	4	1.5	DR	4	1.5	DR
				24	4	1.5	DR	4	0.75	DR

For SI: 1 inch = 25.4 mm; 1 pound per square foot (psf) = 0.0479 kPa.
DR = design required
oc = on center

a. Wood framing and furring shall be Spruce-Pine-Fir or any wood species with a specific gravity of 0.42 or greater in accordance with AWC NDS.
b. Nail fasteners shall comply with ASTM F1667, except nail length shall be permitted to exceed ASTM F1667 standard lengths.
c. Where the required cladding fastener penetration into wood material exceeds ¾ inch and is not more than 1½ inches, a minimum 2× wood furring or an approved design shall be used.
d. Foam sheathing shall have a minimum compressive strength of 15 psi in accordance with ASTM C578 or ASTM C1289.
e. Furring shall be spaced not more than 24 inches on center, in a vertical or horizontal orientation. In a vertical orientation, furring shall be located over wall studs and attached with the required fastener spacing. In a horizontal orientation, the indicated 8 inch and 12 inch fastener spacing in furring shall be achieved by use of two fasteners into studs at 16 inches and 24 inches on center, respectively.

Exceptions:

1. Where the cladding manufacturer has provided approved installation instructions for application over foam sheathing and connection to a masonry or concrete substrate, those requirements shall apply.
2. For exterior insulation and finish systems, refer to Section R703.9.
3. For anchored masonry or stone veneer installed over foam sheathing, refer to Section R703.7.

(*Section R703.16 not shown for brevity.*)

CHANGE SIGNIFICANCE: Section R703.15 provides attachment provisions for exterior wall covering assemblies that include foam plastic insulation and are applied to wood framing members. Section R703.16 provides attachment provisions for exterior wall covering assemblies that include foam plastic insulation applied to cold-formed steel studs with wood or steel sheathing. Section R703.17 contains provisions for cladding attachment over foam sheathing to concrete or masonry walls.

The new sections provide requirements for attachment of furring over foam sheathing to resist wind loading—an application that was not addressed previously in the IRC. During high winds, failures have repeatedly occurred of cladding attached over foam sheathing to gable end walls and first- and second-story walls. In an effort to reduce failures of the sheathing attachment, prescriptive connections that consider foam sheathing have been added to the IRC.

Calculations were completed to determine the wall cladding resistance to wind forces. In the first application, furring was assumed to be placed beneath the cladding and over the foam sheathing to improve siding durability. In the second, cladding is placed directly over foam sheathing with attachments passing through the foam and embedding in wood or cold-formed steel framing. The wind pressure limits are based on the weaker capacity of either fastener withdrawal or furring bending strength, where applicable. From these calculations, minimum attachment requirements are calculated. These attachment calculations include a maximum thickness of foam sheathing.

Sections R703.15 and R703.16 for light-frame construction give prescriptive fastening requirements for cladding materials installed over foam sheathing to ensure adequate performance. The proposed cladding attachment requirements and foam sheathing thickness limits are based on calculations verified by test data to control cladding connection movement to no more than 0.015-inch slip under cladding weight or dead load.

A prescriptive solution for attachment of cladding to masonry or concrete walls through foam sheathing has not been added to the IRC. Prescriptive "off-the-shelf" solutions with standardized types of concrete or masonry fasteners have not been developed. In fact, many fasteners best suited for this application are proprietary and engineered design is required. Section R703.17 requires engineering analysis of cladding connections through foam sheathing to masonry or concrete. As an exception, this engineered design may be done by the manufacturer with an approved prescriptive solution supplied in an evaluation report or installation instructions to the builder.

Tables R802.4, R802.5

Ceiling Joist and Rafter Tables

CHANGE TYPE: Modification

CHANGE SUMMARY: Changes to Southern Pine, Douglas Fir-Larch, and Hemlock Fir capacities have changed the maximum spans for lumber in the ceiling joist and rafter span tables of the *International Residential Code*.

2015 CODE:

TABLE R802.4(1) Ceiling Joist Spans for Common Lumber Species (Uninhabitable attics without storage, live load = 10 psf, L/Δ = 240)

Ceiling Joist Spacing (inches)	Species and Grade		Dead Load = 5 psf			
			2 × 4	2 × 6	2 × 8	2 × 10
			Maximum ceiling joist spans			
			(feet - inches)	(feet - inches)	(feet - inches)	(feet - inches)
16	Douglas fir-larch	SS	11-11	18-9	24-8	Note a
	Douglas fir-larch	#1	11-6	18-1	23-10	Note a
	Douglas fir-larch	#2	11-3	17-8	<u>23-4</u>	Note a
	Douglas fir-larch	#3	<u>9-7</u>	<u>14-1</u>	<u>17-10</u>	<u>21-9</u>
	Hem-fir	SS	11-3	17-8	23-4	Note a
	Hem-fir	#1	11-0	17-4	22-10	Note a
	Hem-fir	#2	10-6	16-6	21-9	Note a
	Hem-fir	#3	9-5	13-9	17-5	21-3
	Southern pine	SS	11-9	18-5	24-3	Note a
	Southern pine	#1	<u>11-3</u>	<u>17-8</u>	<u>23-10</u>	Note a
	Southern pine	#2	<u>10-9</u>	<u>16-11</u>	<u>21-7</u>	<u>25-7</u>
	Southern pine	#3	<u>8-9</u>	<u>12-11</u>	<u>16-3</u>	<u>19-9</u>
	Spruce-pine-fir	SS	11-0	17-4	22-10	Note a
	Spruce-pine-fir	#1	10-9	16-11	22-4	Note a
	Spruce-pine-fir	#2	10-9	16-11	22-4	Note a
	Spruce-pine-fir	#3	9-5	13-9	17-5	21-3

(Portions of table not shown for brevity and clarity.)

CHANGE SIGNIFICANCE: New design values exist for most widths and grades of visually graded Southern Pine lumber. These design values became effective on June 1, 2013. The American Lumber Standards Committee (ALSC) approved the new design values as published in Southern Pine Inspection Bureau *Supplement No. 13* to the *2002 Standard Grading Rules for Southern Pine Lumber*. These values are a result of two years of testing current lumber inventory available on the market to see what changes, if any, had occurred in the strength of the Southern Pine.

Meanwhile, for Douglas Fir-Larch and Hemlock Fir, testing done in the 1990s slightly increased design values for bending. Revised design values for Select Structural, #2, and #3 grades of Douglas Fir-Larch and #1 grade of Hemlock Fir all increased by 25 psi. Testing to check current stock has validated design values set in the 1990s. Although these values were updated in the wood standards, span tables incorporated into the 2000 *International Building Code* (IBC) and 2000 IRC were based on span tables predating the revised design values from the 1990s. These tables are updated with the 2015 IRC.

The 2015 IRC span tables will now be in agreement with wood standard span tables with the revisions for Southern Pine, Douglas Fir-Larch, and Hemlock Fir. For Southern Pine, the changes reflect shorter spans. For Douglas Fir-Larch and Hemlock Fir, the changes result in slightly longer spans.

The new design values apply only to new construction. The integrity of existing structures designed and built using the design values meeting the applicable building codes in effect at the time of permitting is not a concern.

Example—Ceiling Joint Spans

#1 Uninhabitable attic with limited storage

LL = 20 psf
DL = 10 psf
2×10 joists
16″ o.c. spacing
SP #2

	2012	2015
Maximum Span Allowed	20′-9″	18′-1″

The SP #2 span length is significantly reduced from the 2012 IRC span length.

Note: An SP #1 joist will span about the same length in the 2015 IRC Table R802.4(1) or R802.4(2) as the SP #2 did in the tables in the 2012 IRC.

#2 Uninhabitable attic without storage

LL = 10 psf
DL = 5 psf
2×8 joists
24″ o.c. spacing
DFL #2

	2012	2015
Maximum Span Allowed	18′-9″	19′-1″

The span has increased about 2 inches which is the typical increase in the table. Some cells for Douglas fir and Hemlock fir have not changed. Others increased by 1–2 inches.

R806.1

Attic Ventilation

Attic ventilation

CHANGE TYPE: Deletion

CHANGE SUMMARY: The 2012 IRC exception allowing the building official to waive ventilation requirements due to atmospheric or climatic conditions has been deleted.

2015 CODE: R806.1 Ventilation Required. Enclosed attics and enclosed rafter spaces formed where ceilings are applied directly to the underside of roof rafters shall have cross ventilation for each separate space by ventilating openings protected against the entrance of rain or snow. Ventilation openings shall have a least dimension of 1/16 inch (1.6 mm) minimum and 1/4 inch (6.4 mm) maximum. Ventilation openings having a least dimension larger than 1/4 inch (6.4 mm) shall be provided with corrosion-resistant wire cloth screening, hardware cloth, or similar material with openings having a least dimension of 1/16 inch (1.6 mm) minimum and 1/4 inch (6.4 mm) maximum. Openings in roof framing members shall conform to the requirements of Section R802.7. Required ventilation openings shall open directly to the outside air.

> ~~**Exception:** Attic ventilation shall not be required when determined not necessary by the code official due to atmospheric or climatic conditions.~~

CHANGE SIGNIFICANCE: With recent revisions to the IRC roof ventilation requirements, and changes in the 2015 *International Building Code*, both codes now contain specific details on vented and unvented attics, with requirements related to use of vapor retarders and climate-specific instructions on use of air-impermeable insulation. As the former exception was based on climatic conditions with no direction to the building official on matters related to construction methods or details, it has been deleted. As always, the building official has the authority to accept alternative materials, design, and methods of construction in accordance with Section R104.11.

Table R806.5

Insulation for Condensation Control in Unvented Attics

CHANGE TYPE: Modification

CHANGE SUMMARY: For unvented attics and unvented rafter spaces, Table R806.5 has a new footnote allowing calculation of insulation thickness when the insulation is placed above the structural roof sheathing.

2015 CODE:

TABLE R806.5 Insulation for Condensation Control

Climate Zone	Minimum Rigid Board on Air-Impermeable Insulation *R*-Value[a,b]
2B and 3B tile roof only	0 (none required)
1, 2A, 2B, 3A, 3B, 3C	R-5
4C	R-10
4A, 4B	R-15
5	R-20
6	R-25
7	R-30
8	R-35

a. Contributes to but does not supersede the requirements in Section N1102.
b. Alternatively, sufficient continuous insulation shall be installed directly above the structural roof sheathing to maintain the monthly average temperature of the underside of the structural roof sheathing above 45°F (7°C). For calculation purposes, an interior air temperature of 68°F (20°C) is assumed and the exterior air temperature is assumed to be the monthly average outside air temperature of the three coldest months.

CHANGE SIGNIFICANCE: Section R806.5 provides three options for installing insulation at the roof line for unvented attics and unvented rafter spaces: air-impermeable insulation (typically foam plastic) installed directly below the roof sheathing, a combination of air-impermeable and air-permeable insulation installed below the roof sheathing, and air-impermeable insulation (rigid board or sheet insulation) installed above the structural roof sheathing. The minimum *R*-value for the rigid board or sheet insulation is determined from Table R806.5 based on climate zone to prevent condensation on the underside of the structural roof sheathing. The balance of the required insulation is accomplished with air-permeable insulation installed below the roof sheathing. The *R*-values in Table R806.5 are based on a total R-49 roof/ceiling insulation in Climate Zones 4, 5, 6, 7, and 8 and R-38 insulation in Climate Zones 2 and 3. Footnote b provides a calculation procedure to determine necessary rigid board or air-impermeable insulation *R*-values for roof assemblies that do not meet the requirements of Table R806.5. The footnote is consistent with similar language in 2015 IBC Section 1203.3.

R905.1.1

Underlayment

CHANGE TYPE: Modification

CHANGE SUMMARY: The multiple code provisions placed in the 2012 IRC for underlayment have been combined into Section R905.1.1, with three tables listing underlayment type, application, and attachment. Sections on ice barriers from the 2012 IRC are reorganized and combined into Section R905.1.2.

2015 CODE: **R905.1.1 Underlayment.** Underlayment for asphalt shingles, clay and concrete tile, metal roof shingles, mineral-surfaced roll roofing, slate and slate-type shingles, wood shingles, wood shakes, and metal roof panels shall conform to the applicable standards listed in this chapter. Underlayment materials required to comply with ASTM D226, D1970, D4869, and D6757 shall bear a label indicating compliance to the standard designation and, if applicable, type classification indicated in Table R905.1.1(1). Underlayment shall be applied in accordance with Table R905.1.1(2). Underlayment shall be attached in accordance with Table R905.1.1(3).

Exceptions:

1. As an alternative, self-adhering polymer modified bitumen underlayment complying with ASTM D1970 installed in accordance with both the underlayment manufacturer's and roof covering manufacturer's installation instructions for the deck material, roof ventilation configuration, and climate exposure for the roof covering to be installed, shall be permitted.

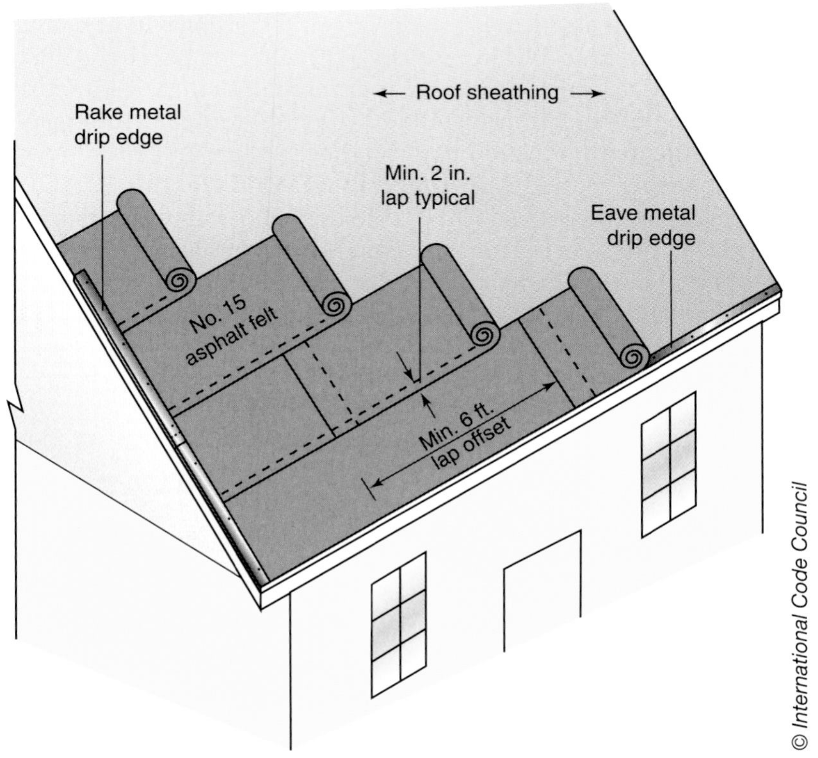

Underlayment

2. As an alternative, a minimum 4-inch (102 mm) wide strip of self-adhering polymer-modified bitumen membrane complying with ASTM D 1970 installed in accordance with the manufacturer's installation instructions for the deck material, shall be applied over all joints in the roof decking. An approved underlayment for the applicable roof covering for maximum ultimate design wind speeds, V_{ult}, less than 140 mph (63 m/s) shall be applied over the entire roof over the 4-inch (102 mm) wide membrane strips.

R905.1.2 Ice Barriers. In areas where there has been a history of ice forming along the eaves causing a backup of water as designated in Table R301.2(1), an ice barrier shall be installed for asphalt shingles, metal roof shingles, mineral-surfaced roll roofing, slate and slate-type shingles, wood shingles and wood shakes. The ice barrier shall consist of not fewer than two layers of underlayment cemented together or a self-adhering polymer modified bitumen sheet shall be used in place of normal underlayment and extend from the lowest edges of all roof surfaces to a point at least 24 inches (610 mm) inside the exterior wall line of the building. On roofs with slope equal to or greater than 8 units vertical in 12 units horizontal, the ice barrier shall also be applied not less than 36 inches (914 mm) measured along with the roof slope from the eave edge of the building.

Exception: Detached accessory structures that contain no conditioned floor area.

TABLE R905.1.1(1) Underlayment Types

Roof Covering	Section	Maximum Ultimate Design Wind Speed, $V_{ult} < 140$ mph	Maximum Ultimate Design Wind Speed, $V_{ult} \geq 140$ mph
Asphalt shingles	R905.2	ASTM D 226 Type I; ASTM D 4869 Type I, II, III, or IV; ASTM D 6757	ASTM D 226 Type II; ASTM D 4869 Type IV; ASTM D 6757
Clay and concrete tile	R905.3	ASTM D 226 Type II; ASTM D 2626 Type I; ASTM D 6380 Class M mineral surfaced roll roofing	ASTM D 226 Type II; ASTM D 2626 Type I; ASTM D 6380 Class M mineral surfaced roll roofing
Metal roof shingles	R905.4	ASTM D 226 Type I or II; ASTM D 4869 Type I, II, III, or IV	ASTM D 226 Type II; ASTM D 4869 Type IV
Mineral-surfaced roll roofing	R905.5	ASTM D 226 Type I or II; ASTM D 4869 Type I, II, III, or IV	ASTM D 226 Type II; ASTM D 4869 Type IV
Slate and slate-type shingles	R905.6	ASTM D 226 Type I; ASTM D 4869 Type I, II, III, or IV	ASTM D 226 Type II; ASTM D 4869 Type IV
Wood shingles	R905.7	ASTM D 226 Type I or II; ASTM D 4869 Type I, II, III, or IV	ASTM D 226 Type II; ASTM D 4869 Type IV
Wood shakes	R905.8	ASTM D 226 Type I or II; ASTM D 4869 Type I, II, III, or IV	ASTM D 226 Type II; ASTM D 4869 Type IV
Metal panels	R905.10	Manufacturer's instructions	ASTM D 226 Type II; ASTM D 4869 Type IV

R905.1.1 continues

R905.1.1 continued

TABLE R905.1.1(2) Underlayment Application

Roof Covering	Section	Maximum Ultimate Design Wind Speed, V_{ult} < 140 mph	Maximum Ultimate Design Wind Speed, $V_{ult} \geq$ 140 mph
Asphalt shingles	R905.2	For roof slopes from two units vertical in 12 units horizontal (2:12), up to four units vertical in 12 units horizontal (4:12), underlayment shall be two layers applied in the following manner. Apply a 19-inch strip of underlayment felt parallel to and starting at the eaves. Starting at the eave, apply 36-inch-wide sheets of underlayment, overlapping successive sheets 19 inches. Distortions in the underlayment shall not interfere with the ability of the shingles to seal. For roof slopes of four units vertical in 12 units horizontal (4:12) or greater, underlayment shall be one layer applied in the following manner. Underlayment shall be applied shingle fashion, parallel to and starting from the eave and lapped 2 inches. Distortions in the underlayment shall not interfere with the ability of the shingles to seal. End laps shall be 4 inches and shall be offset by 6 feet.	Same as Maximum Ultimate Design Wind Speeds, V_{ult} < 140 mph except all laps shall be a minimum of 4 inches.
Clay and concrete tile	R905.3	For roof slopes from two and one-half units vertical in 12 units horizontal (2 ½:12), up to four units vertical in 12 units horizontal (4:12), underlayment shall be a minimum of two layers applied as follows: starting at the eave, apply a 19-inch strip of underlayment parallel with the eave. Starting at the eave, apply 36-inch-wide strips of underlayment felt, overlapping successive sheets 19 inches. For roof slopes of four units vertical in 12 units horizontal (4:12) or greater, underlayment shall be a minimum of one layer of underlayment felt applied shingle fashion, parallel to and starting from the eaves and lapped 2 inches. End laps shall be 4 inches and shall be offset by 6 feet.	Same as Maximum Ultimate Design Wind Speeds, V_{ult} < 140 mph except all laps shall be a minimum of 4 inches.
Metal roof shingles Mineral-surfaced roll roofing Slate and slate-type shingles Wood shingles Wood shakes Metal panels	R905.4 R905.5 R905.6 R905.7 R905.8 R905.10	Apply in accordance with the manufacturer's installation instructions.	For roof slopes from two units vertical in 12 units horizontal (2:12), up to four units vertical in 12 units horizontal (4:12), underlayment shall be two layers applied in the following manner: apply a 19-inch strip of underlayment felt parallel to and starting at the eaves. Starting at the eave, apply 36-inch-wide sheets of underlayment, overlapping successive sheets 19 inches, and fastened sufficiently to hold in place. For roof slopes of four units vertical in 12 units horizontal (4:12) or greater, underlayment shall be one layer applied in the following manner: underlayment shall be applied shingle fashion, parallel to and starting from the eave and lapped 4 inches. End laps shall be 4 inches and shall be offset by 6 feet.

TABLE R905.1.1(3) Underlayment Attachment

Roof Covering	Section	Maximum Ultimate Design Wind Speed, $V_{ult} \leq 140$ mph	Maximum Ultimate Design Wind Speed, $V_{ult} \geq 140$ mph
Asphalt shingles	R905.2	Fastened sufficiently to hold in place	The underlayment shall be attached with corrosion-resistant fasteners in a grid pattern of 12 inches between side laps with a 6-inch spacing at the side laps. Underlayment shall be attached using metal or plastic cap nails or cap staples with nominal cap diameter of not less than 1 inch. Metal caps shall have a thickness not less than 32-gauge sheet metal. Power-driven metal caps shall have a minimum thickness of 0.010 inch. Minimum thickness of the outside edge of plastic caps shall be 0.035 inch. The cap-nail shank shall be not less than 0.083 inch for ring shank cap nails and 0.091 inch for smooth shank cap nails. Staple gage shall be not less than 21 gage. Cap-nail shank and cap staple legs shall have a length sufficient to penetrate through the roof sheathing or not less than ¾ inch into the roof sheathing.
Clay and concrete tile	R905.3		
Metal roof shingles	R905.4	Manufacturer's installation instructions.	The underlayment shall be attached with corrosion resistant fasteners in a grid pattern of 12 inches between side laps with a 6-inch spacing at the side laps. Underlayment shall be attached using metal or plastic cap nails or cap staples with nominal cap diameter of not less than 1 inch. Metal caps shall have a thickness of at least 32-gauge sheet metal. Power-driven metal caps shall have a minimum thickness of 0.010 inch. Minimum thickness of the outside edge of plastic caps shall be 0.035 inch. The cap-nail shank shall be not less than 0.083 inch for ring shank cap nails and 0.091 inch for smooth shank cap nails. Staple gage shall be not less than 21 gage. Cap-nail shank and cap staple legs shall have a length sufficient to penetrate through the roof sheathing or not less than ¾ inch into the roof sheathing.
Mineral-surfaced roll roofing	R905.5		
Slate and slate-type shingles	R905.6		
Wood Shingles	R905.7		
Wood shakes	R905.8		
Metal panels	R905.10		

CHANGE SIGNIFICANCE: This code change reorganizes the underlayment provisions contained within the IRC. In the 2012 IRC, underlayment provisions were specified individually for each type of roof covering. Many of the roof covering provisions contained identical requirements for underlayment type, application, and attachment. This change relocates the underlayment requirements for roof covering to a single section at the beginning of Section R905. There are three new tables that address underlayment type (Table R905.1.1[1]), application (Table R905.1.1[2]), and attachment (Table R905.1.1[3]) for each roof covering in the IRC that requires underlayment. Consolidating the underlayment requirements into a single section makes provisions easier to locate and highlights key differences between requirements for underlayment for different types of roof coverings.

For metal roof panels in areas with wind speeds of 140 mph or greater, ASTM D4869 Type IV underlayment is added as an approved underlayment.

R905.7.5

Wood Shingle Application

CHANGE TYPE: Modification

CHANGE SUMMARY: The minimum requirements for application of wood shingles are expanded. Fastener type is clarified and a new table lists minimum sizes for box nails. Labeling requirements for fastener packaging have also been added.

2015 CODE: R905.7.5 Application. Wood shingles shall be installed in accordance with this chapter and the manufacturer's installation instructions. Wood shingles shall be laid with a side lap not less than 1½ inch (38 mm) between joints in courses, and two joints shall not be in direct alignment in any three adjacent courses. Spacing between shingles shall be not less than ¼ inch to ⅜ inch (6.4 mm to 9.5 mm). Weather exposures for wood shingles shall not exceed those set in Table R905.7.5(1). Fasteners for untreated (naturally durable) wood shingles shall be ~~corrosion resistant with a minimum penetration of ½ inch (13 mm) into the sheathing. For sheathing less than ½ inch (13 mm) in thickness, the fasteners shall extend through the sheathing.~~ box nails in accordance with Table R905.7.5(2). Nails shall be stainless steel Type 304 or Type 316 or hot-dipped galvanized, with a coating weight of ASTM A153 Class D (1.0 oz/ft^2). Alternatively, two 16-gage stainless steel Type 304 or Type 316 staples with crown widths $^7/_{16}$ inch (11.1 mm) minimum, ¾ inch (19.1 mm) maximum shall be used. Fasteners installed within 15 miles (24 km) of salt water coastal areas shall be stainless steel Type 316. Fasteners for fire retardant-treated shingles in accordance with Section R902 or pressure-impregnated-preservative-treated shingles of naturally durable wood in accordance with AWPA U1 shall be stainless steel Type 316. All fasteners shall have a minimum penetration into the sheathing of ¾ inch (19.1 mm). For sheathing less than ~~½ inch~~ ¾ inch (19.1 mm) thickness, each fastener shall ~~extend~~ penetrate through the sheathing. Wood shingles shall be attached to the roof with two fasteners per shingle, positioned ~~no more than ¾ inch from each edge and no more than 1 inch (25 mm) above the exposure line.~~ in accordance with the manufacturer's installation instructions. Fastener packaging shall bear a label indicating the appropriate grade material or coating weight.

TABLE R905.7.5(2) Nail Requirements for Wood Shakes and Wood Shingles

Shakes	Nail Type and Minimum Length	Minimum Head Size	Minimum Shank Diameter
18″ Straight-Split	5d Box 1¾″	0.19″	0.08″
18″ and 24″ Handsplit and Resawn	6d Box 2″	0.19″	0.0915″
24″ Tapersplit	5d Box 1¾″	0.19″	0.08″
18″ and 24″ Tapersawn	6d Box 2″	0.19″	0.0915″
Shingles	**Nail Type and Minimum Length**	**Minimum Head Size**	**Minimum Shank Diameter**
16″ and 18″	3d Box 1¼″	0.19″	0.08″
24″	4d Box 1½″	0.19″	0.08″

CHANGE SIGNIFICANCE: This change clarifies the fasteners required for attachment of wood shingles. The type of fastener to be used is determined by environmental factors and product types. For naturally durable wood shingles, two types of stainless steel nail and hot-dipped galvanized box nails are approved. Alternatively, 16 gauge Type 304 or 316 stainless steel staples may also be used. For fasteners installed on dwellings within 15 miles of a saltwater coastal area, only Type 316 stainless steel nails are allowed. Fasteners of fire-retardant-treated wood must be Type 316 stainless steel nails.

Minimum sheathing penetration is ¾-inch or all the way through sheathing in order to attach the product to hold it in place and prevent withdrawal of the fastener. Additionally, shingles may not be applied with vertical edges tight together. Tightly applied shingles do not leave room for expansion, causing fish-mouthing, cupping, and curling. Wood shingles must be attached in accordance with manufacturers' directions.

R905.8.6

Wood Shake Application

CHANGE TYPE: Modification

CHANGE SUMMARY: The minimum requirements for application of wood shakes are expanded. Fastener type is clarified, and a new table lists minimum sizes for box nails. Labeling requirements for fastener packaging have also been added.

2015 CODE: **R905.8.6 Application.** Wood shakes shall be installed in accordance with this chapter and the manufacturer's installation instructions. Wood shakes shall be laid with a side lap not less than 1½ inch (38 mm) between joints in adjacent courses. Spacing between shakes in the same course shall be ³⁄₈ inch to ⁵⁄₈ inch (9.5 mm to 15.9 mm) ~~for shakes and~~ <u>including</u> tapersawn shakes ~~of naturally durable wood shall be ³⁄₈ inch to ⁵⁄₈ inch (9.5 mm to 15.9 mm) for preservative-treated taper sawn shakes~~. Weather exposures for wood shakes shall not exceed those set in Table R905.8.6. Fasteners for <u>untreated (naturally durable)</u> wood shakes shall be ~~corrosion resistant with a minimum penetration of ½ inch (12.7 mm) into the sheathing. For sheathing less than ½ inch (13 mm) thick, the fasteners shall extend through the sheathing.~~ <u>box nails in accordance with Table R905.7.5(2). Nails shall be stainless steel Type 304 or Type 316 or hot-dipped galvanized, with a coating weight of ASTM A153 Class D (1.0 oz/ft^2).</u> Alternatively, <u>two 16-gage Type 304 or Type 316 stainless steel staples, with crowns width ⁷⁄₁₆ inch (11.1 mm) minimum, ¾ inch (19.1 mm) maximum, shall be used. Fasteners installed within 15 miles (24 km) of salt water coastal areas shall be stainless steel Type 316.</u> Wood shakes shall be attached to the roof with two fasteners per shake positioned ~~no more than 1 inch (25 mm) no more than 2 inches (25 mm) above the exposure line.~~ <u>in accordance with the manufacturer's installation instructions. Fasteners for fire-retardant-treated (as defined in Section R902) shakes or pressure-impregnated-preservative-treated shakes of naturally durable</u>

© Tom Grundy/Shutterstock.com.

Wood shake roofing

wood in accordance with AWPA U1 shall be stainless steel Type 316. All fasteners shall have a minimum penetration into the sheathing of ¾ inch (19.1 mm). Where the sheathing is less than ¾ inch (19.1 mm) thick, each fastener shall penetrate through the sheathing. Fastener packaging shall bear a label indicating the appropriate grade material or coating weight.

CHANGE SIGNIFICANCE: This change clarifies the fasteners required for attachment of wood shakes. The type of fastener to be used is determined by environmental factors and product types. For naturally durable wood, two types of stainless steel nail and hot-dipped galvanized box nails are approved. Alternatively, 16 gauge Type 304 or 316 stainless steel staples may also be used. For fasteners installed on dwellings within 15 miles of a saltwater coastal area, only Type 316 stainless steel nails are allowed. Fasteners of fire-retardant treated wood must be Type 316 stainless steel nails. See significant change article R905.7.5 for Table R905.7.5(2) listing minimum box nail sizes.

Minimum sheathing penetration is ¾-inch or all the way through sheathing in order to attach the product to hold it in place and prevent withdrawal of the fastener. Shakes may not be applied with vertical edges tight together. Tightly applied shakes do not leave room for expansion, causing fish-mouthing, cupping, and curling. Wood shakes must be attached in accordance with manufacturers' directions.

R905.16

Photovoltaic Shingles

CHANGE TYPE: Modification

CHANGE SUMMARY: Additional requirements and limits for photovoltaic shingles have been added to Section R905.16.

2015 CODE: R905.16 Photovoltaic ~~Modules/~~Shingles. The installation of photovoltaic ~~modules/~~shingles shall comply with the provisions of this section, Section R324 and NFPA 70.

R905.16.1 Deck Requirements. Photovoltaic shingles shall be applied to a solid or closely fitted deck, except where the roof covering is specifically designed to be applied over spaced sheathing.

R905.16.2 Deck Slope. Photovoltaic shingles shall be used only on roof slopes of two units vertical in 12 units horizontal (2:12) or greater.

R905.16.3 Underlayment. Unless otherwise noted, required underlayment shall conform to ASTM D4869 or ASTM D6757.

R905.16.4 Underlayment Application. Underlayment shall be applied shingle fashion, parallel to and starting from the eave, lapped 2 inches (51 mm) and fastened sufficiently to hold in place.

R905.16.4.1 Ice Barrier. In areas where there has been a history of ice forming along the eaves causing a backup of water as designated in Table R301.2(1), an ice barrier that consists of not less than two layers of underlayment cemented together or of a self-adhering polymer-modified

Photo Courtesy of Beldon Roofing Company

Photovoltaic shingle

bitumen sheet, shall be used in lieu of normal underlayment and extend from the lowest edges of all roof surfaces to a point not less than 24 inches (610 mm) inside the exterior wall line of the building.

> **Exception:** Detached accessory structures that contain no conditioned floor area.

R905.16.4.2 Underlayment and High Winds. Underlayment applied in areas subject to high winds [above 140 mph (63 m/s) in accordance with Figure R301.2(4)A] shall be applied with corrosion-resistant fasteners in accordance with the manufacturer's installation instructions. Fasteners are to be applied along the overlap not farther apart than 36 inches (914 mm) on center.

Underlayment installed where the ultimate design wind speed equals or exceeds 150 mph (67 m/s) shall comply with ASTM D4869 Type IV, or ASTM D6757. The underlayment shall be attached in a grid pattern of 12 inches (305 mm) between side laps with a 6-inch (152 mm) spacing at the side laps. Underlayment shall be applied as required for asphalt shingles in accordance with Table R905.1.1(2). Underlayment shall be attached using metal or plastic cap nails with a head diameter of not less than 1 inch (25.4 mm) with a thickness of not less than 32-gage sheet metal. The cap-nail shank shall be not less than 12-gage (0.105 inches) with a length to penetrate through the roof sheathing or not less than ¾ inch (19 mm) into the roof sheathing.

> **Exception:** As an alternative, adhered underlayment complying with ASTM D1970 shall be permitted.

CHANGE SIGNIFICANCE: Section R905.16, Photovoltaic Shingles, is expanded. The section now contains requirements for roof decks, minimum roof deck slope, underlayment, underlayment application, ice barrier, and underlayment for high-wind areas. The new requirements are consistent with similar attributes for other non-flat, shingle-type roof coverings. Reference to NFPA 70 and Section R324 for photovoltaic solar energy systems is added.

The word "modules" is deleted from the section title because it is not defined in the code for photovoltaic applications. "Photovoltaic shingles" is now the descriptor for this application. Additionally, the section mirrors the information and format found in the 2015 *International Building Code* for photovoltaic shingles.

R907

Rooftop-Mounted Photovoltaic Systems

CHANGE TYPE: Addition

CHANGE SUMMARY: This code provision describes the requirements and limits of rooftop-mounted photovoltaic systems.

2015 CODE: **R907.1 Rooftop-Mounted Photovoltaic Systems.** Rooftop-mounted photovoltaic panels or modules shall be installed in accordance with this section, Section R324 and NFPA 70.

R907.2 Wind Resistance. Rooftop-mounted photovoltaic panel or modules systems shall be installed to resist the component and cladding loads specified in Table R301.2(2), adjusted for height and exposure in accordance with Table R301.2(3).

R907.3 Fire Classification. Rooftop-mounted photovoltaic panels or modules shall have the same fire classification as the roof assembly required in Section R902.

R907.4 Installation. Rooftop-mounted photovoltaic panels or modules shall be installed in accordance with the manufacturer's installation instructions.

R907.5 Photovoltaic Panels and Modules. Rooftop-mounted photovoltaic panels and modules shall be listed and labeled in accordance with UL 1703 and shall be installed in accordance with the manufacturer's printed instructions.

CHANGE SIGNIFICANCE: Specific requirements applicable to rooftop-mounted photovoltaic panels and modules are added. These provisions complement the existing requirements for photovoltaic solar energy systems in Section R324. The new section also references requirements in NFPA 70. Panels and modules must be listed and labeled to meet the requirements of UL 1703.

Requirements for resistance of component and cladding loads and minimum fire classification are added. Installation is in accordance with the manufacturer's directions. These requirements mirror provisions in the 2015 *International Building Code.*

Photo Courtesy of Peter Kulczyk

Rooftop-mounted photovoltaic system

PART **4**

Energy Conservation

Chapter 11

■ **Chapter 11** Energy Efficiency

The IRC energy provisions are extracted from the residential provisions of the *International Energy Conservation Code* (IECC) and editorially revised to conform to the scope and application of the IRC. The section numbers appearing in parentheses after each IRC section number are the section numbers of the corresponding text in the IECC. The IECC Residential Provisions and Chapter 11 of the IRC provide for the effective use and conservation of energy in new residential buildings by regulating the building envelope, mechanical systems, electrical systems, and service water heating systems. IRC Section N1101 establishes climate zones for geographical locations as the basis for determining thermal envelope requirements for conserving energy. The various elements of the building thermal envelope are covered in Section N1102 and include specific insulation, fenestration, and air-leakage requirements for improving energy efficiency. Section N1103 primarily is concerned with mechanical system controls, insulation and sealing of ductwork, equipment sizing, and mandatory mechanical ventilation systems. The insulation of mechanical and service hot water piping systems is also covered in the mechanical systems provisions of Section N1103. Energy-efficient lighting is covered in Section N1104. Alternative compliance provisions appear in Sections N1105 and N1106. The new Sections N1107 through N1111 address the application of the energy provisions for work performed on existing buildings. ■

N1101.13

Compliance Paths

N1101.14

Permanent Energy Certificate

N1102.1.3

R-Value Computation—Insulated Siding

N1102.2.4

Access Hatches and Doors

N1102.2.7, TABLE N1102.1.2

R-Value Reduction for Walls with Partial Structural Sheathing

N1102.2.8, TABLE N1102.4.1.1

Floor Framing Cavity Insulation

TABLE N1102.4.1.1

Insulation at Wall Corners and Headers

N1102.4.2, TABLE N1102.4.1.1

Wood-Burning Fireplace Doors

N1103.3

Duct Sealing and Testing

N1103.5

Heated Water Circulation and Temperature Maintenance Systems

N1101.13

Compliance Paths

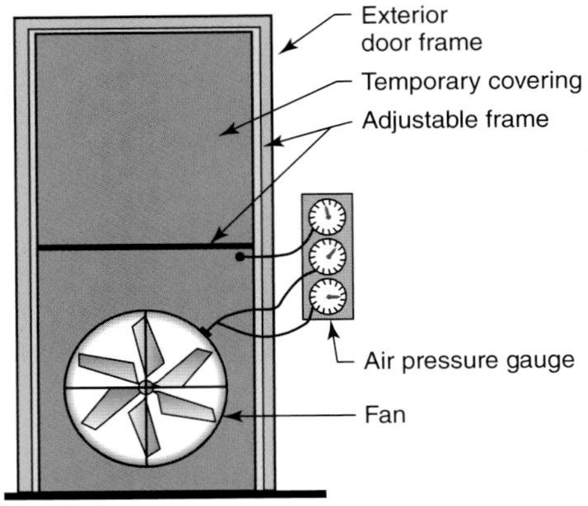

© International Code Council

Blower door test: Air-leakage provisions, including testing of the building thermal envelope, are mandatory.

CHANGE TYPE: Modification

CHANGE SUMMARY: The compliance paths in the energy provisions have been clarified. The mandatory provisions combined with either the prescriptive provisions or the performance provisions are deemed to comply with the code.

2015 CODE: ~~N1101.15~~ N1101.13 (R401.2) Compliance. Projects shall comply with ~~Sections identified as "mandatory" and with either sections identified as "prescriptive" or the performance approach in Section N1105.~~ one of the following:

1. Sections N1101.14 through N1104.
2. Section N1105 and the provisions of Sections N1101.14 through N1104 labeled "Mandatory."
3. An energy rating index (ERI) approach in Section N1106.

CHANGE SIGNIFICANCE: A number of sections in the energy provisions are labeled as "mandatory." There are no trade-offs or equivalency provisions for these sections. For example, Section N1101.14 requires a permanent energy certificate to be posted in the building. Likewise, the code mandates the limits and testing requirements for the air-leakage provisions related to the building thermal envelope in Section N1102.4. Section N1103.3.3 sets the circumstances requiring duct testing and mandates the methods for conducting that testing. These mandatory provisions must be complied with in pursuing any path for satisfying the energy efficiency requirements in the code. Sections labeled as "prescriptive" offer clear rules to follow to gain compliance with the code. The prescriptive rules are easiest to follow and the most commonly used in combination with the mandatory provisions for achieving compliance. However, the designer or builder may choose alternative performance methods to demonstrate compliance rather than follow the prescriptive provisions. The code offers a simulated performance alternative in Section N1105 and an energy rating index compliance alternative in new Section N1106. Both of these methods employ approved compliance software to generate compliance reports that demonstrate equivalency with the prescriptive provisions for conserving energy.

The various compliance paths offered in Section N1101.13 are consistent with the intent statement in Section N1101.2 that the energy provisions intend to provide flexibility to permit the use of innovative approaches and techniques to achieve the objective for effective use and conservation of energy. Section N1101.3 authorizes the building official to accept specific computer software and calculation methods to meet the intent of the code. Likewise, the building official is authorized to approve a national, state, or local energy efficiency program as exceeding the energy efficiency requirements of the IRC and therefore in compliance with the code.

CHANGE TYPE: Modification

CHANGE SUMMARY: The code now requires the permanent energy certificate to be placed on a wall in proximity to the furnace, in a utility room, or in another approved location inside the building.

N1101.14

Permanent Energy Certificate

2015 CODE: ~~N1101.16~~ N1101.14 (R401.3) Certificate (Mandatory). A permanent certificate shall be completed by the builder or registered design professional and posted on ~~or in the electrical distribution panel~~ a wall in the space where the furnace is located, a utility room, or an approved location inside the building ~~by the builder or registered design professional~~. Where located on an electrical panel, the certificate shall not cover or obstruct the visibility of the circuit directory label, service disconnect label, or other required labels. The certificate shall list the predominant *R*-values of insulation installed in or on ceiling/roof, walls, foundation (slab, basement wall, crawlspace wall, and/or floor), and ducts outside conditioned spaces; *U*-factors for fenestration and the solar heat gain coefficient (SHGC) of fenestration, and the results from any required duct system and building envelope air leakage testing done on the building. Where there is more than one value for each component, the certificate shall list the value covering the largest area. The certificate shall list the types and efficiencies of heating, cooling, and service water heating equipment. Where a gas-fired unvented room heater, electric furnace,

N1101.14 continues

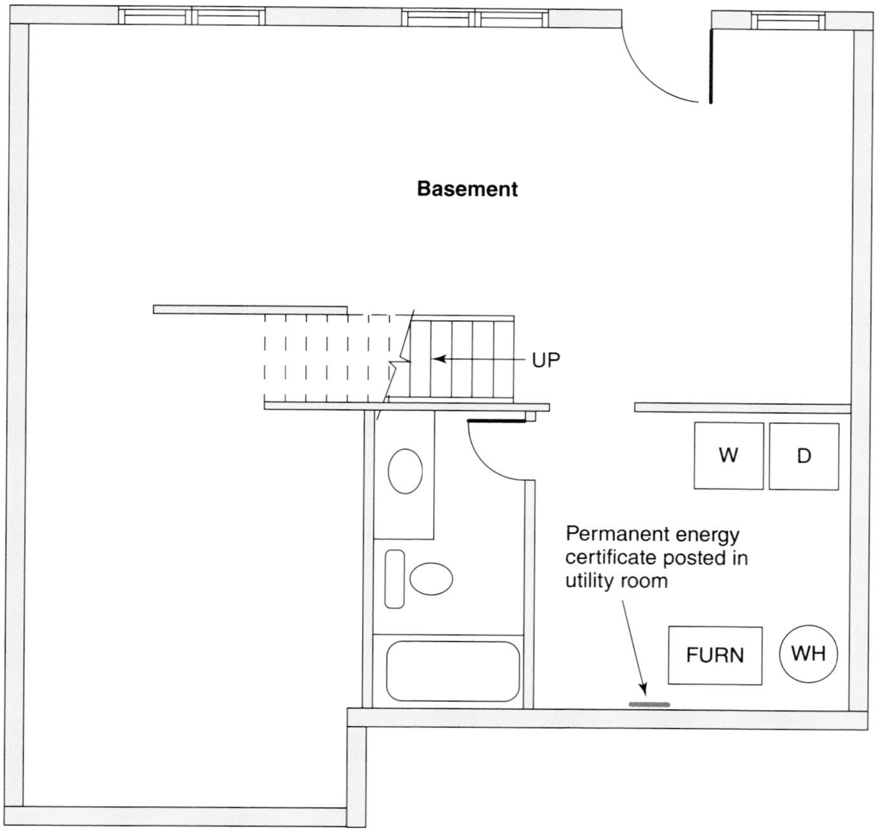

A permanent energy certificate must be posted inside the building.

N1101.14 continued

Energy Efficiency Certificate	
Insulation Rating	
Ceiling/roof	
Walls	
Floors	
Ducts	
Air-leakage Test Results	
Blower door	Duct testing
Fenestration Rating	
Window	
Opaque door	
Skylight	
Equipment Performance	
Heating system	
Cooling system	
Water heater	
Designer/builder	

Permanent energy certificate

or baseboard electric heater is installed in the residence, the certificate shall list "gas-fired unvented room heater," "electric furnace," or "baseboard electric heater," as appropriate. An efficiency shall not be listed for gas-fired unvented room heaters, electric furnaces, or electric baseboard heaters.

CHANGE SIGNIFICANCE: The IRC requires the builder or registered design professional to complete an energy efficiency certificate for each dwelling and to post the certificate for permanent display. The certificate lists the installed insulation and fenestration values, and the results of air-leakage tests. The certificate must also list the type and efficiency of installed heating, cooling, and water heating equipment. Because electric furnaces, baseboard heaters, and unvented gas-fired heaters may not provide the lowest energy consumption when compared to other methods of comfort heating and their energy efficiency ratings may be misleading, the code requires such appliances to be individually listed on the certificate without an efficiency designation.

Previously, the code required the permanent certificate to be affixed to the electrical service panel in a way that did not cover the service directory or other required information governed by the electrical code or the listing of the panel. In some cases, particularly in certain regions of the United States, the electrical service panel is located on the outside of a building and exposed to the weather. There has been a concern that a certificate placed in an outdoor location will be destroyed in a short period of time. The IRC has not permitted such an installation because it prescribes a permanent certificate. Other concerns described the lack of visibility of a certificate installed inside an electrical panel and the distraction of having extra information in or on the panel that is not related to the required safety information that is part of the electrical installation. The 2015 IRC addresses these concerns by requiring that the permanent energy certificate be installed inside the building. There are two prescribed locations—in a space that contains the furnace or a utility room. The intent is that the certificate be located in proximity to the furnace or other mechanical equipment where it will be visible to the homeowner. As an alternative, the code allows other locations provided that they are inside the building and approved by the building official. If approved, a certificate posted in or on an electrical service panel is still allowed by the code, provided it is inside the building and it does not interfere with the visibility of the required labels of the electrical equipment.

N1102.1.3

R-Value Computation—Insulated Siding

CHANGE TYPE: Modification

CHANGE SUMMARY: The code now allows insulated siding to be used in the calculation for satisfying the wall insulation *R*-value. The labeled *R*-value for the siding must be reduced by R-0.6 for calculation purposes.

2015 CODE: N1102.1.3 (R402.1.3) *R*-value Computation. Insulation material used in layers, such as framing cavity insulation ~~and insulating sheathing~~ or continuous insulation, shall be summed to compute the corresponding component *R*-value. The manufacturer's settled *R*-value shall be used for blown insulation. Computed *R*-values shall not include an *R*-value for other building materials or air films. Where insulated siding is used for the purpose of complying with the continuous insulation requirements of Table N1102.1.2, the manufacturer's labeled *R*-value for insulated siding shall be reduced by R-0.6.

N1101.6 (R202) Defined Terms.

INSULATED SIDING. A type of continuous insulation with manufacturer-installed insulating material as an integral part of the cladding product having a minimum *R*-value of R-2.

N1102.1.3 continues

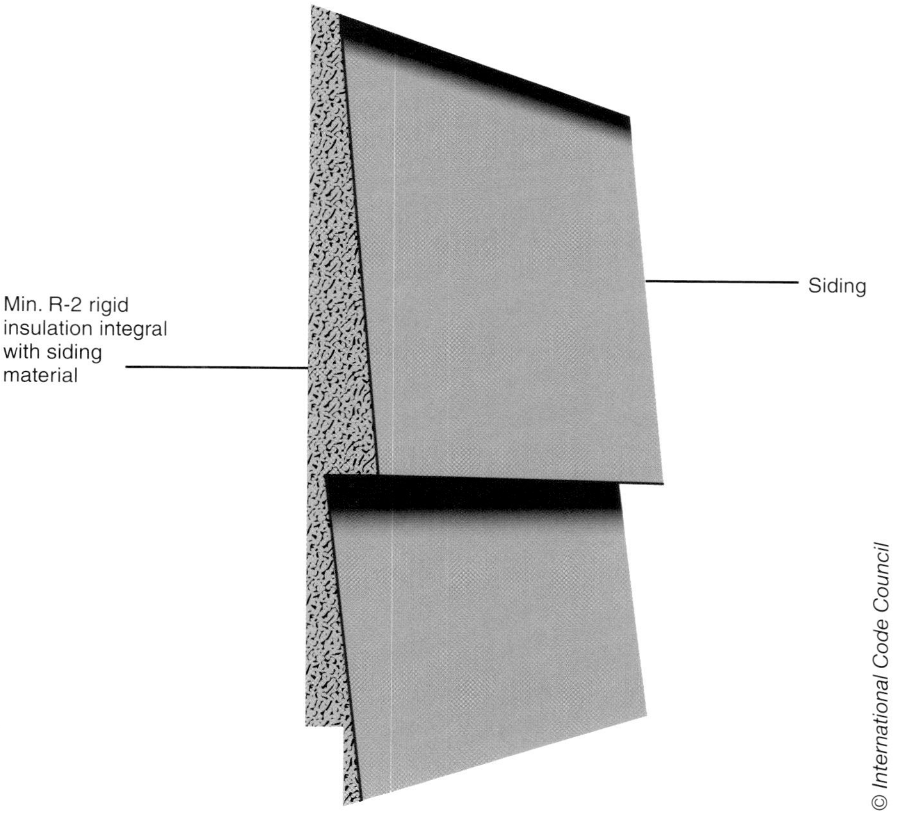

Insulated siding

N1102.1.3 continued

CONTINUOUS INSULATION (ci). Insulating material that is continuous across all structural members without thermal bridges other than fasteners and service openings. It is installed on the interior or exterior or is integral to any opaque surface of the building envelope.

SECTION R202
DEFINITIONS

INSULATED VINYL SIDING. A vinyl cladding product with manufacturer-installed foam plastic insulating material as an integral part of the cladding product, having a thermal resistance of not less than R-2.

CHANGE SIGNIFICANCE: The prescriptive provisions setting the minimum thermal resistance (*R*-value) for insulation of the building thermal envelope are based on the installed *R*-value of the insulation only. Typically, insulation is either cavity insulation or continuous insulation, or a combination. In some climate zones the code prescribes a combination. For example, for Climate Zones 6 through 8, the code prescribes a minimum R-20 cavity insulation combined with a minimum R-5 continuous insulation for walls of the building thermal envelope. Because continuous insulation covers all structural members, there are no thermal gaps at the wall studs, plates, or headers and the insulation is more effective at conserving energy and reducing air leakage when compared to cavity insulation. The improved efficiency is recognized in the code-prescribed wall *R*-values in Climate Zones 3 through 5, where a combination of R-13 cavity insulation plus R-5 continuous insulation is shown as an option that is equivalent to R-20 cavity insulation.

For the prescriptive building thermal envelope provisions, the code does not allow a component or assembly approach for satisfying the *R*-value requirements. Only the insulation *R*-value counts in satisfying the requirement. No credit is given for interior or exterior finishes, air barriers, water-resistive barriers, or other components or materials of the thermal envelope. For that reason, siding has been excluded from this calculation for wall *R*-value. The 2015 IRC adds a definition for insulated siding, which is siding with rigid foam insulation that is laminated or permanently attached to the siding material by the manufacturer. The minimum thermal resistance of this product is R-2. A new definition for insulated siding considers the siding to be continuous insulation and permits it to contribute to the wall insulation calculation. A reduction of R-0.6 must be applied to the manufacturer's labeled *R*-value. Insulated siding labeled as having a thermal resistance of R-3.6 (based on testing to ASTM C1363) receives an R-3.0 for calculation purposes in complying with the insulation provisions.

CHANGE TYPE: Clarification

N1102.2.4
Access Hatches and Doors

CHANGE SUMMARY: Vertical doors that access unconditioned attics and crawl spaces do not require an *R*-value to match the required wall insulation. Such doors must comply with the fenestration *U*-factor requirements of Table N1102.1.2.

2015 CODE: N1102.2.4 (R402.2.4) Access Hatches and Doors. Access doors from conditioned spaces to unconditioned spaces such as attics and crawl spaces shall be weatherstripped and insulated to a level equivalent to the insulation on the surrounding surfaces. Access shall be provided to all equipment that prevents damaging or compressing the insulation. A wood-framed or equivalent baffle or retainer is required to be provided when loose fill insulation is installed, the purpose of which is to prevent the loose fill insulation from spilling into the living space when the attic access is opened, and to provide a permanent means of maintaining the installed *R*-value of the loose fill insulation.

> **Exception:** Vertical doors that provide access from conditioned to unconditioned spaces shall be permitted to meet the fenestration requirements of Table N1102.1.2 based on the applicable climate zone specified in Section N1101.7.

N1102.2.4 continues

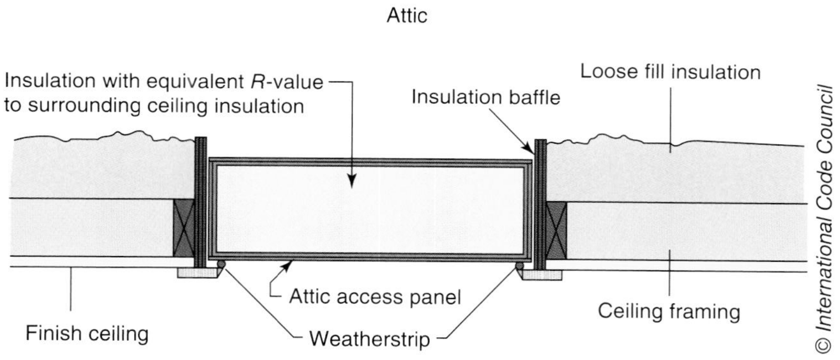

Horizontal attic access hatch

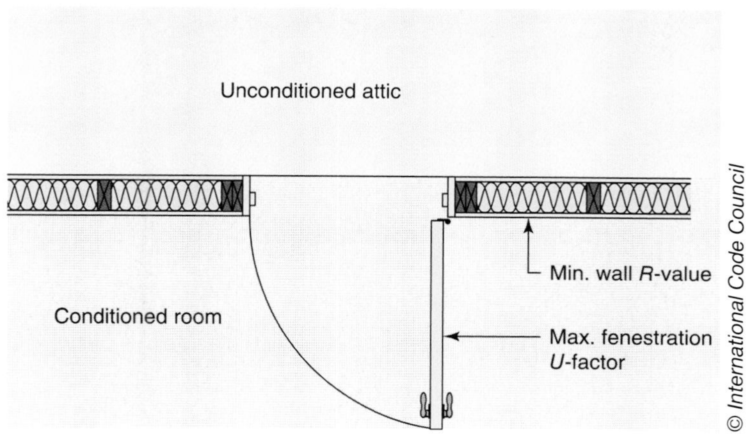

Vertical attic access door

N1102.2.4 continued

CHANGE SIGNIFICANCE: Provisions introduced in the 2009 edition of the IRC clarified that the building thermal envelope requirements apply to hatches and doors that access unconditioned areas such as attics and crawl spaces. The primary intent was to regulate typical attic access hatches that are installed in the ceiling of the dwelling unit to gain entry to the unconditioned attic. In addition to meeting the *R*-value of the area surrounding the access hatch, the code calls for weatherstripping of the hatch to reduce air leakage and infiltration. The code also provides for a baffle or barrier to retain loose fill insulation at the access opening and to maintain the installed *R*-value of the insulation. The provision for a baffle seems to imply that the section applies to horizontal access hatches in ceilings. However, the section title includes doors, and many code users understood that the insulation *R*-value applied to a vertical door used to provide access from conditioned to unconditioned space. A typical installation would be a door in a wall separating a second-floor room from the unconditioned attic of a single-story portion of the building.

The 2015 IRC specifically excludes vertical access doors from the requirement for meeting the required *R*-value of the surrounding wall. Proponents of this change reasoned that such a requirement conflicts with the provisions that apply to exterior doors. Both an access door and an exterior door serve to move from a conditioned space to an unconditioned space, define a portion of the building thermal envelope, and should have the same minimum thermal resistance properties. Table N1102.1.2 requires fenestration, which by definition includes doors, to comply with the maximum *U*-factor. For example, in Climate Zone 5, the table sets a maximum *U*-factor of 0.32 for an exterior door. The provisions for wood frame wall insulation would require a minimum *R*-value of R-20, which is much more stringent than the maximum *U*-factor and would be difficult to achieve with a standard exterior door. The 2015 IRC clarifies that the fenestration *U*-factor from the table applies to a vertical door used for access to an unconditioned space such as an attic. For a horizontal access hatch to the attic, the insulation for the hatch must meet the minimum ceiling *R*-value, which for Climate Zone 5 is R-49.

N1102.2.7, Table N1102.1.2

R-Value Reduction for Walls with Partial Structural Sheathing

CHANGE TYPE: Clarification

CHANGE SUMMARY: The allowed *R*-value reduction for portions of walls with structural sheathing and requiring continuous insulation has been moved from footnote h of Table N1102.1.2 and placed in a new section to clarify the application.

2015 CODE: **N1102.2.7 (R402.2.7) Walls with Partial Structural Sheathing.** Where Section N1102.1.2 would require continuous insulation on exterior walls and structural sheathing covers 40 percent or less of the gross area of all exterior walls, the continuous insulation *R*-value shall be permitted to be reduced by an amount necessary to result in a consistent total sheathing thickness, but not more than R-3, on areas of

N1102.2.7, Table N1102.1.2 continues

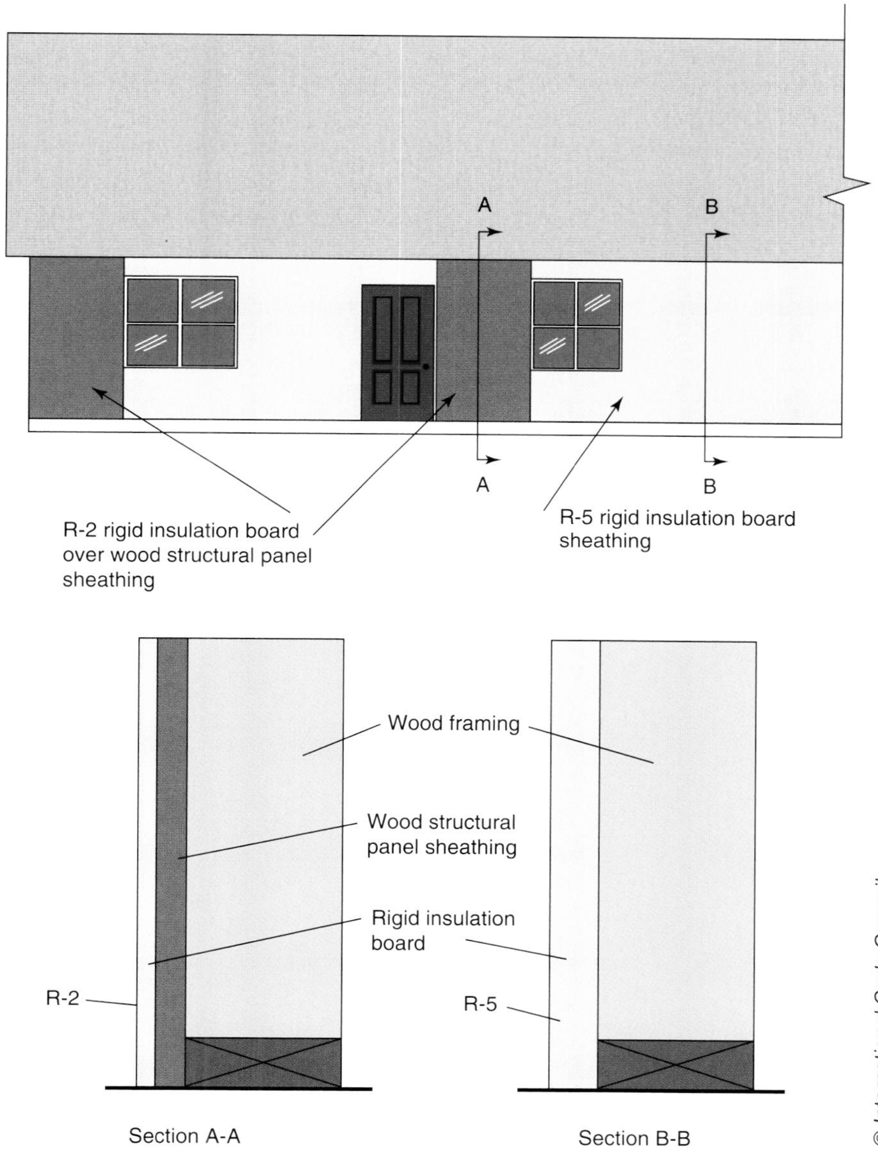

R-value reduction for walls with partial structural sheathing

N1102.2.7, Table N1102.1.2 continued

the walls covered by structural sheathing. This reduction shall not apply to the *U*-factor alternative approach in Section N1102.1.4 and the total UA alternative in Section N1102.1.5.

TABLE ~~N1102.1.1~~ N1102.1.2 (R402.1.2) Insulation and Fenestration Requirements by Component

(Portions of table not shown remain unchanged.)

h. The first value is cavity insulation, the second value is continuous insulation, so "13 + 5" means R-13 cavity insulation plus R-5 continuous insulation. ~~If structural sheathing covers 40 percent or less of the exterior, continuous insulation *R*-value shall be permitted to be reduced by no more than R-3 in the locations where structural sheathing is used to maintain a consistent total sheathing thickness.~~

(Footnotes not shown remain unchanged.)

CHANGE SIGNIFICANCE: New section N1102.2.7 clarifies the provisions for reducing the required *R*-value for portions of walls having structural sheathing and requiring continuous insulation. Previously, this reduction was covered in footnote h of Table N1102.1.1 (now Table N1102.1.2) that sets values for insulation and fenestration of the various components of the building thermal envelope. There is no change to the technical requirements related to the reduction in *R*-value. Moving the relevant text out of the footnote and into a separate code section allows for a more thorough description of the provision that is easier to locate and improves understanding of the code. Structural sheathing refers to wood structural panels or structural fiberboard, or other similar products used to comply with the wood wall bracing requirements. In the northernmost Climate Zones 6 through 8, Table N1102.1.2 requires wall cavity insulation with a value of not less than R-20 plus continuous insulation with a value of not less than R-5 (alternatively, cavity insulation can be R-13 when the continuous insulation is R-10). Continuous insulation refers to rigid foam plastic insulation that covers the wall framing and cavities so there is no thermal gap at the studs and plates as there is with cavity insulation. Where structural sheathing covers an area not greater than 40 percent of the gross area of exterior walls, the rating of the continuous foam plastic insulation can be reduced by as much as R-3. This means that a cavity insulation of R-20 plus continuous insulation of R-2 (R-5 – R-3 = R-2) at the locations having structural sheathing satisfies the code requirement.

N1102.2.8, Table N1102.4.1.1

Floor Framing Cavity Insulation

CHANGE TYPE: Modification

CHANGE SUMMARY: The code now permits an air space above required insulation installed in a floor framing cavity above unconditioned space. Table N1102.4.1.1 has been reformatted into three columns to separate the air barrier requirements from the insulation requirements.

2015 CODE: ~~**N1102.2.7**~~ **N1102.2.8 (R402.2.8) Floors.** Floor framing cavity insulation shall be installed to maintain permanent contact with the underside of the subfloor decking.

> **Exception:** The floor framing cavity insulation shall be permitted to be in contact with the topside of sheathing or continuous insulation installed on the bottom side of floor framing where combined with

N1102.2.8, Table N1102.4.1.1 continues

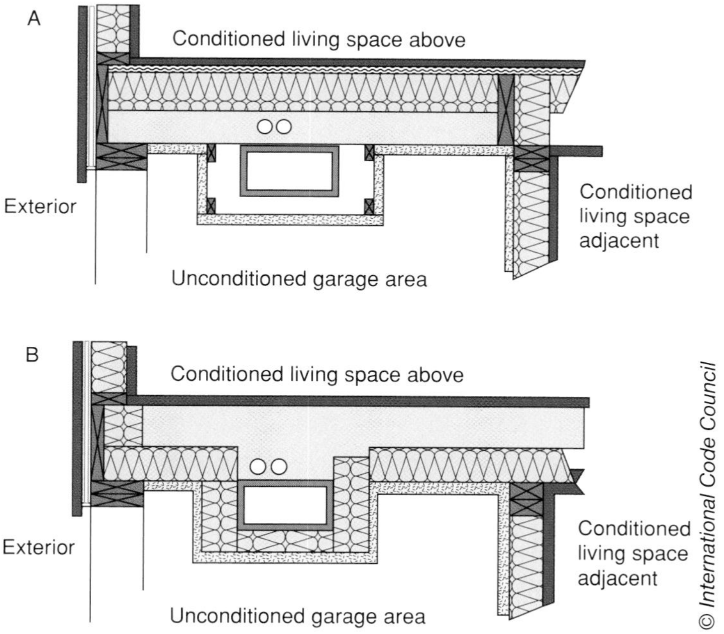

Two options for floor insulation above unconditioned space

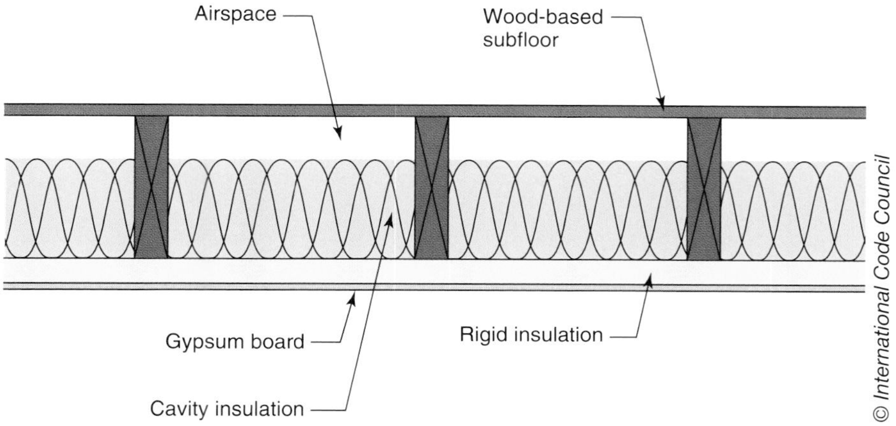

Option for floor insulation above rigid insulation

N1102.2.8, Table N1102.4.1.1 continued

TABLE N1102.4.1.1 (402.4.1.1) Air Barrier and Insulation Installation

Component	Air Barrier Criteria	Insulation Installation Criteria
Floors (including above garage and cantilevered floors)	The air barrier shall be installed at any exposed edge of insulation	Floor framing cavity insulation shall be installed to maintain permanent contact with the underside of subfloor decking, or floor framing cavity insulation shall be permitted to be in contact with the topside of sheathing, or continuous insulation installed on the underside of floor framing; and extends from the bottom to the top of all perimeter floor framing members.

(Portions of table not shown for brevity and clarity.)

insulation that meets or exceeds the minimum wood frame wall *R*-value in Table N1102.1.2 and that extends from the bottom to the top of all perimeter floor framing members.

CHANGE SIGNIFICANCE: Previously, the code required insulation that was installed in a floor framing cavity to be in contact with the underside of the floor sheathing. The code still permits that as one option, but adds another option to have an air space between the floor sheathing and the top of the cavity insulation. In this case, the cavity insulation is in direct contact with the topside of the sheathing or continuous insulation installed on the underside of the floor framing and is combined with perimeter insulation that meets or exceeds the *R*-value requirements for walls. This second option leads to fewer cold spots and does not increase heat loss. It also facilitates ductwork, piping and wiring to be enclosed within the thermal envelope.

Previously, Table N1102.4.1.1 contained only two columns. The first column described the component of construction under consideration and the second column prescribed both the air barrier and insulation installation criteria. The 2015 IRC reformats the table to place the air barrier and insulation requirements in separate columns. The reformatting does not change the technical requirements but intends to clarify the application of the table and reduce confusion by code users.

Table N1102.4.1.1

Insulation at Wall Corners and Headers

CHANGE TYPE: Clarification

CHANGE SUMMARY: Insulation requirements at framed wall corners and headers only apply when there is space to install insulation. The minimum insulation thermal resistance is R-3 per inch of insulation.

2015 CODE:

TABLE N1102.4.1.1 (402.4.1.1) Air Barrier and Insulation Installation

Component	Air Barrier Criteria	Insulation Installation Criteria
Walls	The junction of the foundation and sill plate shall be sealed. The junction of the top plate and the top of exterior walls shall be sealed. Knee walls shall be sealed.	Cavities within corners and headers of frame walls shall be insulated by completely filling the cavity with a material having a thermal resistance of R-3 per inch minimum. Exterior thermal envelope insulation for framed walls shall be installed in substantial contact and continuous alignment with the air barrier.

(Portions of table not shown for brevity and clarity.)

Table N1102.4.1.1 continues

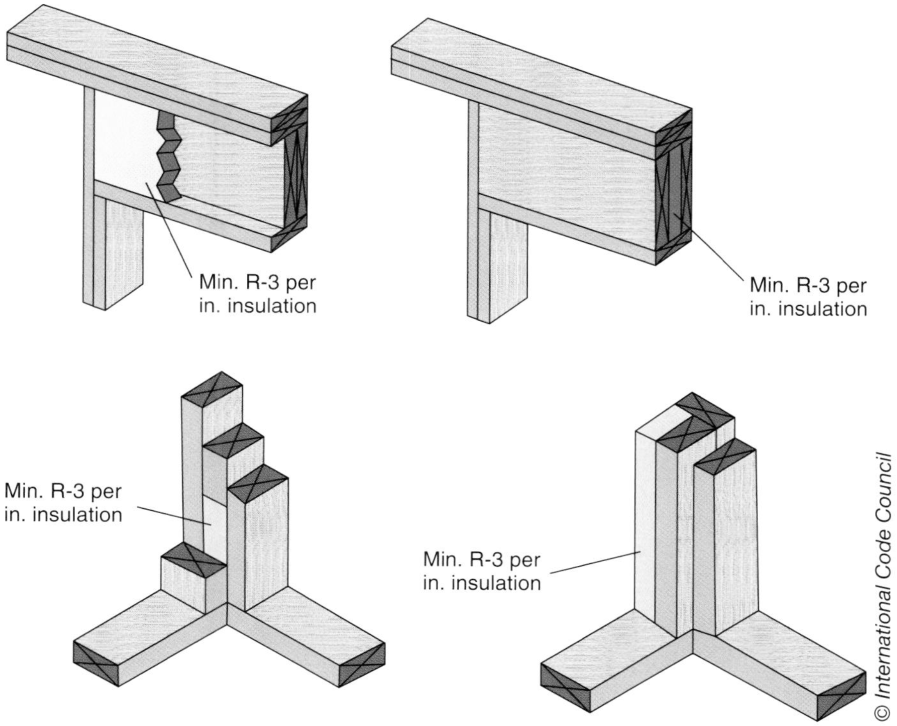

Insulation required to fill space at headers and exterior wall corners

Table N1102.4.1.1 continued

CHANGE SIGNIFICANCE: The code is now more specific as to the required amount of insulation at framed wall corners and headers. The intent of the change is to clarify when insulation is required and the minimum thermal resistance value of the insulation. Proponents of this change reasoned that some headers and wall corners are solid, and there is no air space within which to install insulation. In this case, the code does not intend that insulation be installed or that the solid header or solid corner must meet a certain *R*-value. For example, a two-ply header of nominal 2-inch-thick lumber with a continuous ½-inch wood structural panel sandwiched between the 2× members measures 3½ inches in thickness and is the same thickness as a 2 × 4 wall. In this case, no insulation is required. When space is available, insulation must be installed. The code requires a minimum thermal resistance of R-3 per inch. Most insulation, including fiberglass and rigid foam plastic, meets or exceeds the value of R-3 per inch.

N1102.4.2, Table N1102.4.1.1

Wood-Burning Fireplace Doors

CHANGE TYPE: Modification

CHANGE SUMMARY: Doors on wood-burning fireplaces must be listed for the application. The requirement for gasketed doors on fireplaces has been removed.

2015 CODE: N1102.4.2 (R402.4.2) Fireplaces. New wood-burning fireplaces shall have tight-fitting flue dampers or doors, and outdoor combustion air. Where using tight-fitting doors on factory-built fireplaces listed and labeled in accordance with UL 127, the doors shall be tested and listed for the fireplace. Where using tight-fitting doors on masonry fireplaces, the doors shall be listed and labeled in accordance with UL 907.

N1102.4.2, Table N1102.4.1.1 continues

TABLE N1102.4.1.1 (R402.4.1.1) Air Barrier and Insulation Installation

Component	Air Barrier Criteria	Insulation Installation Criteria
~~Fireplace~~	~~An air barrier shall be installed on fireplace walls. Fireplaces shall have gasketed doors.~~	

(Portions of table not shown for brevity and clarity.)

© Photographee.eu/Shutterstock.com.

Fireplace doors must be listed for the application.

N1102.4.2, Table N1102.4.1.1 continued

CHANGE SIGNIFICANCE: Fireplaces can potentially be major sources of air leakage because of the natural drafting of the chimney. The provisions for doors and dampers on fireplaces intend to reduce air leakage when the fireplace is not in use and improve energy efficiency. In previous code editions, Section N1102.4.2 has required a tight-fitting damper on new wood-burning fireplaces. The general insulation and air barrier requirements in Table N1102.4.1.1 required all fireplaces to have gasketed doors. The combination of a tight-fitting damper, gasketed door, and outdoor combustion air introduced directly into the firebox greatly improves the energy efficiency of this popular feature of homes. However, most factory-built fireplaces are not listed for use with gasketed doors. Factory-built fireplaces must be tested, listed, and labeled in accordance with UL 127 and installed in accordance with the conditions of the listing per IRC Section R1004. Installing a gasketed door that was not listed for use on a factory-built fireplace would be a violation of the listing and could cause a safety hazard if the door was closed while a fire was burning. The requirement for a gasketed door in Table N1102.4.1.1 has been removed and a requirement for a door on all new wood-burning fireplaces has been added to Section N1102.4.2. The code now stipulates that tight-fitting doors installed on wood-burning fireplaces must be listed to the applicable standard.

N1103.3
Duct Sealing and Testing

CHANGE TYPE: Modification

CHANGE SUMMARY: The duct sealing and testing provisions have been reorganized to clarify the application. The maximum duct leakage rates are now prescriptive rather than mandatory provisions to accommodate design flexibility.

2015 CODE: **~~N1103.2~~ N1103.3 (R403.3) Ducts.** Ducts and air handlers shall be in accordance with Sections N1103.3.1 through N1103.3.5.

N1103.3.2 (R403.3.1) Sealing (Mandatory). Ducts, air handlers, and filter boxes shall be sealed. Joints and seams shall comply with either the *International Mechanical Code* or Section M1601.4.1 of this code, as applicable.

Exceptions:

1. Air-impermeable spray foam products shall be permitted to be applied without additional joint seals.
~~2. Where a duct connection is made that is partially inaccessible, three screws or rivets shall be equally spaced on the exposed portion of the joint so as to prevent a hinge effect.~~
~~3~~2. For ducts having a static pressure classification of less than 2 inches of water column (500 Pa), additional closure systems shall not be required for ~~C~~continuously welded joints and seams, and locking-type ~~longitudinal~~ joints and seams of other than the snap-lock and button-lock types ~~in ducts operating at static pressures less than 2 inches of water column (500 Pa) pressure classification shall not require additional closure systems~~.

~~Duct tightness shall be verified by either of the following:~~

~~1. Postconstruction test: Total leakage shall be less than or equal to 4 cfm (113.3 L/min) per 100 square feet (9.29 m^2) of conditioned floor area when tested at a pressure differential of~~

N1103.3 continues

© International Code Council

Duct seal-mastic

N1103.3 continued

~~0.1 inches w.g. (25 Pa) across the entire system, including the manufacturer's air handler enclosure. All register boots shall be taped or otherwise sealed during the test.~~

2. ~~Rough-in test: Total leakage shall be less than or equal to 4 cfm (113.3 L/min) per 100 square feet (9.29 m^2) of conditioned floor area when tested at a pressure differential of 0.1 inches w.g. (25 Pa) across the system, including the manufacturer's air handler enclosure. All registers shall be taped or otherwise sealed during the test. If the air handler is not installed at the time of the test, total leakage shall be less than or equal to 3 cfm (85 L/min) per 100 square feet (9.29 m^2) of conditioned floor area.~~

Exception: ~~The total leakage test is not required for ducts and air handlers located entirely within the building thermal envelope.~~

N1103.3.3 (R403.3.3) Duct Testing (Mandatory). Ducts shall be pressure tested to determine air leakage by one of the following methods:

1. Rough-in test: Total leakage shall be measured with a pressure differential of 0.1 inches w.g. (25 Pa) across the system, including the manufacturer's air handler enclosure if installed at the time of the test. All registers shall be taped or otherwise sealed during the test.
2. Postconstruction test: Total leakage shall be measured with a pressure differential of 0.1 inches w.g. (25 Pa) across the entire system, including the manufacturer's air handler enclosure. Registers shall be taped or otherwise sealed during the test.

Exception: A duct air leakage test shall not be required where the ducts and air handlers are located entirely within the building thermal envelope.

A written report of the results of the test shall be signed by the party conducting the test and provided to the building official.

N1103.3.4 (R403.3.4) Duct Leakage (Prescriptive). The total leakage of the ducts, where measured in accordance with Section N1103.3.3, shall be as follows:

1. Rough-in test: The total leakage shall be less than or equal to 4 cubic feet per minute (113.3 L/min) per 100 square feet (9.29 m^2) of conditioned floor area where the air handler is installed at the time of the test. Where the air handler is not installed at the time of the test, the total leakage shall be less than or equal to 3 cubic feet per minute (85 L/min) per 100 square feet (9.29 m^2) of conditioned floor area.
2. Postconstruction test: The total leakage shall be less than or equal to 4 cubic feet per minute (113.3 L/min) per 100 square feet (9.29 m^2) of conditioned floor area.

~~N1103.2.3~~ N1103.3.5 (R403.3.5) Building Cavities (Mandatory). Building framing cavities shall not be used as ducts or plenums.

(Portions of Section N1103.3 are not shown for brevity and clarity.)

CHANGE SIGNIFICANCE: The changes to Section N1103.3 regarding ducts are largely editorial and a reorganization to improve understanding and application of the provisions. The significant change places the duct pressure testing methods and the maximum air-leakage rates in separate sections. The requirements for testing and the testing methods utilized remain as mandatory provisions. The limits on air leakage are now prescriptive provisions rather than mandatory. This means that alternative performance measures may be used to provide equivalency in satisfying the code requirements for air-leakage rates. This flexibility is particularly important in the case of an unexpected test failure where the air-leakage rate exceeded the limits of Section N1103.3.4.

N1103.5

Heated Water Circulation and Temperature Maintenance Systems

CHANGE TYPE: Modification

CHANGE SUMMARY: The code now requires automatic controls to maintain hot water temperature for heated water circulation systems and for heat trace temperature maintenance systems when such systems are installed. To save energy, continuously operating circulation pumps are no longer permitted. Heat trace systems must comply with one of the referenced standards.

2015 CODE: **~~N1103.4~~ <u>N1103.5</u> (R403.5) Service Hot Water Systems.** Energy conservation measures for service hot water systems shall be in accordance with Sections N1103.5.1 and N1103.5.4.

N1103.5.1 (R403.5.1) ~~Circulating Hot~~ <u>Heated</u> Water Circulation and Temperature Maintenance Systems (Mandatory). ~~Circulating hot water systems shall be provided with an automatic or readily accessible manual switch that can turn off the hot-water circulating pump when the system is not in use.~~ <u>Heated water circulation systems shall be in accordance with Section N1103.5.1.1. Heat trace temperature maintenance systems shall be in accordance with Section N1103.5.1.2. Automatic controls, temperature sensors and pumps shall be accessible. Manual controls shall be readily accessible.</u>

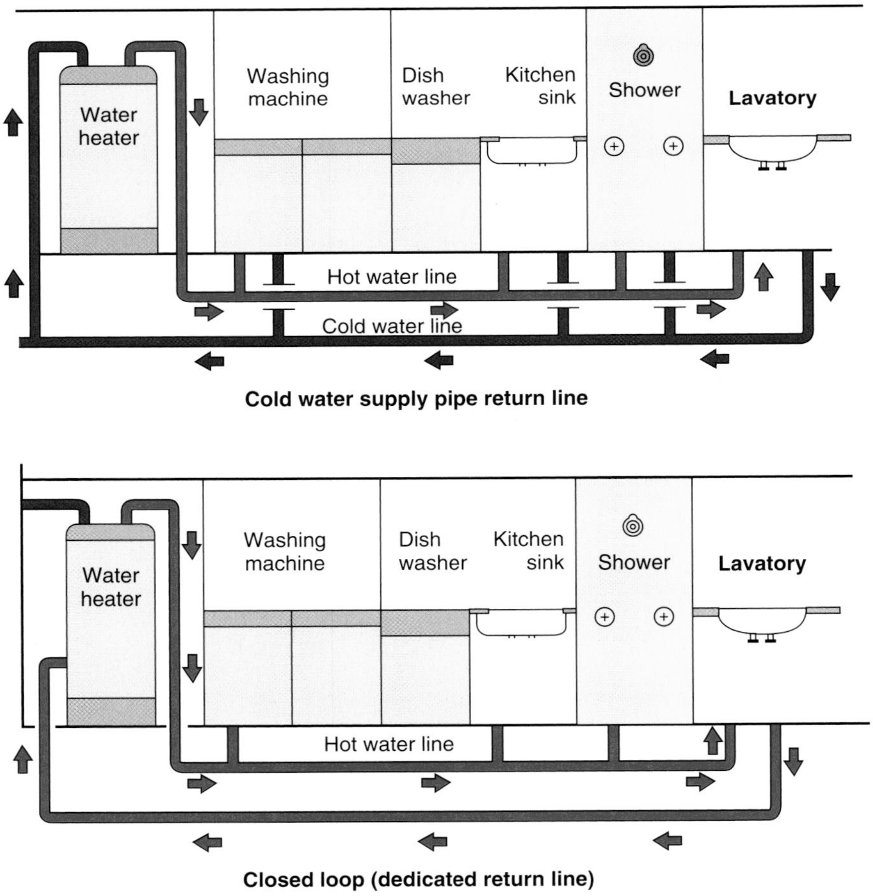

Automatic demand heated water circulation systems

N1103.5.1.1 (R403.5.1.1) Circulation Systems. Heated water circulation systems shall be provided with a circulation pump. The system return pipe shall be a dedicated return pipe or a cold water supply pipe. Gravity and thermo-syphon circulation systems shall be prohibited. Controls for circulating hot water system pumps shall start the pump based on the identification of a demand for hot water within the occupancy. The controls shall automatically turn off the pump when the water in the circulation loop is at the desired temperature and when there is no demand for hot water.

N1103.5.1.2 (R403.5.1.2) Heat Trace Systems. Electric heat trace systems shall comply with IEEE 515.1 or UL 515. Controls for such systems shall automatically adjust the energy input to the heat tracing to maintain the desired water temperature in the piping in accordance with the times when heated water is used in the occupancy.

(Portions of Section N1103.5 are not shown for brevity and clarity.)

CHANGE SIGNIFICANCE: Service water heating supplies hot water to the dwelling unit for purposes other than comfort heating. This is the dwelling's hot water supply for bathing, washing, kitchen sink, laundry, and similar uses. Heated water circulation and temperature maintenance systems are not required, but when installed they must meet the mandatory requirements of Section N1103.5. These systems use circulation pumps or heat trace components to maintain the desired temperature of hot water for the convenience of the user and to conserve water that would otherwise be drawn until hot water reached the fixture outlet. The previous language only addressed circulation systems, not heat trace temperature maintenance systems, and only required an automatic switch or a readily accessible manual switch to turn off the circulating pump when the system was not in use. In addition, the language permitted a continuously operating circulation pump, which is not the most energy-efficient system. There have been no provisions for the more efficient demand-activated circulation systems. There also was no requirement that heat trace components be suitable for the application.

The new provisions in Section N1103.5 do not permit a continuously operating circulating pump. The pump must operate on automatic controls activated when the hot water in the system falls below the desired temperature or when there is a demand for hot water. Pipe insulation is required for hot water circulation systems and the water in the circulation piping can stay hot for an extended time depending on the diameter of the piping. Because the pump only operates intermittently when needed, demand-activated circulation is significantly more energy efficient than a continuously operating heated water circulation system.

A heat trace system is the other energy-efficient means for maintaining the desired temperature in the service hot water system. The code requires heat trace systems to comply with one of the referenced standards and to have automatic controls to conserve energy. As with circulation systems, piping in a heat trace system requires pipe insulation.

PART 5
Mechanical
Chapters 12 through 23

- **Chapter 12** Mechanical Administration No changes addressed
- **Chapter 13** General Mechanical System Requirements No changes addressed
- **Chapter 14** Heating and Cooling Equipment No changes addressed
- **Chapter 15** Exhaust Systems
- **Chapter 16** Duct Systems
- **Chapter 17** Combustion Air No changes addressed
- **Chapter 18** Chimneys and Vents No changes addressed
- **Chapter 19** Special Appliances, Equipment and Systems No changes addressed
- **Chapter 20** Boilers and Water Heaters No changes addressed
- **Chapter 21** Hydronic Piping No changes addressed
- **Chapter 22** Special Piping and Storage Systems No changes addressed
- **Chapter 23** Solar Systems No changes addressed

As a comprehensive code that applies to all aspects of residential construction, the IRC contains provisions for the mechanical, fuel gas, plumbing and electrical systems of the building. These systems are covered in their respective parts of the IRC beginning with Part 5. This part contains administrative provisions unique to the application and enforcement of regulations governing mechanical systems, as well as the technical provisions related to system design and installation. Chapter 13 provides the general requirements for all mechanical systems and addresses the listing and labeling of appliances, types of fuel used, access to appliances, clearance to combustibles and other related issues. The remainder of Part 5 deals with requirements for specific mechanical systems related to heating and cooling, exhaust, ventilation, ducts, vents, boilers and hydronic piping. The last two chapters of Part 5 contain provisions specific to fuel oil piping and storage, and solar energy systems. ■

M1502.4.4, M1502.4.5
Dryer Exhaust Duct Power Ventilators

M1502.4.6
Dryer Duct Length Identification

M1503.4
Makeup Air for Range Hoods

M1506.2
Exhaust Duct Length

M1601.1.1, TABLE M1601.1.1, M1601.2
Above-Ground Duct Systems

M1601.4
Duct Installation

M1602
Return Air

M1502.4.4, M1502.4.5

Dryer Exhaust Duct Power Ventilators

CHANGE TYPE: Addition

CHANGE SUMMARY: The code now recognizes the use of dryer exhaust duct power ventilators (DEDPVs) to increase the allowable exhaust duct length for clothes dryers.

2015 CODE: **M1502.4.4 Dryer Exhaust Duct Power Ventilators.** Domestic dryer exhaust duct power ventilators shall conform to UL 705 for use in dryer exhaust duct systems. The dryer exhaust duct power ventilator shall be installed in accordance with the manufacturer's instructions.

~~M1502.4.4~~ M1502.4.5 Duct Length. The maximum allowable exhaust duct length shall be determined by one of the methods specified in ~~Section M1502.4.4.1 or M1502.4.4.2~~ Sections M1502.4.5.1 through M1502.4.5.3.

~~M1502.4.4.1~~ M1502.4.5.1 Specified Length. *(No change to text.)*

~~M1502.4.4.2~~ M1502.4.5.2 Manufacturer's Instructions. *(No change to text.)*

M1502.4.5.3 Dryer Exhaust Duct Power Ventilator. The maximum length of the exhaust duct shall be determined in accordance with the manufacturer's instructions for the dryer exhaust duct power ventilator.

CHANGE SIGNIFICANCE: The code limits the length of clothes dryer exhaust ducts to protect against potential fire hazards and to ensure that dryers efficiently discharge warm, moist air to the outdoors. Allowable

M1502.4.4, M1502.4.5 continues

Dryer exhaust duct power ventilator (DEDPV)

M1502.4.4, M1502.4.5 continued

length is based on the airflow capacity of modern dryers. Elbow fittings reduce the allowable length, resulting in a calculated "equivalent length" based on the additional resistance to airflow for each fitting. In addition to lint buildup, excessive duct length creates moisture and maintenance problems and increases drying times causing the dryer to be inefficient and waste energy.

Previous editions of the code did not recognize dryer exhaust duct power ventilators (DEDPVs) as an option for clothes dryer installations. DEDPVs are typically referred to as "dryer booster fans" in the marketplace, because they "boost" or increase the airflow of the dryer discharge. Greater airflow increases the distance that the discharge air can be effectively pushed to the outdoors. Prior to the 2015 IRC, the two options for determining the maximum exhaust duct length were to comply with the prescriptive limit of 35 feet, a conservative average for modern dryers, or to follow the clothes dryer manufacturer's instructions for length limits. If the desired location did not fall within those limits, the designer or builder was left to relocate the dryer to reduce the length of the exhaust duct. Another possible solution was to make application to the building official requesting approval to install a DEDPV under the alternative materials, design, and methods of construction provisions in Section R104.11. The 2015 IRC now specifically allows DEDPVs in clothes dryer exhaust systems to increase the equivalent length of duct.

DEDPVs are listed to a revised version of UL 705 that now contains tests and construction requirements that are specific to these devices. DEDPVs have been around for years, but until recently were not listed to a national consensus standard that was specific to these devices. The UL 705 standard contains requirements for the construction, testing and installation of DEDPVs and requires them to be equipped with features such as interlocks, limit controls, monitoring controls and enunciator devices to make certain that the dryers or dryer operators are aware of the operating status of the DEDPVs. The maximum length of the dryer exhaust duct is determined based on the manufacturer's instructions for the DEDPV.

M1502.4.6

Dryer Duct Length Identification

CHANGE TYPE: Modification

CHANGE SUMMARY: A permanent label identifying the concealed length of the dryer exhaust duct is no longer required where the equivalent duct length does not exceed 35 feet. For the dryer exhaust duct exceeding 35 feet, a label or tag is required whether the duct is concealed or not.

2015 CODE: **~~M1502.4.5~~ M1502.4.6 Length Identification.** Where the exhaust duct equivalent length exceeds 35 feet (10 668 mm) ~~is concealed within the building construction~~, the equivalent length of the exhaust duct shall be identified on a permanent label or tag. The label or tag shall be located within 6 feet (1829 mm) of the exhaust duct connection.

CHANGE SIGNIFICANCE: The provisions for identifying the equivalent length of dryer exhaust duct first appeared in the 2009 IRC. The code has since required a permanent label or tag installed within 6 feet of the dryer when the duct was concealed behind finish materials. The purpose

M1502.4.6 continues

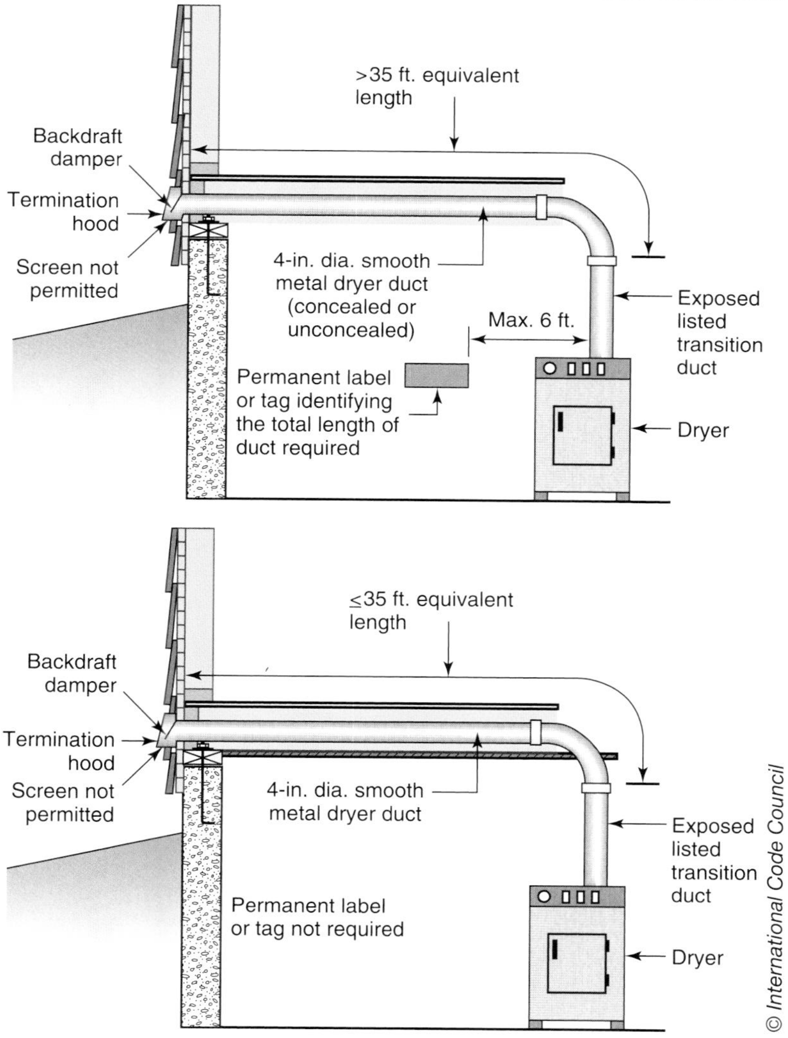

A permanent label or tag is only required when the equivalent length of the dryer exhaust duct exceeds 35 feet.

M1502.4.6 continued

was to alert occupants of the length of the concealed duct so they could make an informed decision to install a dryer with adequate airflow capacity. This provision recognizes that homes change hands and many dryers may be installed over the building's lifetime. The primary concern was aimed at exhaust duct systems that were based on the dryer manufacturer's instructions at the time of construction. A given dryer might have a capacity much greater than the specified length of 35 feet, the default value when the manufacturer and model of the dryer is unknown. This change to the 2015 IRC recognizes that there is no concern if the exhaust duct does not exceed 35 feet in equivalent length and the permanent label in this case provides no benefit to the owner. In addition, the proponents reasoned that the purpose of the permanent sign is to notify the owners and installers that the dryer duct length is exceptional and any installed dryer must be compatible with that duct of exceptional length. Therefore, the criterion for providing signage only when the duct is concealed has been removed. The code now requires a permanent label or tag when the equivalent length of the dryer exhaust duct exceeds 35 feet, whether or not the duct is concealed within construction.

CHANGE TYPE: Modification

M1503.4

Makeup Air for Range Hoods

CHANGE SUMMARY: Automatic operation of a mechanical damper is no longer required for supplying makeup air for kitchen exhaust systems exceeding a rating of 400 cubic feet per minute (cfm). Transfer openings are permitted to obtain makeup air from rooms other than the kitchen.

2015 CODE: **M1503.4 Makeup Air Required.** Exhaust hood systems capable of exhausting in excess of 400 cubic feet per minute (0.19 m^3/s) shall be mechanically or naturally provided with makeup air at a rate approximately equal to the exhaust air rate. Such makeup air systems shall be equipped with ~~a means of closure and shall be automatically controlled to start and operate simultaneously with the exhaust system~~ not less than one damper. Each damper shall be a gravity damper or an electrically operated damper that automatically opens when the exhaust system operates. Dampers shall be accessible for inspection, service,

M1503.4 continues

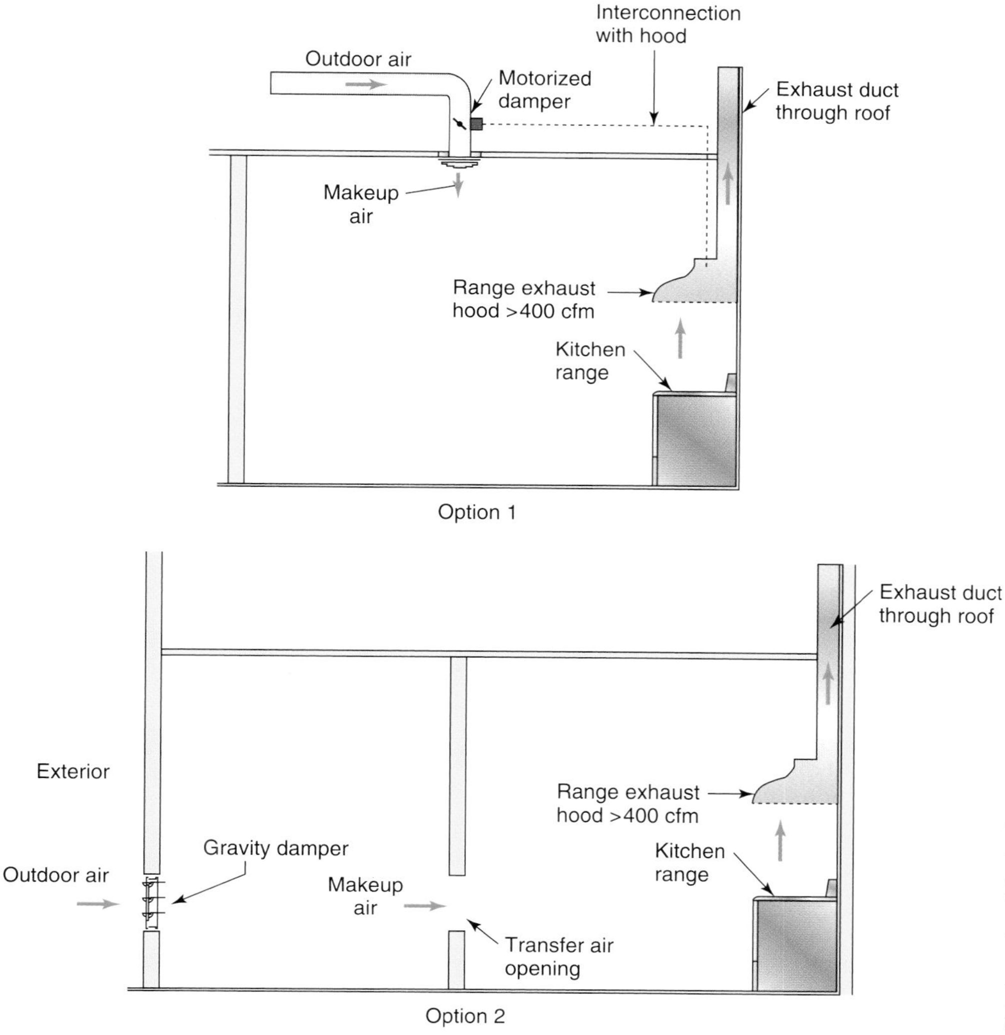

Required makeup air for kitchen exhaust hoods exceeding 400 cfm

M1503.4 continued

repair and replacement without removing permanent construction or any other ducts not connected to the damper being inspected, serviced, repaired or replaced.

M1503.4.1 Location. Kitchen exhaust makeup air shall be discharged into the same room in which the exhaust system is located or into rooms or duct systems that communicate through one or more permanent openings with the room in which such exhaust system is located. Such permanent openings shall have a net cross-sectional area not less than the required area of the makeup air supply openings.

CHANGE SIGNIFICANCE: The 2009 IRC introduced provisions for makeup air for high-velocity kitchen exhaust hoods that were capable of an airflow rate exceeding 400 cfm. Although this size of kitchen hood equipment is unusual in residential construction, the concern is that kitchens in modern homes are getting larger and oversized equipment is becoming more popular. With tighter building thermal envelopes for energy conservation, the high rate of exhaust requires outside makeup air to prevent negative pressure and the adverse effects on other appliances and systems. The code previously required an automatic damper that was interlocked with the exhaust hood so that outdoor makeup air was provided any time the hood fan was in operation.

The 2015 IRC provides more flexibility in achieving adequate makeup air for high-velocity kitchen exhaust fans. The outdoor makeup air can be obtained either mechanically or naturally. In either case, a damper is required to provide a means of closure, reduce air leakage, and conserve energy. Electrically operated dampers must still be interlocked to automatically open when the exhaust system operates. The other option is to provide a gravity damper that opens in response to pressure differentials created when the exhaust fan operates. The gravity damper is balanced to close when exhaust fan operation ceases. Proponents of this change offered that allowing a gravity damper is compatible with other similar applications in the IRC and that the residential code does not require automatic motorized dampers elsewhere. Proponents also stated that a gravity damper has the added benefit of equalizing depressurization in the house during the operation of other equipment such as bath fans and clothes dryers. Both types of damper—gravity and motorized—require maintenance and may need to be replaced at some time. Therefore, the code requires the dampers to be accessible.

The code also clarifies that the source of makeup air may be from a room or space other than the kitchen where the range hood is located. When outdoor air is introduced into another room through a mechanical damper or gravity damper, permanent openings of adequate size are required between the rooms for the makeup air to pass though. This provision recognizes that homeowners have valid reasons for not wanting the opening in the kitchen. Locating the opening in another room or bringing the makeup air in through the duct system allows the unconditioned air to mix and temper. This is beneficial in both the heating and cooling seasons in various climate zones. Proponents of this change reasoned that requiring the outside air opening in the kitchen created the possibility that it would be covered or otherwise disabled due to the discomfort of introducing unconditioned air to the kitchen.

M1506.2

Exhaust Duct Length

CHANGE TYPE: Addition

CHANGE SUMMARY: The code establishes maximum exhaust duct lengths based on duct diameter, type of duct and the exhaust fan airflow rating.

2015 CODE:

SECTION M1506
EXHAUST DUCTS AND EXHAUST OPENINGS

M1506.1 Ducts Construction. Where exhaust duct construction is not specified in this chapter, construction shall comply with Chapter 16.

M1506.2. Duct Length. The length of exhaust and supply ducts used with ventilating equipment shall not exceed the lengths determined in accordance with Table M1506.2.

> **Exception:** Duct length shall not be limited where the duct system complies with the manufacturer's design criteria or where the flow rate of the installed ventilating equipment is verified by the installer or approved third party using a flow hood, flow grid or other airflow measuring device.

M1506.23 Exhaust Openings. *(No change to text.)*

M1506.2 continues

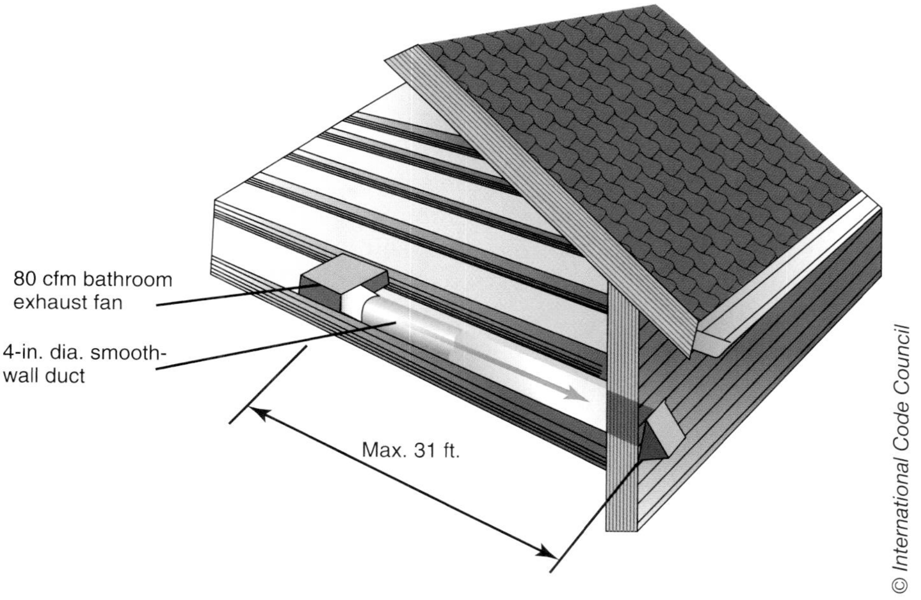

Maximum length of exhaust duct is based on fan rating and type and diameter of duct.

M1506.2 continued

TABLE M1506.2 Duct Length

Duct Type	Flex Duct								Smooth-Wall Duct							
Fan airflow rating (CFM @ 0.25 inch wc[a])	50	80	100	125	150	200	250	300	50	80	100	125	150	200	250	300
Diameter[b] (inches)	Maximum length[c, d, e] (feet)															
3	X	X	X	X	X	X	X	X	5	X	X	X	X	X	X	X
4	56	4	X	X	X	X	X	X	114	31	10	X	X	X	X	X
5	NL	81	42	16	2	X	X	X	NL	152	91	51	28	4	X	X
6	NL	NL	158	91	55	18	1	X	NL	NL	NL	168	112	53	25	9
7	NL	NL	NL	NL	161	78	40	19	NL	NL	NL	NL	NL	148	88	54
8 and above	NL	NL	NL	NL	NL	189	111	69	NL	NL	NL	NL	NL	NL	198	133

a. Fan airflow rating shall be in accordance with ANSI/AMCA 210-ANSI/ASHRAE 51.
b. For non-circular ducts, calculate the diameter as four times the cross-sectional area divided by the perimeter.
c. This table assumes that elbows are not used. Fifteen feet (5 m) of allowable duct length shall be deducted for each elbow installed in the duct run.
d. NL = no limit on duct length of this size.
e. X = not allowed. Any length of duct of this size with assumed turns and fittings will exceed the rated pressure drop.

CHANGE SIGNIFICANCE: The 2015 IRC introduces a prescriptive table for sizing exhaust ducts. The table is taken from ASHRAE 62.2-2010, addendum F. The intent is to match duct size (diameter and maximum length) to the airflow capacity of the exhaust fan to ensure the exhaust system operates efficiently and at the intended designed airflow rate. For residential buildings regulated by the IRC, mechanical exhaust is typically local exhaust for bathrooms and kitchens, and may also be part of a whole-house mechanical ventilation system. Local exhaust is defined as an exhaust system that uses one or more fans to exhaust air from a specific room or rooms within a dwelling. Although the section title is "Exhaust Ducts and Exhaust Openings," the new Section M1506.2 also mentions supply ducts used for ventilating equipment. The intent is that the supply ducts for introducing outside air into a whole-house mechanical ventilation system are to be sized in accordance with the prescriptive values in Table M1506.2.

The maximum duct length in the table is based on three variables: duct diameter, duct type (flexible or smooth-wall) and the fan airflow rating. Smooth-wall ducts provide less resistance to airflow and the allowable duct lengths are greater than those for flexible ducts that have greater resistance to airflow. Footnote b provides the calculation method for converting the tabular values to apply to rectangular ducts. Footnote c establishes a 15-foot reduction in allowable duct length for each elbow installed. The code requires that the fan flow rate be verified by the manufacturer in accordance with ANSI/AMCA 210-ANSI/ASHRAE 51 or be field verified by the installer or an approved third party. The intent is to provide a minimum level of quality assurance for the installation of ventilation fans. Confirmation that a ventilation fan's flow rate is in compliance with ANSI/AMCA 210-ANSI/ASHRAE 51 is typically based on a Home Ventilating Institute (HVI) sticker in the fan housing.

CHANGE TYPE: Modification

CHANGE SUMMARY: The list of duct system requirements has been revised to reference the applicable standards and delete redundant language. The table for material thickness of metal ducts was replaced with what is currently consistent with the SMACNA sheet metal construction standard.

2015 CODE: **M1601.1.1 Above-Ground Duct Systems.** Above-ground duct systems shall conform to the following:

1. Equipment connected to duct systems shall be designed to limit discharge air temperature to ~~a maximum of~~ not greater than 250°F (121°C).
2. Factory-made ~~air~~ ducts shall be ~~constructed of Class 0 or Class 1 materials as designated in Table M1601.1.1(1).~~ listed and labeled in accordance with UL 181 and installed in accordance with the manufacturer's instructions.
3. Fibrous glass duct construction shall conform to the SMACNA *Fibrous Glass Duct Construction Standards* or NAIMA *Fibrous Glass Duct Construction Standards.*
4. Field-fabricated and shop-fabricated metal and flexible duct constructions shall conform to the SMACNA HVAC *Duct Construction Standards - Metal and Flexible* except as allowed by ~~Minimum thicknesses of metal duct material shall be as listed in~~ Table ~~M1601.1.1(2)~~ M1601.1.1. Galvanized steel shall conform to ASTM A 653. ~~Metallic ducts shall be fabricated in accordance with SMACNA Duct Construction Standards Metal and Flexible.~~
5. The use of gypsum products to construct return air ducts or plenums is permitted, provided that the air temperature does not exceed 125°F (52°C) and exposed surfaces are not subject to condensation.

(No change to Items 6 and 7.)

M1601.1.1, Table M1601.1.1, M1601.2 continues

M1601.1.1, Table M1601.1.1, M1601.2

Above-Ground Duct Systems

© International Code Council

Ducts not greater than 14 inches in diameter may be constructed of 30-gage galvanized sheet metal.

~~TABLE M1601.1.1(1)~~
~~Classification of Factory-Made Air Ducts~~

~~Duct Class~~	~~Maximum Flame Spread Index~~
~~0~~	~~0~~
~~1~~	~~25~~

M1601.1.1, Table M1601.1.1, M1601.2 continued

TABLE ~~M1601.1(2)~~ M1601.1.1 Duct Construction Minimum Sheet Metal Thickness for Single Dwelling Units[a] ~~Gages of Metal Ducts and Plenums Used For Heating or Cooling~~

ROUND DUCT DIAMETER (inches)	STATIC PRESSURE ½-inch water gage Thickness (inches)		STATIC PRESSURE 1-inch water gage Thickness (inches)	
	Galvanized	Aluminum	Galvanized	Aluminum
≤12	0.013	0.018	0.013	0.018
12 to 14	0.013	0.018	0.016	0.023
15 to 17	0.016	0.023	0.019	0.027
18	0.016	0.023	0.024	0.034
19 to 20	0.019	0.027	0.024	0.034

RECTANGULAR DUCT DIMENSION (inches)	STATIC PRESSURE ½-inch water gage Thickness (inches)		STATIC PRESSURE 1-inch water gage Thickness (inches)	
	Galvanized	Aluminum	Galvanized	Aluminum
≤8	0.013	0.018	0.013	0.018
9 to 10	0.013	0.018	0.016	0.023
11 to 12	0.016	0.023	0.019	0.027
13 to 16	0.019	0.027	0.019	0.027
17 to 18	0.019	0.027	0.024	0.034
19 to 20	0.024	0.034	0.024	0.034

For SI: 1 inch = 25.4 mm, 1-inch water gage = 249 Pa.

a. Ductwork that exceeds 20 inches by dimension or exceeds a pressure of 1-inch water gage (250 Pa) shall be constructed in accordance with SMACNA *HVAC Duct Construction Standards Metal and Flexible.*

~~Duct size~~	~~Galvanized~~ ~~Minimum Thickness (inches)~~	~~Galvanized~~ ~~Equivalent Galvanized Gage No.~~	~~Aluminum~~ ~~Minimum Thickness (inches)~~
~~Round ducts and enclosed rectangular ducts~~			
~~14 inches or less~~	~~0.0157~~	~~28~~	~~0.0145~~
~~16 and 18 inches~~	~~0.0187~~	~~26~~	~~0.018~~
~~20 inches and over~~	~~0.0236~~	~~24~~	~~0.023~~
~~Exposed rectangular ducts~~			
~~14 inches or~~	~~0.0157~~	~~28~~	~~0.0145~~
~~Over 14[a] inches~~	~~0.0187~~	~~26~~	~~0.018~~

~~For SI: 1 inch = 25.4 mm.~~

~~a. For duct gages and reinforcement requirements at static pressures of 1/2 inch, 1 inch and 2 inches w.g., SMACNA Duct Construction Standard, Tables 2-1; 2-2 and 2-3 shall apply.~~

~~**M1601.2 Factory-Made Ducts.** Factory-made air ducts or duct material shall be approved for the use intended, and shall be installed in accordance with the manufacturer's installation instructions. Each portion of a factory-made air duct system shall bear a listing and label indicating compliance with UL 181 and UL 181A or UL 181B.~~

CHANGE SIGNIFICANCE: Minor revisions to the list of seven requirements for above-ground duct systems clarify the application and bring the information up to date with the referenced standards. The reference to duct classification in Item 2 has been removed, and Table M1601.1.1(1) for classification of factory-made air ducts based on flame spread index has been deleted. These burning classifications are already covered in the referenced UL 181 standard and it is not necessary to repeat them in the code. Item 4 clarifies that the referenced standard, SMACNA HVAC *Duct Construction Standards - Metal and Flexible*, applies to both field-fabricated and shop-fabricated metal and flexible duct construction. Minimum sheet metal thickness is determined by the standard or by revised Table M1601.1.1 (previously Table M1601.1.1[2]).

Previous to the 2009 edition of the IRC, the code permitted a material thickness of 30 gage (0.013 inches) for round metal ducts 14 inches or less in diameter. In the 2009 IRC, the minimum sheet metal thickness was increased to 28 gage (0.0157 inches). The 2015 IRC returns the minimum thickness for 14-inch diameter ducts to 30 gage (0.013 inches). Table M1601.1.1 has replaced Table M1601.1.1(2). The new table expands the number of rows for duct sizes and bases the material thickness on either ½-inch or 1-inch water gage static pressure. The table is consistent with information in SMACNA HVAC *Duct Construction Standards - Metal and Flexible*. Proponents of the change reasoned that there was no demonstrated justification for eliminating 30-gage sheet metal thickness for 14-inch diameter duct from a strength, longevity, functionality, economic or energy standpoint.

M1601.4

Duct Installation

CHANGE TYPE: Modification

CHANGE SUMMARY: Tapes and mastics used to seal sheet metal ducts must be listed to UL 181 B as has been required for sealing flexible ducts. Snap-lock and button-lock seams are no longer exempt from the sealing requirements.

2015 CODE: M1601.4 Installation. Duct installation shall comply with Sections M1601.4.1 through M1601.4.~~9~~10.

M1601.4.1 Joints, Seams and Connections. ~~All~~ Longitudinal and transverse joints, seams and connections in metallic and nonmetallic ducts shall be constructed as specified in SMACNA *HVAC Duct Construction Standards—Metal and Flexible* and NAIMA *Fibrous Glass Duct Construction Standards*. ~~All~~ Joints, longitudinal and transverse seams, and connections in ductwork shall be securely fastened and sealed with welds, gaskets, mastics (adhesives), mastic-plus-embedded-fabric systems, liquid sealants or tapes. Tapes and mastics used to seal fibrous glass ductwork shall be listed and labeled in accordance with UL 181 A and shall be marked "181 A-P" for pressure-sensitive tape, "181 A-M" for mastic or "181 A-H" for heat-sensitive tape.

~~Closure systems~~ Tapes and mastics used to seal metallic and flexible air ducts and flexible air connectors shall comply with UL 181 B and shall be marked "181 B-FX" for pressure-sensitive tape or "181 BM" for mastic. Duct connections to flanges of air distribution system equipment shall be sealed and mechanically fastened. Mechanical fasteners for use with flexible nonmetallic air ducts shall comply with UL 181 B and shall be marked "181 B-C." Crimp joints for round metallic ducts shall have a contact lap of not less than 1 inch (25 mm) and shall be mechanically fastened by means of not less than three sheet-metal screws or rivets equally spaced around the joint.

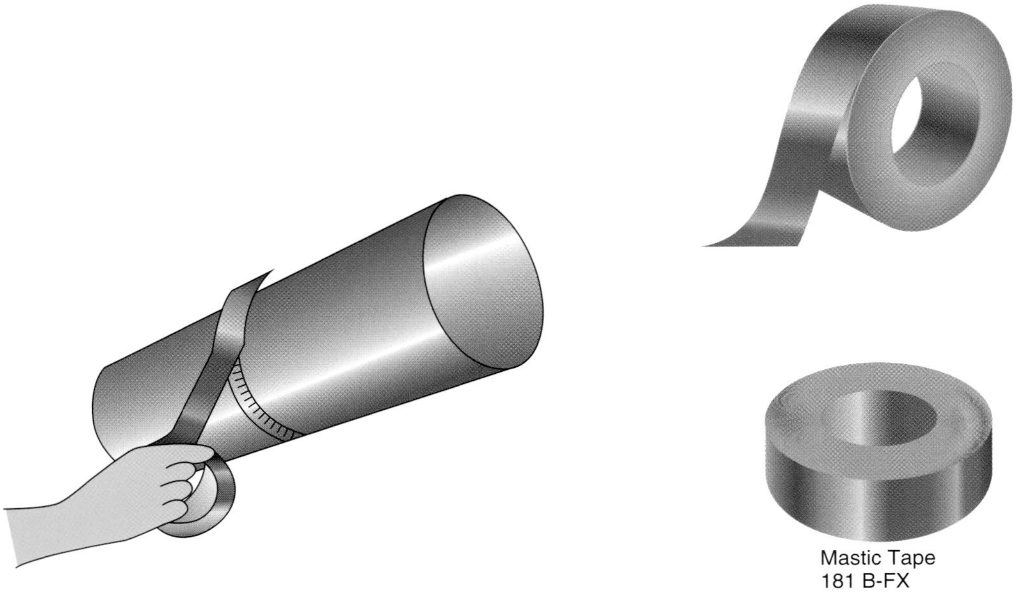

© International Code Council

Tapes and mastics used to seal sheet metal ducts must be listed to UL 181 B.

Closure systems used to seal ~~metal~~ all ductwork shall be installed in accordance with the manufacturers' instructions. ~~Round metallic ducts shall be mechanically fastened by means of at least three sheet metal screws or rivets spaced equally around the joint. Unlisted duct tape shall not be permitted as a sealant on any duct.~~

Exceptions:

1. Spray polyurethane foam shall be permitted to be applied without additional joint seals.
2. Where a duct connection is made that is partially inaccessible, three screws or rivets shall be equally spaced on the exposed portion of the joint so as to prevent a hinge effect.
3. For ducts having a static pressure classification of less than 2 inches of water column (500 Pa), additional closure systems shall not be required for continuously welded joints and seams and locking-type ~~longitudinal~~ joints and seams of other than the snap-lock and button-lock types ~~in ducts operating at static pressures less than 2 inches of water column (500 Pa) pressure classification shall not require additional closure systems~~.

M1601.4.2 Duct Lap. Crimp joints for round and oval metal ducts shall be lapped not less than 1 inch (25 mm) and the male end of the duct shall extend into the adjoining duct in the direction of airflow.

~~M1601.4.3~~ M1601.4.4 Support. ~~Metal ducts shall be supported by 1/2-inch-wide (13 mm) 18-gage metal straps or 12-gage galvanized wire at intervals not exceeding 10 feet (3048 mm) or other approved means. Nonmetallic~~ Factory made ducts listed in accordance with UL 181 shall be supported in accordance with the manufacturer's installation instructions. Field- and shop-fabricated fibrous glass ducts shall be supported in accordance with the SMACNA *Fibrous Glass Duct Construction Standards* or the NAIMA *Fibrous Glass Duct Construction Standards*. Field- and shop-fabricated metal and flexible ducts shall be supported in accordance with the SMACNA HVAC *Duct Construction Standards—Metal and Flexible*.

(No significant changes to portions of Section M1601.4 not shown.)

CHANGE SIGNIFICANCE: Section M1601.4.1, Joints, Seams and Connections, has been revised to remove redundant language and to clarify the sealing requirements and applicable standards for various types of ducts. The UL 181 A standard is specific to fibrous glass duct systems and UL 181 B is specific to flexible duct systems. There is no closure system listed specifically for metal ducts, but UL 181 B is judged appropriate for sealing of metal ducts. All mastics and tapes used for sealing ductwork must be listed, so the language prohibiting unlisted duct tape is unnecessary and has been removed. The manufacturer's instructions now apply to closure systems for all types of ducts, not just those for metal ducts.

Snap-lock and button-lock types of ducts are no longer exempt from the closure requirements because such types allow considerable air leakage unless sealed. Some manufacturers place a sealant or gasket in the seams of snap- or button-lock ducts, which satisfies the intent of the code

M1601.4 continues

M1601.4 continued

to have a closure (sealing) system for such ducts. Some locking joints are leak-proof, such as mechanically folded seams used for spiral seam ducts, and the code still recognizes this exception.

New Section M1601.4.2 prescribes the appropriate connection of crimp joints for round and oval metal ducts in the direction of airflow. Previously, the code was silent on oval ducts, which are commonly installed in dwellings, and did not address the direction of the lap relative to airflow.

Section M1601.4.4 regarding duct supports has been revised to reference the appropriate SMACNA standards and the manufacturer's instructions as opposed to prescribing a support interval and method of support. The previous 10-foot interval requirement was considered too broad and inappropriate for many sizes and types of ducts. In practice, 18-gage metal straps are not typically used to support residential ducts.

M1602
Return Air

CHANGE TYPE: Modification

CHANGE SUMMARY: The provisions for return air have been simplified and clarified to improve understanding while preserving the intent of keeping contaminants out of the airstream of the heating, ventilation and air-conditioning (HVAC) system. The provisions for outdoor air openings have been removed and the code now references the applicable provisions for outdoor air in Chapter 3.

2015 CODE:

SECTION M1602
RETURN AIR

~~**M1602.1 Return Air.** Return air shall be taken from inside the dwelling. Dilution of return air with outdoor air shall be permitted.~~

~~**M1602.3**~~ **M1602.1** ~~**Inlet Opening Protection**~~ **Outdoor Air Openings.** ~~Outdoor air inlets shall be covered with screens having openings that are not less than 1/4 inch (6.4 mm) and not greater than 1/2 inch (12.7 mm).~~ Outdoor intake openings shall be located in accordance with Section R303.5.1. Opening protection shall be in accordance with Section R303.6.

~~**M1602.2 Prohibited Sources.** Outdoor and return air for a forced-air heating or cooling system shall not be taken from the following locations:~~

M1602.2 Return Air Openings. Return air openings for heating, ventilation and air-conditioning systems shall comply with all of the following:

1. Openings shall not be located less than 10 feet (3048 mm) measured in any direction from an open combustion chamber or draft hood of another appliance located in the same room or space.

M1602 continues

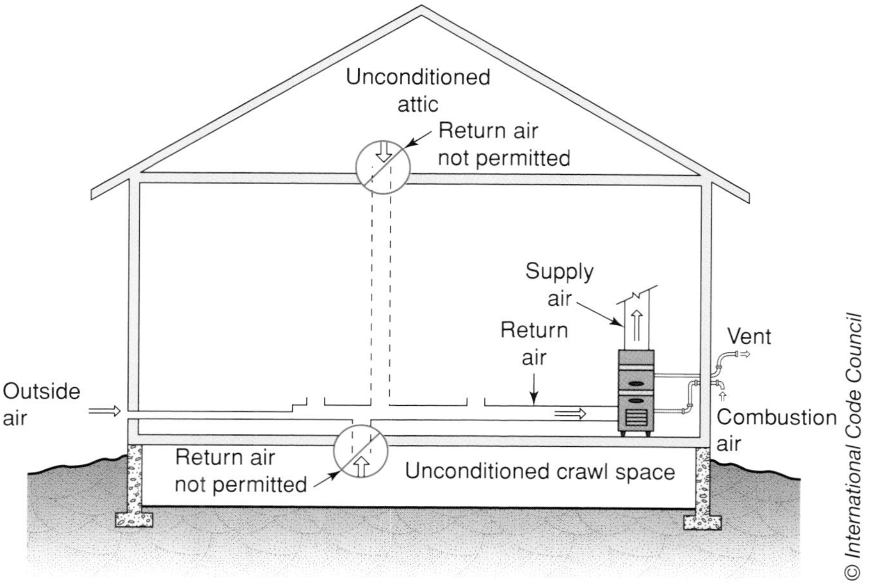

Prohibited sources of return air include unconditioned attics and crawl spaces.

M1602 continued

2. The amount of return air taken from any room or space shall be not greater than the flow rate of supply air delivered to such room or space.
3. Return and transfer openings shall be sized in accordance with the appliance or equipment manufacturers' installation instructions, Manual D or the design of the registered design professional.
4. Return air shall not be taken from a closet, bathroom, toilet room, kitchen, garage, mechanical room, boiler room, furnace room or unconditioned attic.

Exceptions:

1. Taking return air from a kitchen is not prohibited where such return air openings serve the kitchen only, and are located not less than 10 feet (3048 mm) from the cooking appliances.
2. Dedicated forced-air systems serving only ~~a~~ the garage shall not be prohibited from obtaining return air from the garage.
3. Taking return air from an unconditioned crawl space shall not be accomplished through a direct connection to the return side of a forced-air furnace. Transfer openings in the crawl space enclosure shall not be prohibited.
4. Return air from one dwelling unit shall not be discharged into another dwelling unit.

(Portions of deleted text from the 2012 IRC are not shown for brevity and clarity.)

CHANGE SIGNIFICANCE: Section M1602 has been reorganized and simplified to improve understanding and application. Previously, the code listed locations that were prohibited as sources for return air. Item 3 regarding a space that was less than 25 percent of the entire volume served by the system was not well understood and not typically followed. For modern construction this item was considered outdated and has been removed. In its place, the code now simply requires that the amount of return air taken from any room is not greater than the supply air delivered to that room. Item 5 regarding spaces containing fuel-fired appliances including multiple exceptions was also problematic and has been removed in its entirety. The intent of the return air provisions is to keep contaminants out of the airstream being returned to the air handler where the air is then circulated throughout the dwelling unit. The new text accomplishes that goal and captures the intent of the previous provisions.

PART 6

Fuel Gas

Chapter 24

■ **Chapter 24** Fuel Gas

Fuel gas systems are covered in Part 6, including provisions for approved materials as well as the design and installation of fuel gas piping and other system components. The fuel gas provisions of the IRC are taken directly from the *International Fuel Gas Code* (IFGC). In order to make the correlation and coordination of the two codes easier, after each fuel gas section of the IRC the original section of the IFGC is shown in parentheses. The fuel gas portion of the IRC contains its own specific definitions in Section G2403 in addition to the general definitions found in Chapter 2 of the IRC. The text, tables and figures in other sections of Chapter 24 address the technical issues of fuel gas systems, such as appliance installation; materials, sizing, and installation of fuel gas piping systems; piping support; flow controls; connections; combustion air; venting; and other related system requirements. ■

G2404.11

Condensate Pumps

G2411.1.1

Electrical Bonding of Corrugated Stainless Steel Tubing

G2413.2

Maximum Gas Demand

G2414.6

Plastic Pipe, Tubing and Fittings

G2415.5

Fittings in Concealed Locations

G2415.7

Protection of Concealed Piping Against Physical Damage

G2421.2

Medium-Pressure Regulators

G2422.1

Connecting Portable and Movable Appliances

G2426.7.1

Door Clearance to Vent Terminals

G2427.4.1, G2427.6.8.3

Plastic Piping for Appliance Vents

G2427.8

Venting System Termination Location

G2439.4, G2439.7

Clothes Dryer Exhaust Ducts

G2447.2

Prohibited Location of Commercial Cooking Appliances

G2404.11

Condensate Pumps

CHANGE TYPE: Addition

CHANGE SUMMARY: Condensate pumps located in uninhabitable spaces must be connected to the appliance to shut down the equipment in the event of pump failure.

2015 CODE: **G2404.11 (307.6) Condensate Pumps.** Condensate pumps located in uninhabitable spaces, such as attics and crawl spaces, shall be connected to the appliance or equipment served such that when the pump fails, the appliance or equipment will be prevented from operating. Pumps shall be installed in accordance with the manufacturer's instructions.

CHANGE SIGNIFICANCE: Condensate pumps for Category IV condensing appliances are often located in attics and crawl spaces and above ceilings where they are not readily observable. If they fail, the condensate overflow can cause damage to the building components, especially where the overflow will not be noticed immediately. The majority of such pumps are equipped with simple float controls that can be wired in series with the appliance or equipment control circuit. When the pump system fails, the float will rise in the reservoir and open a switch, interrupting power to the appliance before the condensate starts to overflow the reservoir. These float controls are commonly not connected or, in other cases, the pump might not be equipped with an overflow switch. This new code section requires the installation of condensate pumps that have this overflow shutoff capability and requires that the appliance or equipment served be connected to take advantage of that feature. This automatic shutoff feature will prevent water damage to the building in case of pump failure.

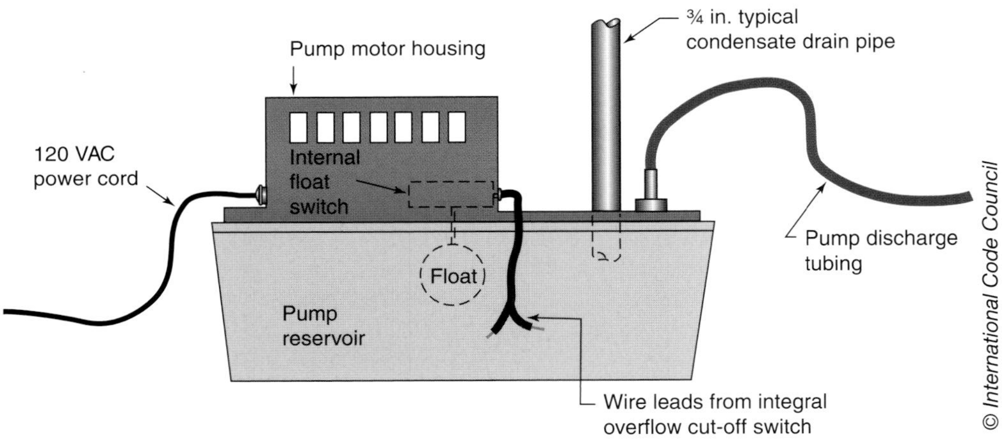

Condensate pumps located in attics and crawl spaces must be connected to the appliance such that when the pump fails the appliance shuts off.

G2411.1.1

Electrical Bonding of Corrugated Stainless Steel Tubing

CHANGE TYPE: Modification

CHANGE SUMMARY: The maximum allowable length of the bonding jumper for corrugated stainless steel tubing (CSST) is 75 feet. Bonding methods must comply with NFPA 70 and devices, such as clamps, must be listed in accordance with UL 467.

2015 CODE: G2411.1.1 (310.1.1) CSST. Corrugated stainless steel tubing (CSST) gas piping systems and piping systems containing one or more segments of CSST shall be bonded to the electrical service grounding electrode system or, where provided, the lightning protection electrode system. ~~The bonding jumper shall connect to a metallic pipe or fitting between the point of delivery and the first downstream CSST fitting. The bonding jumper shall be not smaller than 6 AWG copper wire or equivalent. Gas piping systems that contain one or more segments of CSST shall be bonded in accordance with this section.~~

G2411.1.1 continues

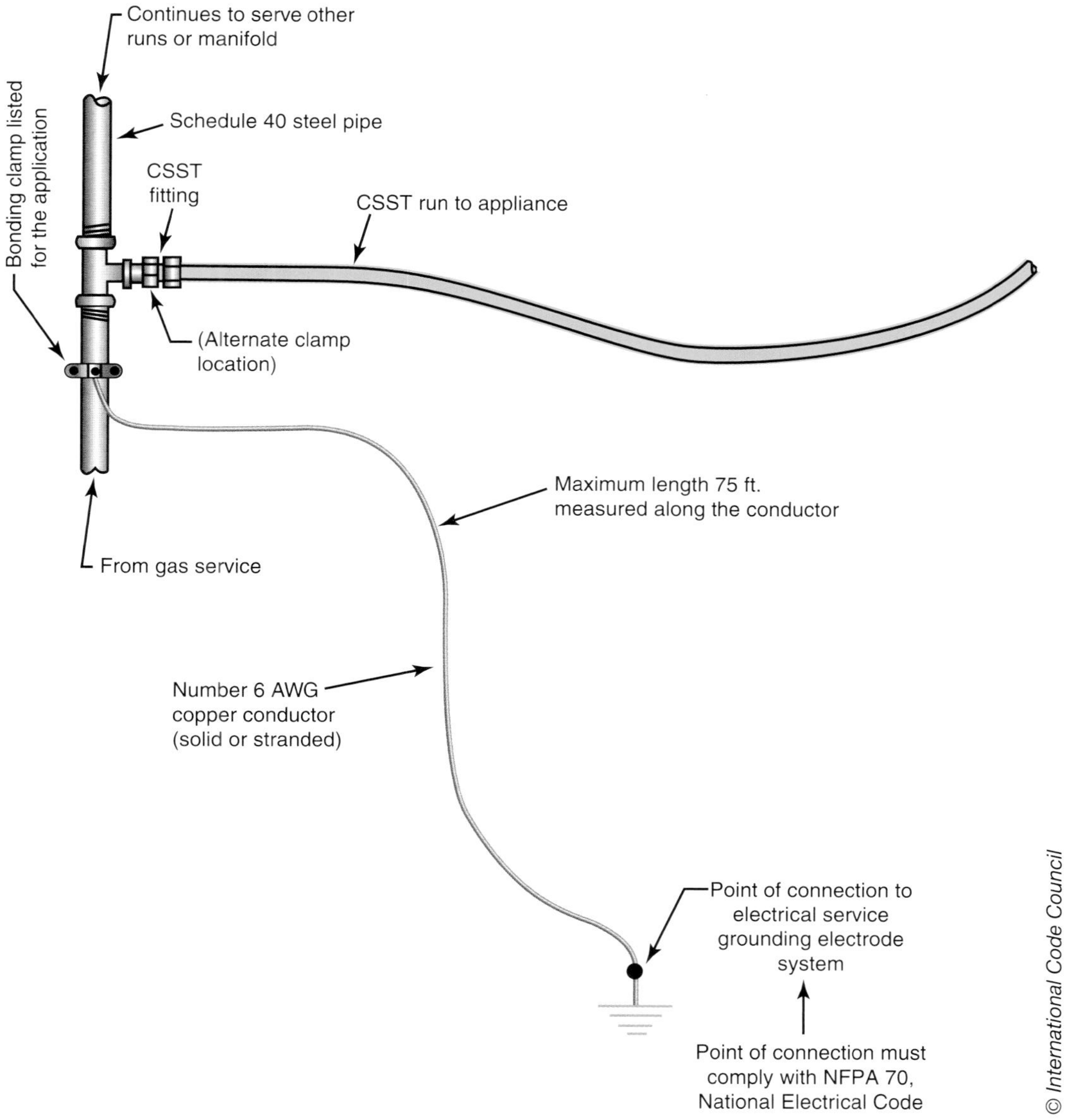

Bonding of corrugated stainless steel tubing (CSST) gas piping

G2411.1.1 continued

G2411.1.1.1 (310.1.1.1) Point of Connection. The bonding jumper shall connect to a metallic pipe, pipe fitting or CSST fitting.

G2411.1.1.2 (310.1.1.2) Size and Material of Jumper. The bonding jumper shall be not smaller than 6 AWG copper wire or equivalent.

G2411.1.1.3 (310.1.1.3) Bonding Jumper Length. The length of the bonding jumper between the connection to a gas piping system and the connection to a grounding electrode system shall not exceed 75 feet (22 860 mm). Any additional grounding electrodes used shall be bonded to the electrical service grounding electrode system or, where provided, the lightning protection grounding electrode system.

G2411.1.1.4 (310.1.1.4) Bonding Connections. Bonding connections shall be in accordance with NFPA 70.

G2411.1.1.5 (310.1.1.5) Connection Devices. Devices used for making the bonding connections shall be listed for the application in accordance with UL 467.

CHANGE SIGNIFICANCE: An electrical bonding jumper becomes less effective as its length increases because of the increasing impedance to electrical flow on the wire. Therefore, shorter lengths improve bonding jumper effectiveness. Extensive testing was performed by the corrugated stainless steel tubing (CSST) industry to determine how well electrical bonding protects the CSST from indirect lightning strikes and lightning-induced currents. The testing concluded that the bonding was effective in preventing perforations in the CSST gas piping under the conditions of the predicted lightning events. The testing determined that the bonding jumper was functionally adequate up to approximately 100 feet in length and there were no data collected to support longer lengths. This suggested the need for a length limit. A length limit of 75 feet was chosen to provide a safety factor and also because it was believed that 75 feet would accommodate the majority of building designs and utility service entrances.

Bonding the CCST to an independent grounding electrode (one that is electrically isolated from the building's grounding electrode system) is prohibited. However, the code does not prevent installation of a supplemental grounding electrode for additional protection. Where such supplemental electrodes are installed, the code requires that they be bonded back to the electrical service grounding electrode system, as is consistent with NFPA 70, *National Electrical Code* (NEC) requirements. The code does not intend to allow the length limit to be circumvented by installing supplemental electrodes. Where supplemental electrodes are installed by choice, the bonding jumper that connects them to the electrical service grounding electrode system is still limited to 75 feet.

The points of connection to the electrical service grounding electrode system, the methods of connection, and the protection of the bonding conductors must be in accordance with NFPA 70 (NEC). The devices, such as clamps, that are used to connect the bonding jumper on both ends must be listed for the application and environment in which they are installed. For example, clamps used outdoors must be listed for exposure to the elements. Some commonly used bonding clamps are suitable only for indoor use and some are suitable for indoor and outdoor use.

G2413.2

Maximum Gas Demand

CHANGE TYPE: Modification

CHANGE SUMMARY: Table G2413.2 and the reference to it were deleted to clarify that the code requires the actual maximum input rating of the appliances to be known and used for gas pipe sizing purposes.

2015 CODE: G2413.2 (402.2) Maximum Gas Demand. The volumetric flow rate of gas to be provided, ~~in cubic feet per hour,~~ shall be the sum of the maximum input ~~calculated using the manufacturer's input ratings~~ of the appliances served ~~adjusted for altitude~~. ~~Where an input rating is not indicated, the gas supplier, appliance manufacturer or a qualified agency shall be contacted, or the rating from Table 402.2 shall be used for estimating the volumetric flow rate of gas to be supplied.~~

The total connected hourly load shall be used as the basis for pipe sizing, assuming that all appliances could be operating at full capacity simultaneously. Where a diversity of load can be established, pipe sizing shall be permitted to be based on such loads.

The volumetric flow rate of gas to be provided shall be adjusted for altitude where the installation is above 2,000 feet (610 m) in elevation.

G2413.2 continues

~~TABLE G2413.2 (402.2) Approximate Gas Input for Typical Appliances~~

~~Appliance~~	~~Input Btu/h (Approx.)~~
~~**Space Heating Units**~~	
~~Hydronic boiler:~~	
~~Single family~~	~~100,000~~
~~Multifamily, per unit~~	~~60,000~~
~~Warm-air furnace:~~	
~~Single family~~	~~100,000~~
~~Multifamily, per unit~~	~~60,000~~

(Portions of deleted table not shown for brevity and clarity.)

G2413.2 continued

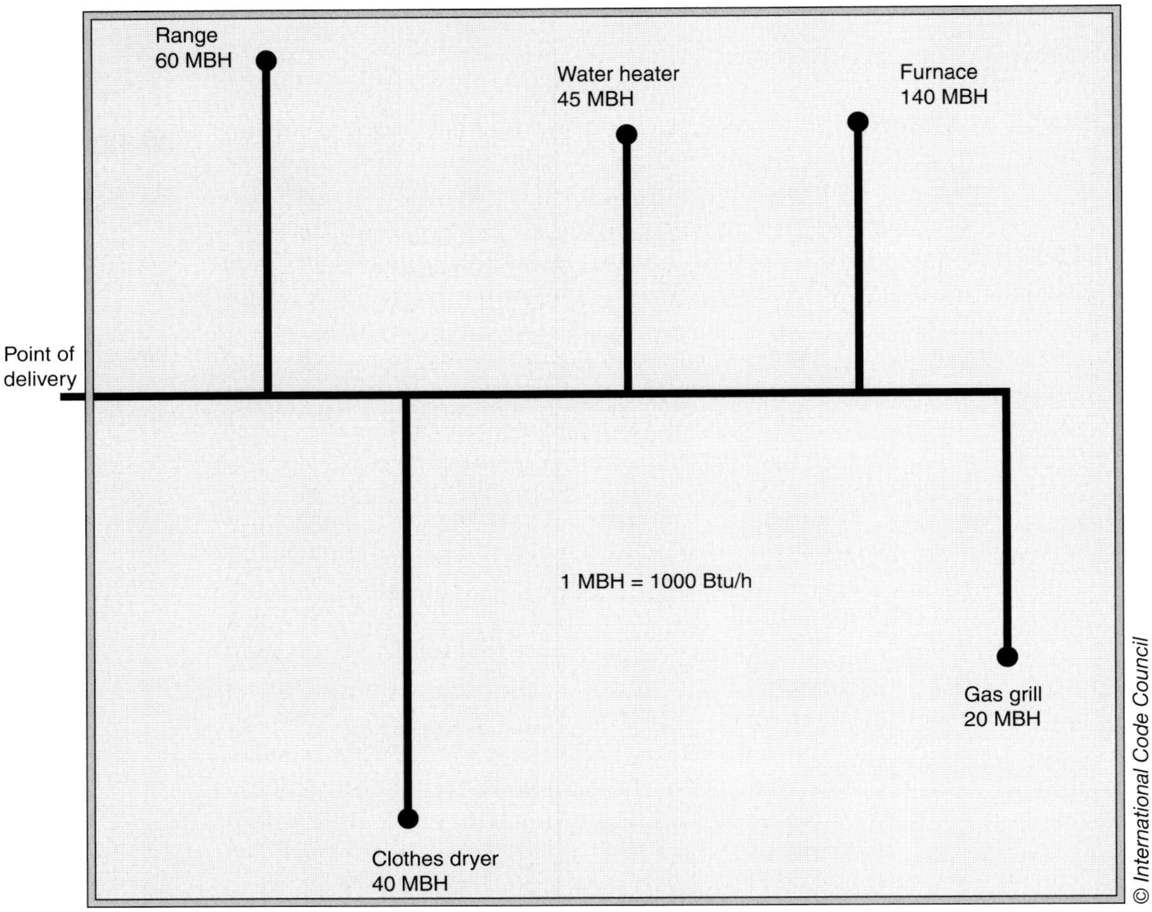

Gas piping size is based on the actual maximum input rating of the appliances

CHANGE SIGNIFICANCE: Table G2413.2 in previous editions of the code provided estimates for determining the total gas demand and, ultimately, the size of the gas piping system. The designer of a gas piping system might have used the table as a starting point in cases where the actual appliance loads were unknown. Then the designer would determine the actual appliance inputs and verify that the design was adequate. The table provided estimates as a placeholder in the piping system design, and the design would have to be verified after the true loads were known. This process carries the risk that the estimate table could be relied upon solely and the piping system might be undersized in some cases. It was felt that such design guidance tables belong in a handbook rather than in a code and the table has been removed. Designers can use estimated loads if they need to, but the code should not encourage the practice. Also, it is difficult for such a table to accurately represent the many different appliances in the marketplace.

CHANGE TYPE: Modification

CHANGE SUMMARY: PVC and CPVC pipe are expressly prohibited materials for supplying fuel gas.

G2414.6

Plastic Pipe, Tubing and Fittings

2015 CODE: G2414.6 (403.6) Plastic Pipe, Tubing and Fittings. Polyethylene plastic pipe, tubing and fittings used to supply fuel gas shall conform to ASTM D 2513. Such pipe shall be marked "Gas" and "ASTM D 2513."

Plastic pipe, tubing and fittings, other than polyethylene, shall be identified and conform to the 2008 edition of ASTM D 2513. Such pipe shall be marked "Gas" and "ASTM D 2513."

Polyvinyl chloride (PVC) and chlorinated polyvinyl chloride (CPVC) plastic pipe, tubing and fittings shall not be used to supply fuel gas.

CHANGE SIGNIFICANCE: The code now references the 2013 edition of ASTM D 2513, which has been revised to address only polyethylene (PE) plastic pipe, tubing and fittings, whereas the 2009 edition addressed all plastic materials. The code had to maintain a reference to the 2008 edition of the standard in order to address plastics other than PE such as polyamide (nylon). It was determined that polyamide pipe is currently used to supply fuel gas; however, PVC and CPVC are not. Further, it was decided that because of the brittle nature of PVC and CPVC, especially at low temperatures, these materials are not suitable for conveying fuel gas. Rather than be silent, the code now prohibits what the marketplace has failed to embrace as a viable material for the conveyance of fuel gas.

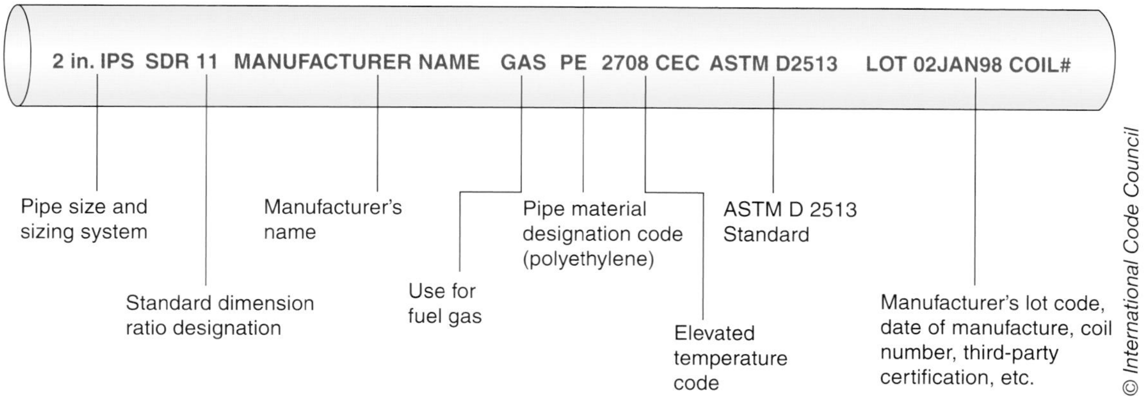

Approved polyethylene gas piping with markings in accordance with the code and ASTM D 2513

G2415.5

Fittings in Concealed Locations

CHANGE TYPE: Clarification

CHANGE SUMMARY: This section retains the basic intent while being completely reorganized to clarify the correct application. Threaded elbows, tees and couplings are now specifically approved for concealed locations as the code always intended. The code now provides the applicable referenced standards for fittings that are listed for concealed locations.

2015 CODE: G2415.5 (404.5) Fittings ~~Piping~~ in Concealed Locations. ~~Portions of a piping~~ Fittings installed in concealed locations shall ~~not have unions, tubing fittings, right and left couplings, bushings, compression couplings and swing joints made by combinations of fittings~~ be limited to the following types:

1. Threaded elbows, tees and couplings
2. Brazed fittings
3. Welded fittings
4. Fittings listed to ANSI LC-1/CSA 6.26 or ANSI LC-4.

~~**Exceptions:**~~

~~1. Tubing joined by brazing.~~

~~2. Fittings listed for use in concealed locations.~~

CHANGE SIGNIFICANCE: Rather than listing what is prohibited and having exceptions, the text of this section was reformatted to state what fittings are allowed in concealed locations. The new text lists the four types of allowed fittings: threaded elbows, tees and couplings; brazed fittings; welded fittings; and proprietary fittings listed to ANSI LC-1 or

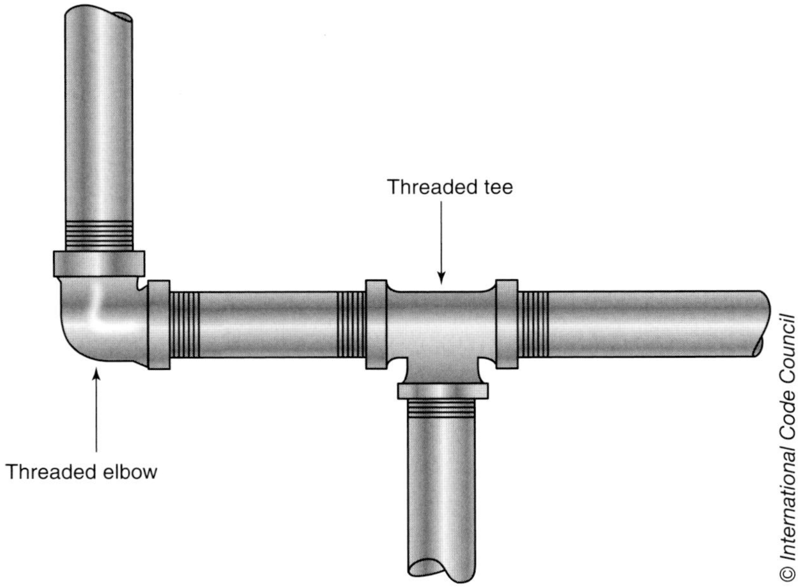

Threaded elbows, tees and couplings for Schedule 40 steel gas piping are permitted in concealed locations.

ANSI LC-4. The fittings allowed for concealment are limited to those four types. By omission, all other types of fittings are prohibited in concealed locations. This section has caused interpretation difficulties, and the new text simply clarifies what has always been the intent.

Note that in future editions of the code, Item 1 will likely be revised to add threaded plugs and caps, as this was revised in ANSI Z223.1. When the code refers to couplings, those fittings are tapered thread couplings, not the straight thread couplings that are commonly found. See Section G2414.9 in the code regarding metallic pipe threads.

G2415.7

Protection of Concealed Piping Against Physical Damage

CHANGE TYPE: Modification

CHANGE SUMMARY: The section on protection of piping has been completely rewritten to address more than just bored holes and notches in structural members. It now addresses piping parallel to framing members and piping within framing members. The new text requires that the protection extend well beyond the edge of members that are bored or notched.

2015 CODE: G2415.7 (404.7) Protection Against Physical Damage. ~~In concealed locations, where piping other than black or galvanized steel is installed through holes or notches in wood studs, joists, rafters or similar members less than 1½ inches (38 mm) from the nearest edge of the member, the pipe shall be protected by shield plates. Protective steel shield plates having a minimum thickness of 0.0575 inch (1.463 mm) (No. 16 gage) shall cover the area of the pipe where the member is notched or bored and shall extend a minimum of 4 inches (102 mm) above sole plates, below top plates and to each side of a stud, joist or rafter.~~ Where piping will be concealed within light-frame construction assemblies, the piping shall be protected against penetration by fasteners in accordance with Sections G2415.7.1 through G2415.7.3.

> **Exception:** Black steel piping and galvanized steel piping shall not be required to be protected.

G2415.7.1 (404.7.1) Piping Through Bored Holes or Notches. Where piping is installed through holes or notches in framing members and the piping is located less than 1½ inches (38 mm) from the framing member face to which wall, ceiling or floor membranes will be attached, the pipe

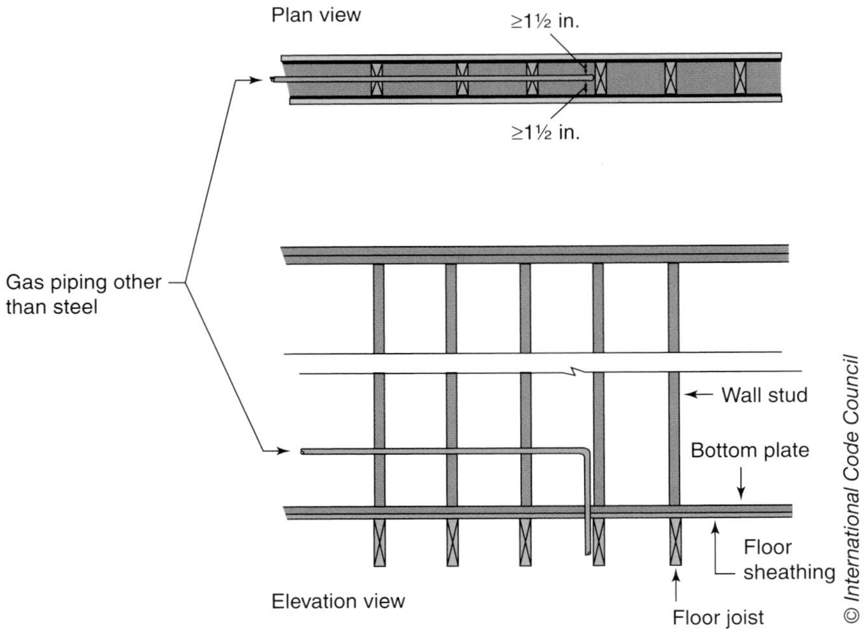

Concealed gas piping with minimum clearance of 1½ inches does not require protection.

shall be protected by shield plates that cover the width of the pipe and the framing member and that extend not less than 4 inches (51 mm) to each side of the framing member. Where the framing member that the piping passes through is a bottom plate, bottom track, top plate or top track, the shield plates shall cover the framing member and extend not less than 4 inches (51 mm) above the bottom framing member and not less than 4 inches (51 mm) below the top framing member.

G2415.7.2 (404.7.2) Piping Installed in Other Locations. Where the piping is located within a framing member and is less than 1½ inches (38 mm) from the framing member face to which wall, ceiling or floor membranes will be attached, the piping shall be protected by shield plates that cover the width and length of the piping. Where the piping is located outside of a framing member and is located less than 1½ inches (38 mm) from the nearest edge of the face of the framing member to which the membrane will be attached, the piping shall be protected by shield plates that cover the width and length of the piping

G2415.7.3 (404.7.3) Shield Plates. Shield plates shall be of steel material having a thickness of not less than 0.0575 inch (1.463 mm) (No. 16 gage).

CHANGE SIGNIFICANCE: Fuel gas tubing in concealed locations is vulnerable to penetration by fasteners used for securing finish materials. When this occurs, a leak may not develop immediately, but may show up years later due to corrosion. For that reason, piping and tubing other than Schedule 40 steel pipe must be protected from penetration by nails and screws where the pipe or tubing is less than 1½ inches from the face of the member where sheathing, membranes or finish materials (typically drywall) will be attached. If the 1½-inch dimension can not be maintained, the code requires the installation of steel shield plates to protect the piping or tubing. This protection is necessary whether the pipe or tube is perpendicular or parallel to the framing member. If a pipe or tube is run inside of a 3½-inch "C" channel of a metal stud parallel to the direction of the stud, it is subject to penetration by screws unless the pipe or tube is ½ inch or less in diameter and located dead center in the stud channel. Where pipes and tubing are attached to and run parallel with the side of a framing member, penetration by nails or screws is also possible if the fastener misses the framing and the pipe or tube is less than 1½ inches from either face of the stud. Extending the protection shield plate 4 inches beyond the edges of the framing member is intended to protect against fasteners that miss the member or that exit the member on an angle. To avoid having protection plates run parallel with a member, the pipe or tube could simply be placed on "standoffs" such that the pipe/tube is not less than 1½ inches from the nearest edge of the member. As always, careful planning of the routing of gas piping and tubing can avoid the need for protection plates.

This section pertains to piping and tubing that is concealed within wood or steel light-frame construction assemblies, which is the same scope as previous editions of the code.

G2421.2

Medium-Pressure Regulators

CHANGE TYPE: Modification

CHANGE SUMMARY: Medium-Pressure (MP) line regulators installed in rigid piping must have a union installed to allow removal of the regulator.

2015 CODE: **G2421.2 (410.2) MP Regulators.** MP pressure regulators shall comply with the following:

1. The MP regulator shall be approved and shall be suitable for the inlet and outlet gas pressures for the application.
2. The MP regulator shall maintain a reduced outlet pressure under lockup (no-flow) conditions.
3. The capacity of the MP regulator, determined by published ratings of its manufacturer, shall be adequate to supply the appliances served.
4. The MP pressure regulator shall be provided with access. Where located indoors, the regulator shall be vented to the outdoors or shall be equipped with a leak-limiting device, in either case complying with Section G2421.3.
5. A tee fitting with one opening capped or plugged shall be installed between the MP regulator and its upstream shutoff valve. Such tee fitting shall be positioned to allow connection of a pressure-measuring instrument and to serve as a sediment trap.
6. A tee fitting with one opening capped or plugged shall be installed not less than 10 pipe diameters downstream of the MP regulator outlet. Such tee fitting shall be positioned to allow connection of a pressure-measuring instrument.
7. <u>Where connected to rigid piping, a union shall be installed within 1 foot (304 mm) of either side of the MP regulator.</u>

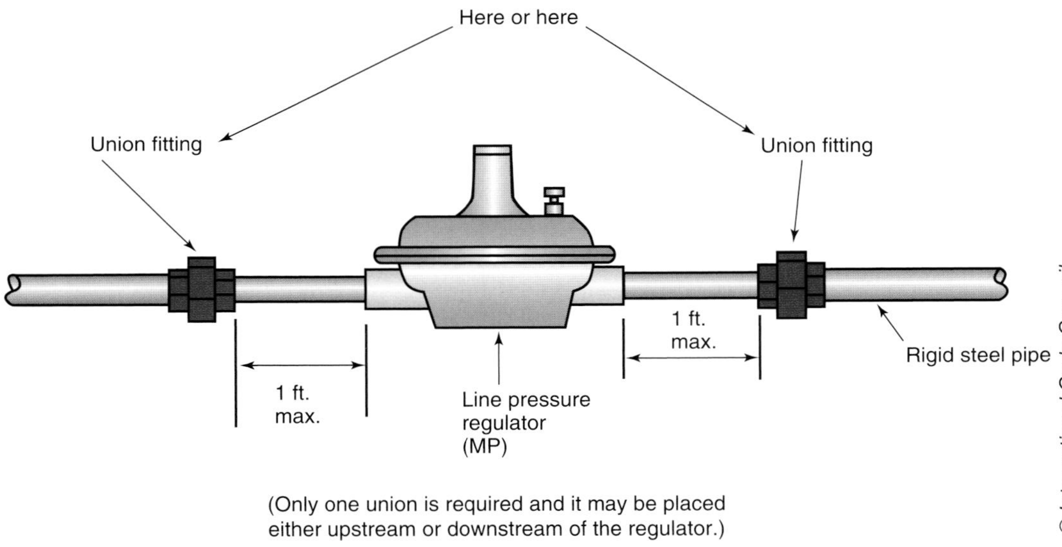

Union required for Medium-Pressure (MP) regulator connected to rigid piping

CHANGE SIGNIFICANCE: MP regulators are line pressure regulators that serve to reduce pressures that are above 0.5 psi and less than or equal to 5 psi, down to some lower pressure. They are typically installed in 2-psi and 5-psi gas distribution systems that serve appliances having a maximum input pressure of 0.5 psi (14 inches water column). If such regulators are installed with steel piping on the inlet and outlet side, it will be impossible to remove the regulator or isolate it without disassembling the piping system for some distance or cutting the piping. To facilitate removal or isolation of the regulator, a union fitting must be placed near the inlet or outlet side of the regulator.

G2422.1

Connecting Portable and Movable Appliances

CHANGE TYPE: Modification

CHANGE SUMMARY: Where portable gas appliances are used outdoors, such as gas grills, fire pits, and patio heaters, the options for connecting to the gas distribution system are practically limited to gas hoses designed for the purpose. Such hoses must comply with ANSI Z21.54.

2015 CODE: G2422.1 (411.1) Connecting Appliances. Appliances shall be connected to the piping system by one of the following:

1. Rigid metallic pipe and fittings.
2. Corrugated stainless steel tubing (CSST) where installed in accordance with the manufacturer's instructions.
3. Listed and labeled appliance connectors in compliance with ANSI Z21.24 and installed in accordance with the manufacturer's instructions and located entirely in the same room as the appliance.
4. Listed and labeled quick-disconnect devices used in conjunction with listed and labeled appliance connectors.

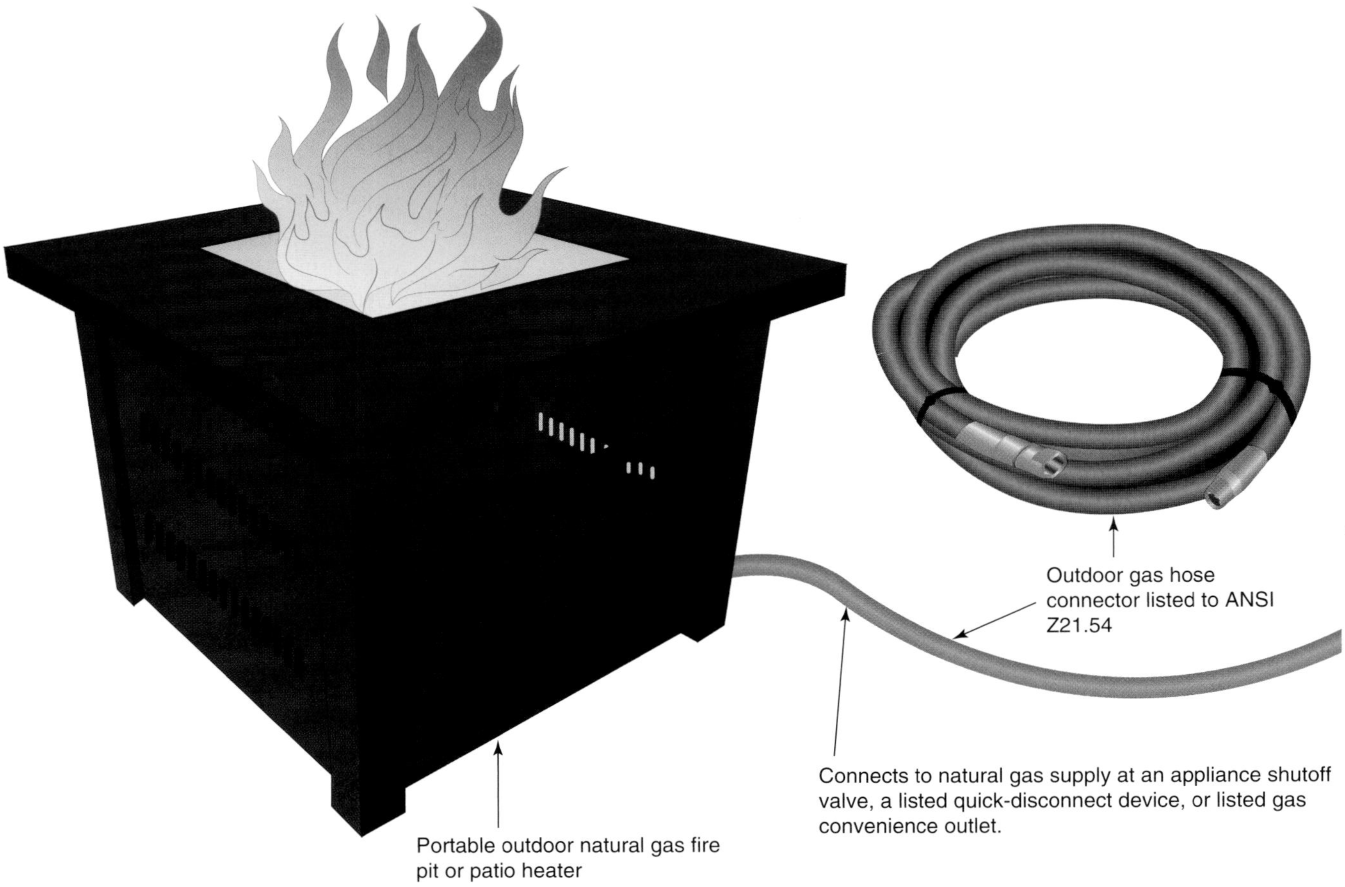

Natural gas hose connector for portable outdoor gas appliances

5. Listed and labeled convenience outlets used in conjunction with listed and labeled appliance connectors.
6. Listed and labeled outdoor appliance connectors in compliance with ANSI Z21.75/CSA 6.27 and installed in accordance with the manufacturer's instructions.
7. Listed outdoor gas hose connectors in compliance with ANSI Z21.54 used to connect portable outdoor appliances. The gas hose connection shall be made only in the outdoor area where the appliance is to be used, and shall be to the gas piping supply at an appliance shutoff valve, a listed quick-disconnect device, or listed gas convenience outlet.

G2422.1.5 (411.1.4) Movable Appliances. Where appliances are equipped with casters or are otherwise subject to periodic movement or relocation for purposes such as routine cleaning and maintenance, such appliances shall be connected to the supply system piping by means of an ~~approved flexible~~ appliance connector ~~designed and labeled for the application~~ listed as complying with ANSI Z21.69 or by means of Item 1 of Section G2422.1. Such flexible connectors shall be installed and protected against physical damage in accordance with the manufacturer's ~~installation~~ instructions.

CHANGE SIGNIFICANCE: Methods 1 through 6 of Section G2422.1 are not designed for connecting portable appliances to the gas distribution piping system. For outdoor portable appliances, new method 7 is the only appropriate option. Outdoor gas hose connectors have to be resistant to mechanical damage, possible heat exposure and the harmful effects of exposure to the weather. Connectors listed to ANSI Z21.54 are evaluated and tested for the particularly harsh environment of outdoor use. The gas hose connector must be located entirely outdoors and must be connected to the gas piping system at a point outdoors. The point of connection to the gas distribution system piping must be through a listed device that allows the hose to be readily disconnected manually or through an appliance shutoff valve. Quick-disconnect devices have safety features such as thermal shutoffs that will close the valve when exposed to high temperatures and interlocking systems that will not allow the hose to be removed until the manual gas valve is closed. The intent of new Item 7 is to address portable outdoor appliance connections and to mandate that such connectors be listed to a specific safety standard.

Movable gas appliances in other than outdoor locations require flexible connectors listed as complying with ANSI Z21.69 installed and protected against physical damage in accordance with the manufacturer's instructions, or they must be connected with rigid metallic piping as referenced in Item 1 of Section G2422.1. Previously, the code only required that movable appliances be connected with approved flexible connectors.

G2426.7.1

Door Clearance to Vent Terminals

CHANGE TYPE: Addition

CHANGE SUMMARY: An appliance vent terminal is not permitted in a location within 12 inches of the arc of a swinging door.

2015 CODE: **G2426.7.1 (502.7.1) Door Swing.** Appliance and equipment vent terminals shall be located such that doors cannot swing within 12 inches (305 mm) horizontally of the vent terminal. Door stops or closers shall not be installed to obtain this clearance.

CHANGE SIGNIFICANCE: Vent terminals for sidewall vented appliances, such as direct-vent gas fireplaces and fireplace heaters, direct-vent room heaters, direct-vent water heaters, furnaces and boilers are sometimes located where a side-swinging door could impact the vent terminal or swing close to the terminal. The results can be damage to the vent terminal, a fire hazard, and interference with the appliance venting and combustion air intake. Another possible scenario is where the door blocks

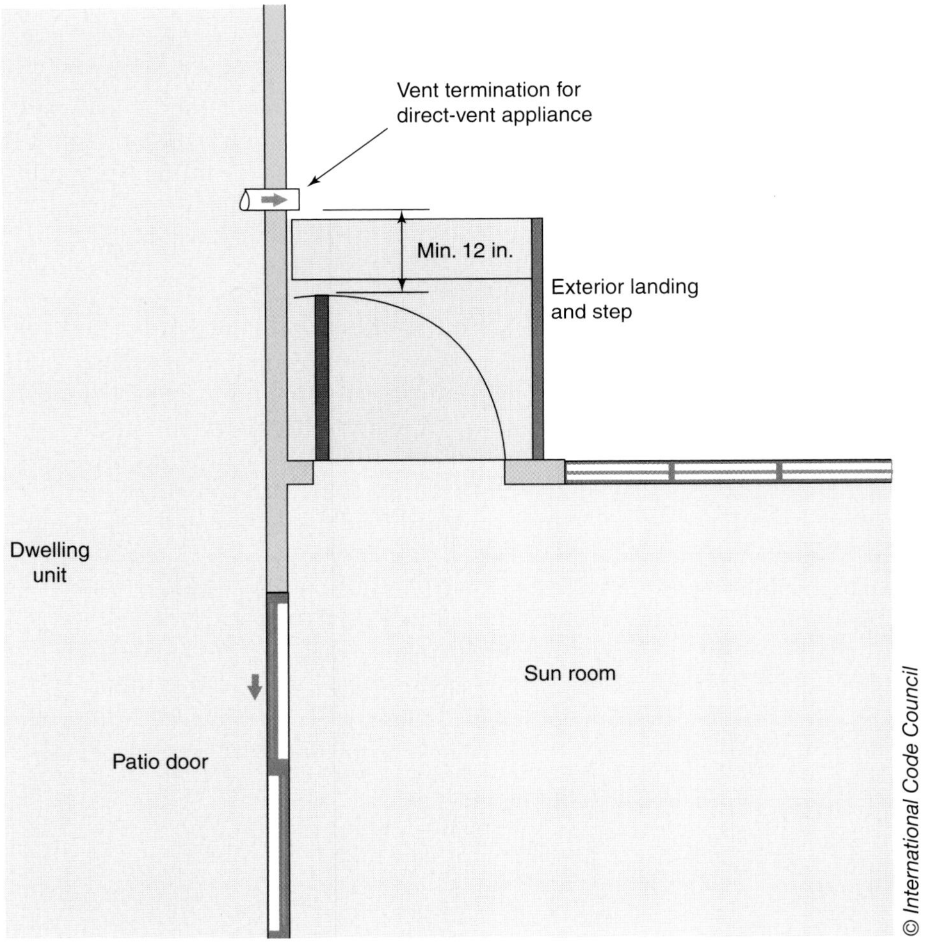

Vent terminals must be located so doors cannot swing within 12 inches to protect against physical damage.

This direct-vent terminal is subject to damage from the swinging door.

or deflects the vent discharge such that the combustion products are pulled back into the combustion air intake resulting in excessive carbon monoxide production, serious appliance malfunction, and sooting. Door stops and closer devices cannot be depended upon because they are easily defeated or removed.

G2427.4.1, G2427.6.8.3

Plastic Piping for Appliance Vents

CHANGE TYPE: Modification

CHANGE SUMMARY: The approval of plastic pipe for venting appliances is no longer a responsibility of the building official and, instead, that responsibility rests with the appliance manufacturer and the appliance listing agency. The code previously addressed only vents, which are defined as listed and labeled factory-made products. The code is no longer silent on the sizing of plastic pipe vents that do not fall under the definition of "vent."

2015 CODE: G2427.4.1 (503.4.1) Plastic Piping. ~~Plastic piping used for venting appliances listed for use with such venting materials shall be approved.~~ Where plastic piping is used to vent an appliance, the appliance shall be listed for use with such venting materials and the appliance manufacturer's installation instructions shall identify the specific plastic piping material.

G2427.6.8.3 (503.6.9.3) Category II, III and IV Appliances. The sizing of gas vents for Category II, III and IV appliances shall be in accordance with the appliance manufacturer's instructions. The sizing of plastic pipe that is specified by the appliance manufacturer as a venting material for Category II, III and IV appliances, shall be in accordance with the manufacturer's instructions.

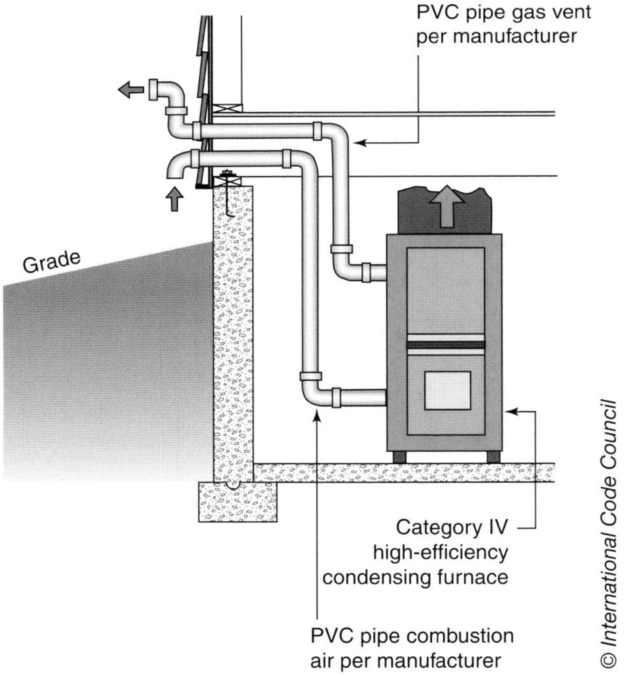

The type of material and sizing of plastic pipe that is specified by the appliance manufacturer as a venting material shall be in accordance with the manufacturer's instructions.

CHANGE SIGNIFICANCE: The previous code text did not actually require that the appliance be listed for use with specific venting system materials, although this was implied by the text and is required in the appliance standards. For appliances vented with plastic piping, the appliance manufacturer's installation instructions must clearly specify what plastic materials are required or allowed for venting an appliance. The installation instructions must be consistent with how the appliance was tested by the listing agency. The product standards for gas appliances contain various testing procedures for plastic venting systems. The appliance manufacturer determines the type of plastic vent that is suitable for venting its product, and the testing and listing agency tests the appliance with that venting system for compliance with the product standards. There must not be any uncertainty about what type of venting system is required for any appliance so that venting system failures can be avoided. Note that the definition of "vent" does not include plastic pipes such as PVC, ABS and CPVC because such pipes are not currently listed as factory-built venting systems. The PVC, ABS and CPVC pipe manufactures do not recommend that their pipe be used for appliance venting because such products are not currently listed for such applications. There are polypropylene venting systems on the market that are listed to UL 1738 as appliance venting systems, and they do fall under the definition of "vent."

Because plastic pipes such PVC, ABS and CPVC plumbing pipes are not listed and labeled as appliance vents (see definition of "vent"), the code was silent on how to size such pipes. The sizing is covered in the appliance manufacturer's instructions, and the code requires compliance with such instructions. For consistency, Section G2427.6.8.3 was modified to address sizing of both listed vents and unlisted materials used as vents.

G2427.8

Venting System Termination Location

CHANGE TYPE: Modification

CHANGE SUMMARY: New text addresses the location of sidewall vent terminals with respect to adjoining buildings. A 10-foot separation is required when a vent discharges in the direction of an opening in an adjacent building.

2015 CODE: G2427.8 (503.8) Venting System Termination Location. The location of venting system terminations shall comply with the following (see Appendix C):

1. A mechanical draft venting system shall terminate ~~at least~~ not less than 3 feet (914 mm) above any forced-air inlet located within 10 feet (3048 mm).

   Exceptions:

   1. This provision shall not apply to the combustion air intake of a direct-vent appliance.
   2. This provision shall not apply to the separation of the integral outdoor air inlet and flue gas discharge of listed outdoor appliances.

2. A mechanical draft venting system, excluding direct-vent appliances, shall terminate ~~at least~~ not less than 4 feet (1219 mm) below, 4 feet (1219 mm) horizontally from, or 1 foot (305 mm) above any door, operable window, or gravity air inlet into any building. The bottom of the vent terminal shall be located ~~at least~~ not less than 12 inches (305 mm) above finished ground level.
3. The vent terminal of a direct-vent appliance with an input of 10,000 Btu per hour (3 kW) or less shall be located ~~at least~~ not less than 6 inches (152 mm) from any air opening into a building.

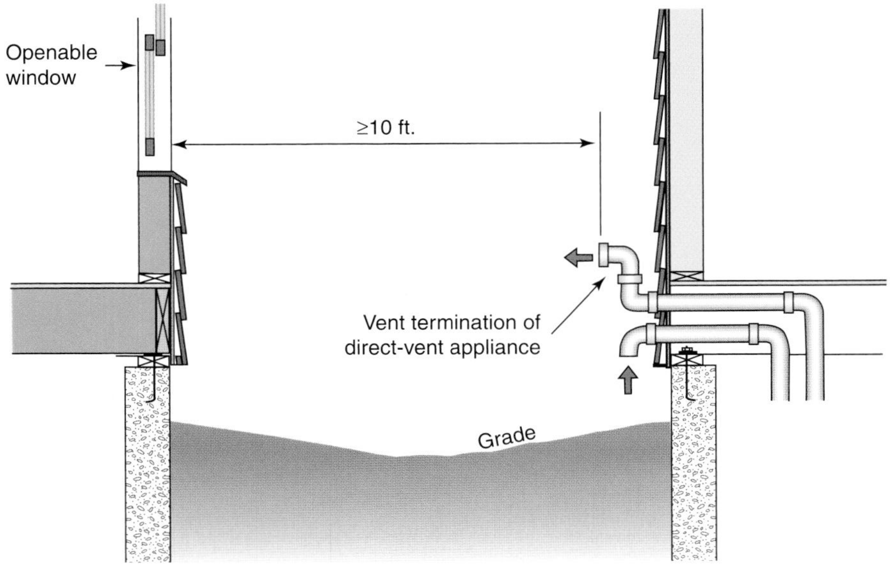

A minimum 10-foot horizontal separation is required between a vent terminal and an opening of an adjacent building.

Such an appliance with an input over 10,000 Btu per hour (3 kW) but not over 50,000 Btu per hour (14.7 kW) shall be installed with a 9-inch (230 mm) vent termination clearance, and an appliance with an input over 50,000 Btu per hour (14.7 kW) shall have ~~at least~~ not less than a 12-inch (305 mm) vent termination clearance. The bottom of the vent terminal and the air intake shall be located ~~at least~~ not less than 12 inches (305 mm) above finished ground level.

4. Through-the-wall vents for Category II and IV appliances and noncategorized condensing appliances shall not terminate over public walkways or over an area where condensate or vapor could create a nuisance or hazard or could be detrimental to the operation of regulators, relief valves or other equipment. Where local experience indicates that condensate is a problem with Category I and III appliances, this provision shall also apply. Drains for condensate shall be installed in accordance with the appliance and vent manufacturers' instructions.
5. Vent systems for Category IV appliances that terminate through an outside wall of a building and discharge flue gases perpendicular to the adjacent wall shall be located not less than 10 feet (3048 mm) horizontally from an operable opening in an adjacent building. This requirement shall not apply to vent terminals that are 2 feet (607 mm) or more above or 25 feet (7620 mm) or more below operable openings.

CHANGE SIGNIFICANCE: The code now addresses a common situation where dwellings are located close to each other and sidewall-vented appliances are installed with the vent terminals directed toward the neighboring home. The concern is that combustion gases will enter the adjacent building through openings in the exterior walls that face the appliance vent terminal. This section applies only to Category IV (condensing) appliances that are sidewall vented with stainless steel or plastic vents.

Computer simulations were conducted as part of a research project and the results indicated that in many scenarios, the combustion products would impinge on the neighboring building. Many factors impact the simulated scenarios, including wind speed and direction, the height of the adjacent buildings and the type of vent terminal (e.g., straight pipe, tee fitting, deflector cap, or directional fitting). If the appliance vent terminal is a straight open-ended pipe, and that pipe is perpendicular to the wall it passes through, it creates a worst-case scenario that the new provision addresses. This scenario is the most common and the most likely to project combustion gases far enough to be a potential danger to the neighbors. The research project suggested that vent terminals that utilize a tee fitting outlet or a deflector cap, or that are directed at some angle downward, are much less likely to create a nuisance or hazard to the neighbors because the combustions gases disperse and lack the velocity to impinge on the adjacent building.

G2439.4, G2439.7

Clothes Dryer Exhaust Ducts

CHANGE TYPE: Modification

CHANGE SUMMARY: New text recognizes the use of dryer exhaust duct power ventilators (DEDPVs) to increase the allowable exhaust duct length for clothes dryers. A permanent label identifying the concealed length of dryer exhaust duct is no longer required where the equivalent duct length does not exceed 35 feet. For dryer exhaust duct exceeding 35 feet, a label or tag is required whether the duct is concealed or not. Instead of prohibiting all duct fasteners such as screws and rivets, the code now limits the penetration of fasteners, where installed.

2015 CODE: G2439.4 (614.5) Dryer Exhaust Duct Power Ventilators. Domestic dryer exhaust duct power ventilators shall be listed and labeled to UL 705 for use in dryer exhaust duct systems. The dryer exhaust duct power ventilator shall be installed in accordance with the manufacturer's instructions.

~~G2439.5 (614.6)~~ G2439.7 (614.8) Domestic Clothes Dryer Exhaust Ducts. Exhaust ducts for domestic clothes dryers shall conform to the requirements of Sections ~~G2439.5.1~~ G2439.7.1 through ~~G2439.5.7~~ G2439.7.6.

~~G2439.5.2 (614.6.2)~~ G2439.7.2 (614.8.2) Duct Installation. Exhaust ducts shall be supported at 4-foot (1219 mm) intervals and secured in place. The insert end of the duct shall extend into the adjoining duct or fitting in the direction of airflow. Ducts shall not be joined with screws or similar fasteners that protrude more than ⅛ inch (3.2 mm) into the inside of the duct.

G2439.7.4.3 (614.8.4.3) Dryer Exhaust Duct Power Ventilator Length. The maximum length of the exhaust duct shall be determined by the dryer exhaust duct power ventilator manufacturer's installation instructions.

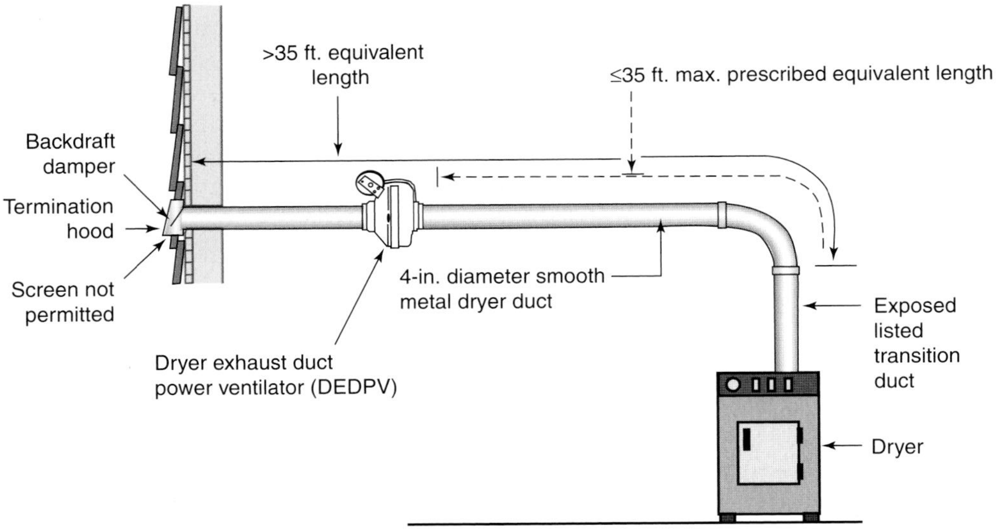

© International Code Council

Installation of a dryer exhaust duct power ventilator (DEDPV) in accordance with the manufacturer's instructions to increase the allowable length of dryer exhaust duct

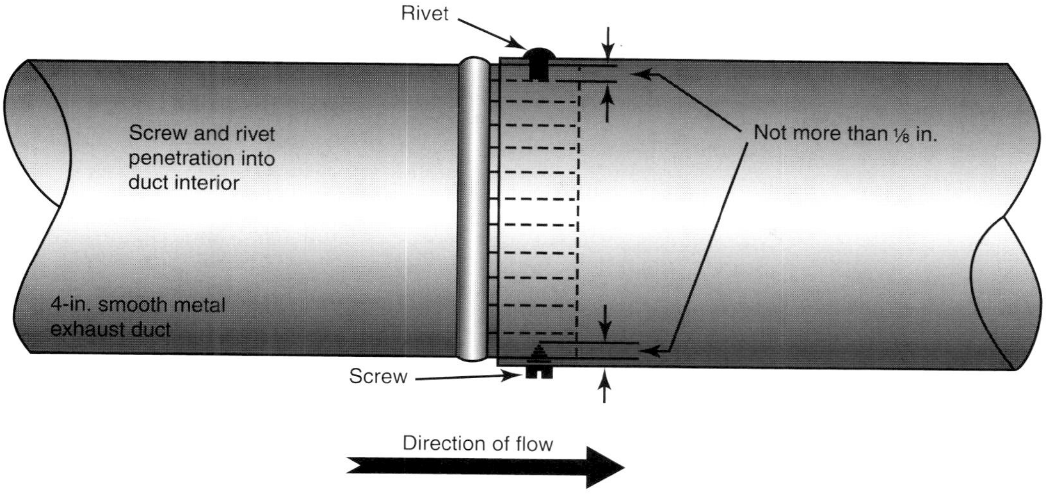

Dryer duct joints must be inserted in the direction of flow and fasteners cannot penetrate more than ⅛ inch

~~G2439.5.6 (614.6.5)~~ G2439.7.5 (614.8.5) Length Identification. Where the exhaust duct ~~is concealed within the building construction~~ equivalent length exceeds 35 feet (10 668 mm), the equivalent length of the exhaust duct shall be identified on a permanent label or tag. The label or tag shall be located within 6 feet (1829 mm) of the exhaust duct connection.

(No significant changes to portions of Section G2439 not shown.)

CHANGE SIGNIFICANCE: The code limits the length of clothes dryer exhaust ducts to protect against potential fire hazards and to ensure that dryers efficiently discharge warm, moist air to the outdoors. Allowable length is based on the airflow capacity of modern dryers. Elbow fittings reduce the allowable length, resulting in a calculated "equivalent length" based on the additional resistance to airflow for each fitting. In addition to lint buildup, excessive duct length creates moisture and maintenance problems, and increases drying times causing the dryer to be inefficient and waste energy.

Previous editions of the code did not recognize dryer exhaust duct power ventilators (DEDPVs) as an option for clothes dryer installations. DEDPVs are typically referred to as "dryer booster fans" in the marketplace, because they "boost" or increase the airflow of the dryer discharge. Greater airflow increases the distance that the discharge air can be effectively pushed to the outdoors. Prior to the 2015 code, the two options for determining the maximum exhaust duct length were to comply with the prescriptive limit of 35 feet, a conservative average for modern dryers, or to follow the clothes dryer manufacturer's instructions for length limits. If the desired location did not fall within those limits, the designer or builder was left to relocate the dryer to reduce the length of the exhaust duct. Another possible solution was to make application to the building official requesting approval to install a DEDPV under the alternative materials, design and methods of construction provisions in Section R104.11. The 2015 IRC now specifically allows DEDPVs in clothes dryer exhaust systems to increase the equivalent length of duct.

G2439.4, G2439.7 continues

G2439.4, G2439.7 continued

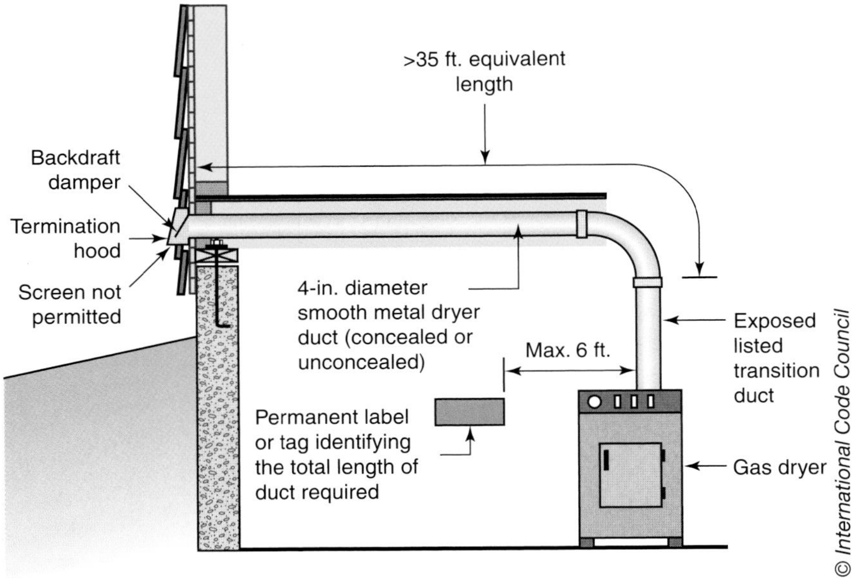

A length identification tag is required if the equivalent duct length exceeds 35 feet.

DEDPVs are listed to a revised version of UL 705 that now contains tests and construction requirements that are specific to these devices. DEDPVs have been around for years, but until recently were not listed to a national consensus standard that was specific to these devices. The UL 705 standard contains requirements for the construction, testing, and installation of DEDPVs and requires them to be equipped with features such as interlocks, limit controls, monitoring controls, and enunciator devices to make certain that the dryers or dryer operators are aware of the operating status of the DEDPVs. The maximum length of the dryer exhaust duct is determined based on the manufacturer's instructions for the DEDPV.

The provisions for identifying the equivalent length of dryer exhaust duct first appeared in the 2009 IRC. The code has since required a permanent label or tag be installed within 6 feet of the dryer when the duct was concealed behind finish materials. The purpose was to alert occupants of the length of concealed duct so they could make an informed decision to install a dryer with adequate airflow capacity. This provision recognizes that homes change hands and many dryers may be installed over the building's lifetime. The primary concern was aimed at exhaust duct systems that were based on the dryer manufacturer's instructions at the time of construction. A given dryer might have a capacity much greater than the specified length of 35 feet, the default value, when the manufacturer and model of the dryer is unknown. This change to the 2015 IRC recognizes that there is no concern if the exhaust duct does not exceed 35 feet in equivalent length and the permanent label in this case provides no benefit to the owner. In addition, the proponents reasoned that the purpose of the permanent sign is to notify the owners and installers that the dryer duct length is exceptional and that any installed dryer must be compatible with that duct of exceptional length. Therefore, the criterion for providing signage only when the duct is concealed has been removed. The code now requires a permanent label or tag when the equivalent length of the dryer exhaust duct exceeds 35 feet, whether or not the duct is concealed within construction.

Section G2439.3 states that fasteners used to join fittings and sections of dryer exhaust duct must not obstruct the airflow. Many times, this was interpreted as a prohibition of screws and rivets. Other times, it was taken to mean that such fasteners must not penetrate too far into the duct. The revision to Section G2439.7.2 makes it clear how Section G2439.3 is to be interpreted. A fastener protrusion of ⅛ inch or less will collect some lint, but it will be insignificant. Actually, smooth duct walls collect lint also. The trade-off for allowing tiny amounts of lint to collect is the improved duct construction. If dryer exhaust ducts are not allowed to be mechanically fastened, the only method to prevent separation of joints is duct tape. Duct tape should never be depended upon as the sole means of securing duct systems. Duct tape is a sealing means, not a fastening means. Now such ducts can be properly and securely fastened and then sealed with tapes or mastics. Note that the IRC Section M1502 requires dryer exhaust ducts to be mechanically fastened and allows the same ⅛-inch maximum penetration.

G2447.2

Prohibited Location of Commercial Cooking Appliances

CHANGE TYPE: Modification

CHANGE SUMMARY: The code does not prohibit the installation of cooking appliances that are listed as both commercial and domestic appliances.

2015 CODE: G2447.2 (623.2) Prohibited Location. Cooking appliances designed, tested, listed and labeled for use in commercial occupancies shall not be installed within dwelling units or within any area where domestic cooking operations occur.

> **Exception:** Appliances that are also listed as domestic cooking appliances.

CHANGE SIGNIFICANCE: Commercial cooking appliances are prohibited in dwelling units and domestic environments because they lack special safety features that domestic appliances must possess. There are appliances built today that are listed as commercial appliances and that are also listed to the domestic appliance standard; therefore, such appliances are allowed in any occupancy. The previous code text would prohibit an appliance listed as a commercial appliance despite the fact that the appliance was dual listed as both commercial and domestic. The code text was revised to eliminate this unintended consequence.

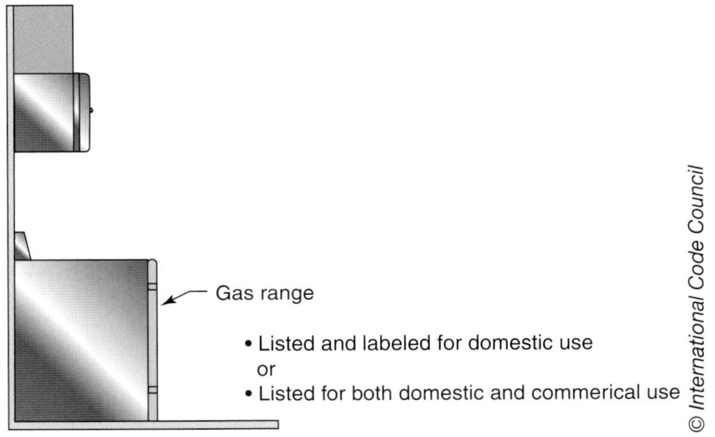

Household cooking appliances

PART **7**

Plumbing

Chapters 25 through 33

- **Chapter 25** Plumbing Administration
- **Chapter 26** General Plumbing Requirements
- **Chapter 27** Plumbing Fixtures
- **Chapter 28** Water Heaters
- **Chapter 29** Water Supply and Distribution
- **Chapter 30** Sanitary Drainage
- **Chapter 31** Vents
- **Chapter 32** Traps
- **Chapter 33** Storm Drainage No changes addressed

Part 7 of the IRC contains provisions for plumbing systems and begins with a chapter on the specific and unique administrative issues related to plumbing code enforcement. Subsequent chapters cover technical subjects for the overall design and installation of plumbing systems in buildings. General plumbing issues such as protection of plumbing systems from damage, piping support, and certification of products are covered in Chapter 26. The other chapters of Part 7 are specific to requirements for plumbing fixtures, water heaters, water supply and distribution, sanitary drainage, vents, traps, and storm drainage. ■

P2502.1, P2503.4

Inspection and Tests for Building Sewers

P2503.5

Drain, Waste, and Vent Systems Testing

P2603.2.1

Protection Against Physical Damage

P2603.3

Protection Against Corrosion

TABLE P2605.1

Piping Support

P2702.1, P2706.1

Waste Receptors

P2717

Dishwashing Machines

P2801

Water Heater Drain Valves and Pans

P2804.6.1

Water Heater Relief Valve Discharge Piping

P2901, P2910 THROUGH P2913

Nonpotable Water Systems

P2905

Heated Water Distribution Systems

continues

P2906.2
Lead Content of Drinking Water Pipe and Fittings

P3003.9
Solvent Cementing of PVC Joints

P3005.2
Cleanouts

P3008.1
Backwater Valves

P3103.1, P3103.2
Vent Terminals

P3201.2
Trap Seal Protection Against Evaporation

CHANGE TYPE: Clarification

P2502.1, P2503.4

Inspection and Tests for Building Sewers

CHANGE SUMMARY: New text clarifies the method for examining existing building sewers and building drains when the entire sanitary drainage system is replaced. Internal examination is required to verify the size, slope, and condition of the existing piping. A new provision prescribes a pressure test for a forced sewer at a test pressure of 5 psi (34.5 kPa) greater than the pump rating.

2015 CODE: P2502.1 Existing Building Sewers and Building Drains. ~~Existing building sewers and drains shall be used in connection with new systems when found by examination and/or test to conform to the requirements prescribed by this document.~~ Where the entire sanitary drainage system of an existing building is replaced, existing building drains under concrete slabs and existing building sewers that will serve the new system shall be internally examined to verify that the piping is sloping in the correct direction, is not broken, is not obstructed and is sized for the drainage load of the new plumbing drainage system to be installed.

P2503.4 Building Sewer Testing. The building sewer shall be tested by insertion of a test plug at the point of connection with the public sewer, ~~and~~ filling the building sewer with water~~, testing with~~ and pressurizing the sewer to not less than ~~a~~ 10-foot (3048-mm) head of water. ~~and be able to maintain such~~ The test pressure shall not decrease during a period of not less than ~~for~~ 15 minutes. The building sewer shall be watertight at all points.

P2502.1, P2503.4 continues

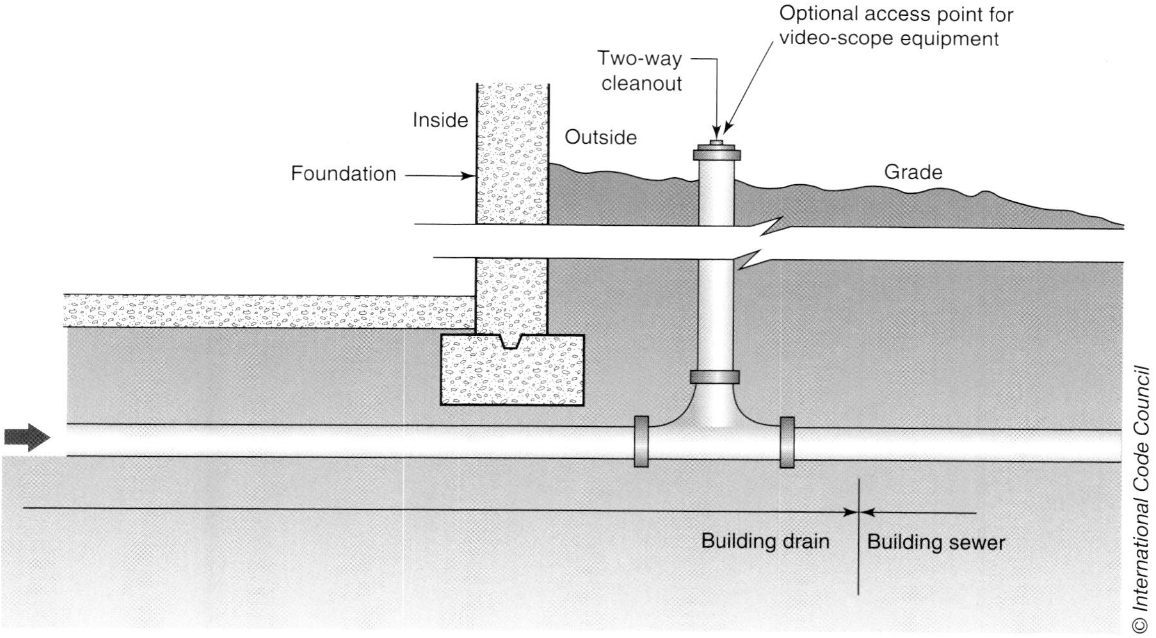

Existing building drains and building sewers require internal examination to verify slope and condition when the entire above-ground sanitary drainage system is replaced.

P2502.1, P2503.4 continued

A forced sewer test shall consist of pressurizing the piping to a pressure of not less than 5 psi (34.5 kPa) greater than the pump rating and maintaining such pressure for not less than 15 minutes. The forced sewer shall be watertight at all points.

CHANGE SIGNIFICANCE: On occasion, an entire plumbing system is replaced except for the below-grade or under-slab sanitary drainage. For example, when a house is substantially damaged by fire, the building may be rebuilt on the existing foundation. In most cases, there is no need to tear out good, serviceable building drains and building sewers for the sake of replacing them with new material. In previous editions of the code, existing sewers and drains required "examination and/or tests" to verify conformance to the code. The language was considered vague and not appropriate for code requirements. The code is now more specific in requiring an internal examination of existing building sewers and building drains when the rest of the sanitary drainage system is replaced. Internal examination is typically accomplished with a video-scope camera without removing or damaging existing piping. The revised provisions specifically state the objective of the internal examination—to verify that the piping is not broken or obstructed, has the proper slope for efficient drainage, and is adequately sized to serve the new sanitary drainage system.

Editorial changes to the first paragraph in Section P2503.4 clarify that the building sewer must be watertight at all points during testing. Although there were proposals to eliminate the 10-foot head pressure and simply test gravity sewers by filling the piping with water, the prevailing consensus was to maintain the existing requirement. The code maintains the 10-foot head pressure test with no drop in pressure for 15 minutes.

A new second paragraph to this section provides criteria for pressure testing of forced sewers. When it is not possible to drain by gravity, a pump is used to force sanitary waste through the piping. The code prescribes pressure testing the piping of forced sewers at a pressure that is 5 psi greater than the pump capacity.

P2503.5

Drain, Waste, and Vent Systems Testing

CHANGE TYPE: Modification

CHANGE SUMMARY: The head pressure for a water test on drain, waste, and vent (DWV) systems has been reduced from 10 feet to 5 feet.

2015 CODE: **P2503.5 ~~DWV~~ Drain, Waste, and Vent Systems Testing.** Rough-in and finished plumbing installations of drain, waste, and vent systems shall be tested in accordance with Sections P2503.5.1 and P2503.5.2.

P2503.5 continues

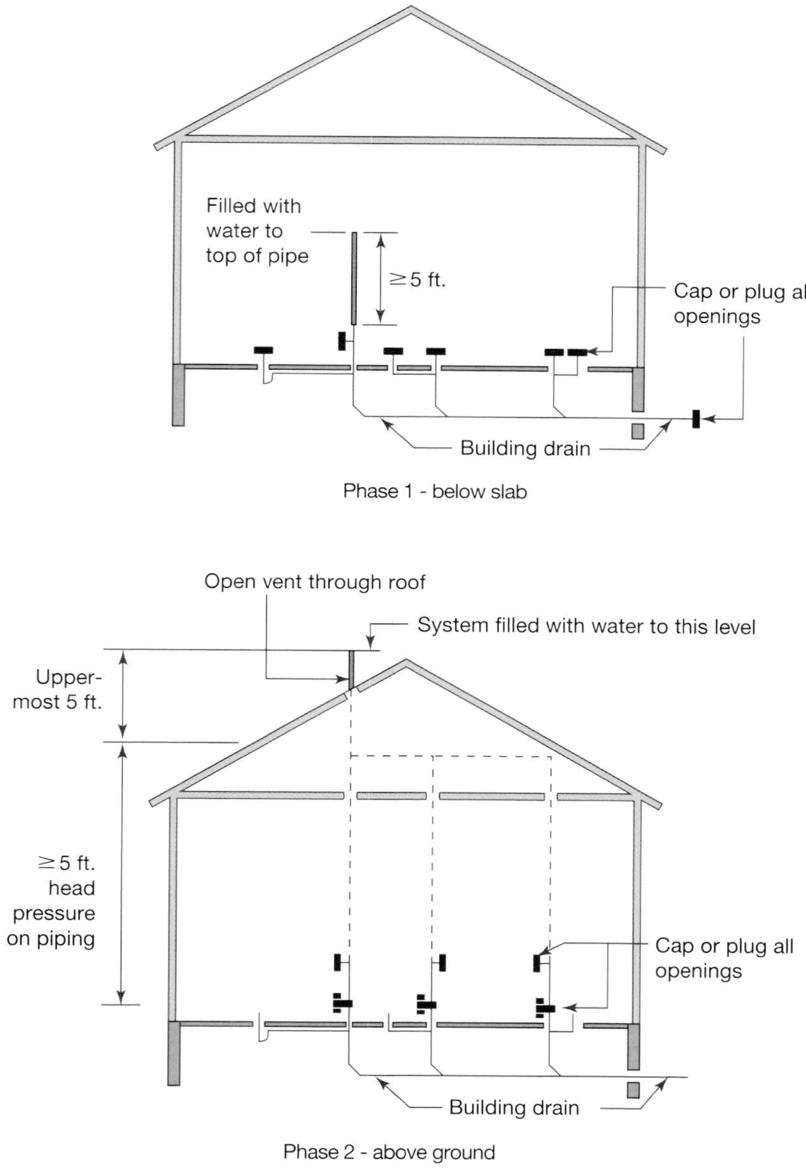

Water test on DWV system

P2503.5 continued

P2503.5.1 Rough Plumbing. DWV systems shall be tested on completion of the rough piping installation by water or for piping systems other than plastic, by air with no evidence of leakage. Either test shall be applied to the drainage system in its entirety or in sections after rough piping has been installed, as follows:

1. Water test. Each section shall be filled with water to a point not less than ~~10~~ 5 feet (~~3048~~ 1524 mm) above the highest fitting connection in that section, or to the highest point in the completed system. Water shall be held in the section under test for a period of 15 minutes. The system shall prove leak free by visual inspection.
2. Air test. The portion under test shall be maintained at a gauge pressure of 5 pounds per square inch (psi) (34 kPa) or 10 inches of mercury column (34 kPa). This pressure shall be held without introduction of additional air for a period of 15 minutes.

P2503.5.2 Finished Plumbing. (*No changes to text.*)

CHANGE SIGNIFICANCE: The code has historically required a 10-foot head pressure for testing drain, waste, and vent (DWV) systems with water. The DWV system is filled with water to a point 10 feet higher than the piping being tested and the piping and joints are visually inspected for any leaks that might develop. The duration of the water test is 15 minutes to ensure that the system is watertight. The top 10 feet of the DWV system, which is typically the highest vent through the roof, is only filled with water to the top of the vent terminal. Adding an additional 10-foot standpipe above the vent terminal would not be easily accomplished and would not provide any benefit because the vent will not carry water and not be under pressure in service.

The 2015 IRC reduces the water head test height from 10 feet to 5 feet. Although the 10-foot head pressure is a long-standing tradition, proponents of this change stated that the actual head pressure is not nearly as critical as the visual nature of the test. They reasoned that a 10-foot (4.34-psi) head test is unlikely to reveal any leaks or defects that would not be detected by a 5-foot (2.17-psi) head test. Testimony offered that some jurisdictions, including the State of Florida, have previously amended the code in favor of the 5-foot head test. Lowering the fill stack to 5 feet enables both the installer and the inspector to visually observe the water level inside the pipe during testing without the use of a ladder.

P2603.2.1

Protection Against Physical Damage

CHANGE TYPE: Modification

CHANGE SUMMARY: For piping installed through bored holes or in notches, the minimum clearance distance from the concealed piping to the edge of the framing member has been reduced from 1½ inches to 1¼ inches. Protection is required for piping installed less than 1¼ inches from the edge of the framing member.

2015 CODE: P2603.2.1 Protection Against Physical Damage. In concealed locations, where piping, other than cast-iron or galvanized steel, is installed through holes or notches in studs, joists, rafters or similar members less than ~~1½~~ 1¼ inches (~~38~~ 31.8 mm) from the nearest edge of the member, the pipe shall be protected by steel shield plates. Such shield plates shall have a thickness of not less than 0.0575 inch (1.463 mm) (No. 16 gage). Such plates shall cover the area of the pipe where the member is notched or bored, and shall extend not less than 2 inches (51 mm) above sole plates and below top plates.

CHANGE SIGNIFICANCE: Plumbing piping other than cast iron or galvanized steel installed through holes in framing members is subject to punctures from fasteners of sheathing or finish materials unless the piping

P2603.2.1 continues

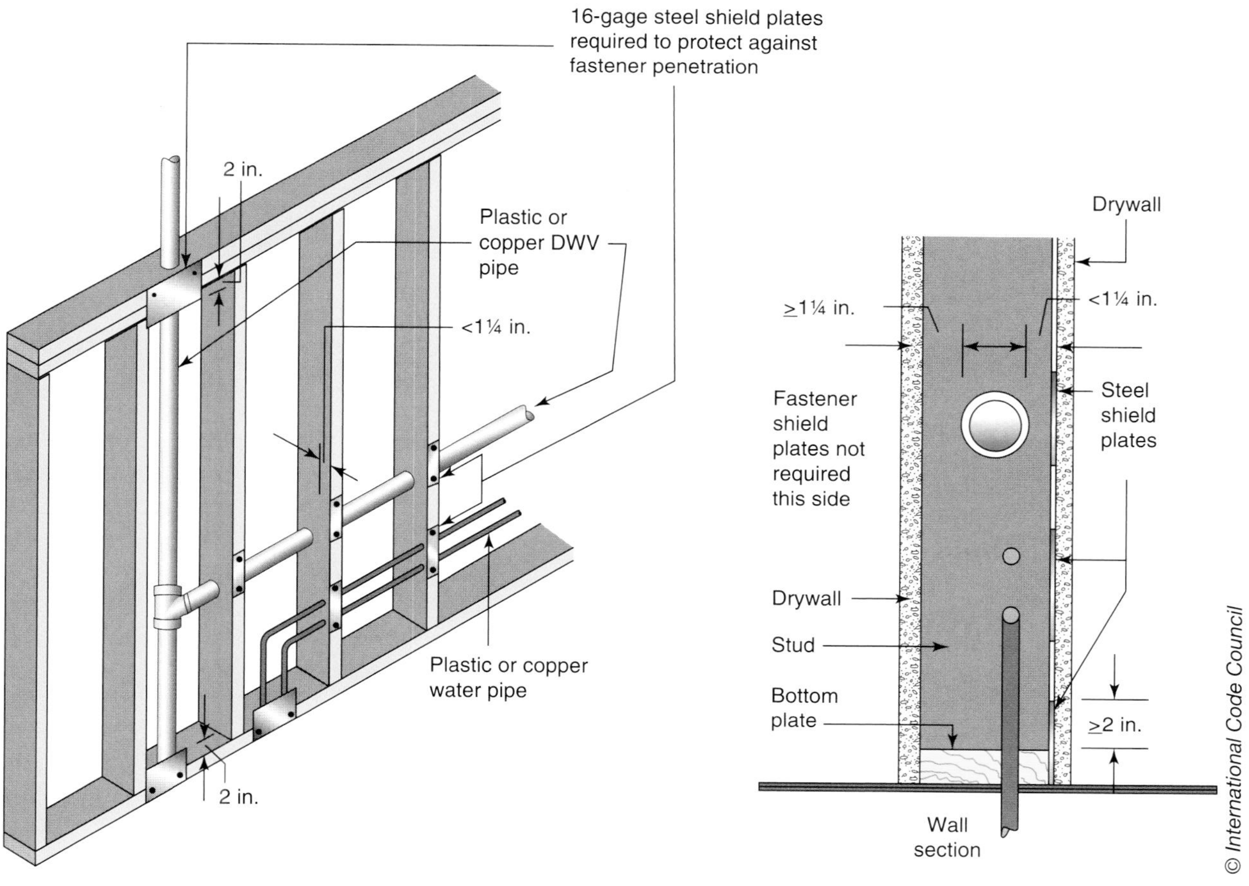

Physical protection of concealed piping

P2603.2.1 continued

is placed a sufficient distance away from the face of the member or protection is provided with steel shield plates applied to the face of the framing member. Similar rules exist for gas piping, hydronic heating and cooling piping, gas vents, clothes dryer ducts, and electrical wiring subject to damage from fasteners. Previously, 1½ inches was considered the safe distance between plumbing piping and the face of a stud, joist or rafter. Piping installed through holes or notches and less than 1½ inches from the edge of the member required protection by steel shield plates. This provision effectively required shield plates for all copper and plastic plumbing piping installed through holes in 2 × 4 plates and studs in conventional wall construction. The actual dimensions of a nominal 2 × 4 are 1½ inches by 3½ inches. Even a ½-inch-diameter (⅝-inch-O.D.) pipe centered in a 2 × 4 stud wall is slightly less than 1½ inches from both edges of the stud or plate and would require a shield plate installed on both sides of the framing member before the application of drywall or sheathing. In order to permit the installation of ½-inch and ¾-inch piping in a 2 × 4 frame wall without steel plate protection, the minimum clearance distance has been reduced to 1¼ inches. This distance is consistent with NFPA 70 *National Electrical Code* (NEC) provisions for the installation of the Type NM nonmetallic cable that is common in residential construction.

CHANGE TYPE: Modification

P2603.3

Protection Against Corrosion

CHANGE SUMMARY: The minimum thickness of sheathing material for protection of piping against corrosion has been reduced from 0.025 inches to 0.008 inches (8 mil). The corrosion protection requirement applies to metallic piping other than cast iron, ductile iron, and galvanized steel that is in direct contact with concrete, masonry or steel framing. Previously, protection was only required for materials passing through walls and floors of these materials. All metallic piping requires corrosion protection when located in corrosive soils.

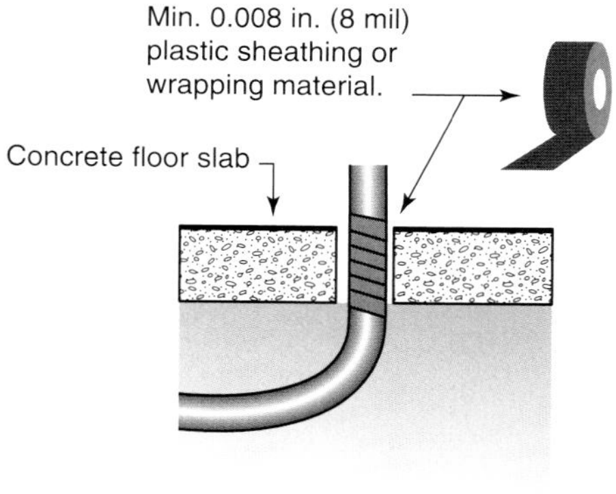

Copper and copper alloy pipe and tubing require protection from corrosion when in contact with concrete, masonry, and steel framing.

2015 CODE: ~~**P2603.3 Breakage and Corrosion.** Pipes passing through concrete or cinder walls and floors, cold-formed steel framing or other corrosive material shall be protected against external corrosion by a protective sheathing or wrapping or other means that will withstand any reaction from lime and acid of concrete, cinder or other corrosive material. Sheathing or wrapping shall allow for movement including expansion and contraction of piping. The wall thickness of material shall be not less than 0.025 inch (0.64 mm).~~

P2603.3 Protection Against Corrosion. Metallic piping, except for cast iron, ductile iron and galvanized steel, shall not be placed in direct contact with steel framing members, concrete or masonry. Metallic piping shall not be placed in direct contact with corrosive soil. Where sheathing is used to prevent direct contact, the sheathing material thickness shall be not less than 0.008 inch (8 mil) (0.203 mm) and shall be made of plastic. Where sheathing protects piping that penetrates concrete or masonry walls or floors, the sheathing shall be installed in a manner that allows movement of the piping within the sheathing.

CHANGE SIGNIFICANCE: The intent of Section P2603.3 is to protect metallic piping from exterior corrosion caused by contact with corrosive materials. Previously, the code required that sheathing or wrapping material used to protect the piping be at least 0.025 inches thick. The proponent of this change submitted that material of this thickness is not commonly stocked by supply houses and is not being installed or required in the field. Much thinner plastic sheathing materials have been used across the country for decades without any reported adverse effects. Cast iron and ductile iron manufacturers recommend for corrosive soil conditions the use of either 0.008-inch-thick low-density polyethylene sheathing or 0.004-inch-thick high-strength cross-laminated polyethylene sheathing. For small metallic pipes such as copper tubing (½ to 1¼ inches) passing through concrete or masonry, plumbing supply houses normally stock 0.004- and 0.006-inch-thick low-density "flat tube" plastic sheathing materials, and that is what is being used in the field. To conform to the most stringent of the recommendations for sheathing materials, the code now prescribes a minimum thickness of 0.008-inch (8-mil) material. The new wording may also change the scope somewhat by requiring corrosion protection for the applicable types of metallic piping (typically copper piping and tubing) that come in contact with concrete, masonry, and steel framing. Previously the code regulated only piping passing

P2603.3 continues

P2603.3 continued

through walls and floors of these materials. Concern was expressed that this language may suggest that some types of metallic piping and tubing may require wrap protection when fastened to the surface of a concrete or masonry foundation wall. Although corrosion protection is not a concern for cast iron, ductile iron, and galvanized steel in contact with masonry, concrete, and steel framing, all metallic piping must be protected from corrosive soils.

This change to the code also intends to clarify the intent for allowing movement of piping that has been wrapped. The previous language "Sheathing or wrapping shall allow for movement including expansion and contraction of piping" was not clear to many code users. Consensus indicates that sheathing or wrapping that protects a pipe passing through concrete or masonry, such as a pipe below a slab coming up through and cast in the slab, should allow for some "give" between the pipe and the concrete or masonry. The new text clarifies the meaning by stating that sheathing shall be installed in a manner that allows movement of the piping within the sheathing for pipes that pass through concrete or masonry.

CHANGE TYPE: Modification

CHANGE SUMMARY: Support spacing requirements for PEX and PE-RT tubing 1¼ inches and greater in diameter have been added to the table. Footnote b of Table P2605.1 clarifies the mid-story guide requirements for some types of vertical pipe 2 inches and smaller in diameter.

Table P2605.1

Piping Support

2015 CODE:

TABLE P2605.1 Piping Support

Piping Material	Maximum Horizontal Spacing (feet)	Maximum Vertical Spacing (feet)
~~Brass Pipe~~	~~10~~	~~10~~
Cross-linked polyethylene (PEX) pipe, 1 inch and smaller	2.67 (32 inches)	10[b]
Cross-linked polyethylene (PEX) pipe, 1¼ inch and larger	4	10[b]
Polyethylene of Raised Temperature (PE-RT) pipe, 1 inch and smaller	2.67 (32 inches)	10[b]
Polyethylene of Raised Temperature (PE-RT) pipe, 1¼ inch and larger	4	10[b]

(Portions of table not shown remain unchanged.)

a. *(No change to text.)*

b. ~~Mid-story guide~~ For sizes 2 inches and smaller, a guide shall be installed midway between required vertical supports. Such guides shall prevent pipe movement in a direction perpendicular to the axis of the pipe.

CHANGE SIGNIFICANCE: Cross-linked polyethylene (PEX) and polyethylene of raised temperature (PE-RT) tubing is being made in larger diameters that are stiffer and require less support. Table P2605.1 now includes support spacing for these materials in pipe sizes 1¼ inches and larger. Horizontal spacing for the larger-diameter piping is 4 feet compared to 2.67 feet for piping 1 inch or less in diameter.

Mid-story guides are required for vertical smaller-diameter flexible piping to restrict the movement of the pipe. When installed vertically, the various types of plastic piping can bow out of line. This side-to-side movement, either parallel or perpendicular to the plane of the wall, typically occurs when the pipe is filled with water (water distribution piping) or is subjected to hot water flow (drainage piping). The bowed piping can come in contact with adjacent piping, fittings or other objects (such as wall coverings) that might cause noise or damage to the piping. If the piping was allowed to bow unrestricted, high bending stresses could occur at the required vertical pipe supports. As a guide allows movement of the pipe in a direction parallel to the pipe axis, the guide is not required to firmly grip the pipe or absolutely prevent any movement whatsoever. For example, in wood frame construction, a mid-story pipe guide could be a horizontal block of wood (between studs) that has a clearance hole for the pipe to pass through. The revision to footnote b better defines mid-story guides and clarifies their purpose to restrain vertical piping from moving sideways at the midpoint between required vertical supports. The guides are required for all types of plastic piping (i.e., PEX, PEX-AL-PEX, PE-RT, ABS, CPVC, PVC, and PP) 2 inches and smaller in diameter.

Table P2605.1 continues

Table P2605.1 continued Brass and bronze are copper alloys and are covered under the copper and copper alloys listed in Table P2605.1. Therefore, brass pipe has been deleted from the table.

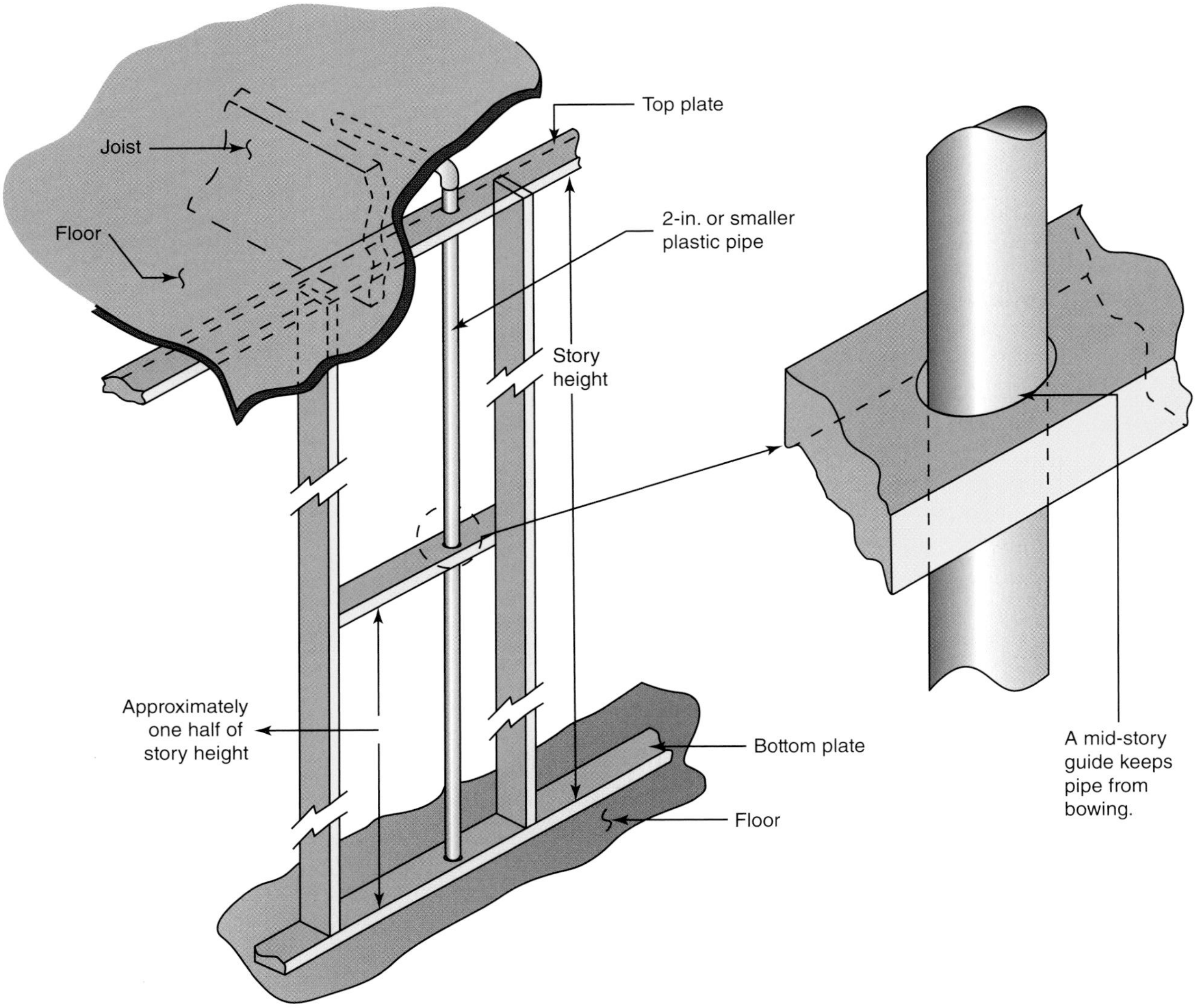

A guide is required at the midpoint between supports for vertical plastic piping 2 inches and smaller.

CHANGE TYPE: Modification

CHANGE SUMMARY: A definition of waste receptor has been added to the code. Waste receptors are now permitted in bathrooms and closets.

P2702.1, P2706.1

Waste Receptors

2015 CODE:

SECTION R202
DEFINITIONS

WASTE RECEPTOR. A floor sink, standpipe, hub drain or a floor drain that receives the discharge of one or more indirect waste pipes.

P2702.1 Plumbing Fixtures. Plumbing fixtures, other than water closets, shall be provided with approved strainers.

> **Exception:** Hub drains receiving only clear water waste and standpipes shall not require strainers.

P2706.1 General. ~~Waste receptors shall be of an approved type. Plumbing fixtures or other receptors receiving the discharge of indirect waste pipes shall be shaped and have a capacity to prevent splashing or flooding and shall be readily accessible for inspection and cleaning. Waste~~

P2702.1, P2706.1 continues

Typical laundry standpipe is an example of a waste receptor.

P2702.1, P2706.1 continued

~~receptors and standpipes shall be trapped and vented and shall connect to the building drainage system.~~ For other than hub drains that receive only clear-water waste and standpipes, ~~A~~a removable strainer or basket shall cover the waste outlet of waste receptors. Waste receptors shall not be installed in ~~ventilated~~ concealed spaces. Waste receptors shall not be installed in ~~bathrooms~~ plenums, attics, crawl spaces, or interstitial spaces above ceilings and below floors. ~~or in any inaccessible or unventilated space such as a closet. Ready access shall be provided to w~~Waste receptors shall be readily accessible.

~~**Exceptions:**~~

~~1. Open hub waste receptors shall be permitted in the form of a hub or pipe extending not less than 1 inch (25 mm) above a water-impervious floor, and are not required to have a strainer.~~

~~2. Clothes washer standpipes shall not be prohibited in bathrooms.~~

P2706.1.1 Hub Drains. Hub drains shall be in the form of a hub or a pipe that extends not less than 1 inch (25mm) above a water-impervious floor.

P2706.1.2 Standpipes. Standpipes shall extend not less than ~~of~~ 18 inches (457 mm) and ~~but~~ not greater than 42 inches (1067 mm) above the trap weir. ~~Access shall be provided to standpipe traps and drains for rodding.~~

P2706.1.2.1 Laundry Tray Connection to Standpipe. Where ~~A~~ a laundry tray waste line ~~is permitted to~~ connects into a standpipe for ~~the~~ an automatic clothes washer drain~~.~~, ~~T~~the standpipe shall extend not less than 30 inches (762 mm) above the standpipe trap weir and shall extend above the flood level rim of the laundry tray. The outlet of the laundry tray shall not be greater than 30 inches (762 mm) horizontally ~~distance~~ from the standpipe trap.

CHANGE SIGNIFICANCE: A definition for "waste receptor" has been added to Chapter 2 to clarify the meaning and give clear direction to the code user. The definition includes only four items—a floor sink, standpipe, hub drain or a floor drain that receives the discharge of one or more indirect waste pipes. Because they are clearly defined, waste receptors do not require approval by the building official. Floor sinks and floor drains are required to comply with standards. Standpipes and hub drains have specific code requirements. Any other receptor that the designer or installer wants to use will have to be approved under Section R104.11 for alternate materials, methods, and equipment.

As defined in Section P2706.1.1, a hub drain is simply a pipe hub or a pipe that extends at least 1 inch above a water-impervious floor, such as concrete. Hub drains that receive only clear water waste and standpipes do not require strainers. There is a low probability that solids will enter these receptors and strainers are not needed. The prohibition against locating waste receptors in bathrooms or closets was deleted. This change recognizes that floor drains, floor sinks or hub drains may be located in closets or bathrooms to receive the condensate from air-conditioning units or the discharge from water heater pan drains or temperature and pressure (T&P) relief valves. Standpipes have specifically been permitted in bathrooms beginning with the 2012 IRC.

The first three sentences of Section P2706.1 have been deleted because they are redundant. Section P2601.2 already covers where waste receptors must be connected, and Section P3201.6 covers the requirement for traps for each fixture. Reference to inaccessible spaces was deleted because all waste receptors must be readily accessible. The term "readily accessible" as defined in Chapter 2 means that access can be gained without the removal of a panel or obstruction. The reference to an unventilated space was unclear and has been deleted.

P2717

Dishwashing Machines

CHANGE TYPE: Modification

CHANGE SUMMARY: The code now references the applicable standards for integral air gaps protecting the potable water supply to dishwashers. The term "food waste disposer" replaces "food waste grinder." Sections P2717.2 and P2717.3 regarding dishwasher discharge to the sink tailpiece or the food waste disposer have been combined into a single Section P2717.2, eliminating redundant language and improving understanding of the provisions.

2015 CODE:

SECTION P2717
DISHWASHING MACHINES

P2717.1 Protection of Water Supply. The water supply ~~for~~ <u>to a</u> dishwasher~~s~~ shall be protected against backflow by an air gap <u>complying with ASME A112.1.3 or A112.1.2 that is installed integrally within the machine</u> or <u>a</u> ~~integral~~ backflow preventer <u>in accordance with Section P2902</u>.

P2717.2 Sink and Dishwasher. ~~A sink and dishwasher are permitted to discharge through a single 1½-inch (38 mm) trap. The discharge pipe from the dishwasher shall be increased to not less than 3/4 inch (19 mm) in diameter and shall be connected with a wye fitting to the sink tailpiece.~~ <u>The combined discharge from a dishwasher and a one- or two-compartment sink, with or without a food waste disposer, shall be served by a trap of not less than 1½ inches (38 mm) in outside diameter.</u> The dishwasher ~~waste line~~ <u>discharge pipe or tubing</u> shall rise <u>to the underside of the counter</u> and be ~~securely~~ fastened <u>or otherwise held in that position</u> before connecting to <u>the head of the food-waste disposer or to a wye fitting in</u> the sink tailpiece.

~~**P2717.3 Sink, Dishwasher and Food Grinder.** The combined discharge from a sink, dishwasher, and waste grinder is permitted to discharge through a single 1 1/2-inch (38 mm) trap. The discharge pipe~~

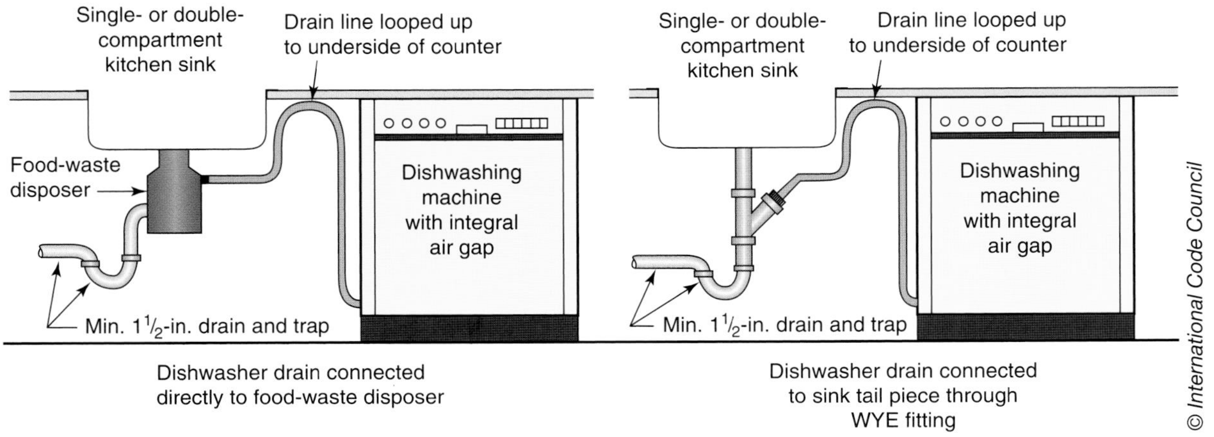

Dishwasher drain

~~from the dishwasher shall be increased to not less than 3/4 inch (19 mm) in diameter and shall connect with a wye fitting between the discharge of the food-waste grinder and the trap inlet or to the head of the food grinder. The dishwasher waste line shall rise and be securely fastened to the underside of the counter before connecting to the sink tail piece or the food grinder.~~

CHANGE SIGNIFICANCE: Revised Section P2717 adds references to applicable standards for the integral backflow protection of the potable water supply serving dishwashing machines. The requirement for dishwashing machines to comply with ASSE 1006 (covering the requirement for an internal air gap on the water supply) was removed from the 2012 code because the standard was withdrawn by ASSE and the machine manufacturers were no longer indicating compliance with that standard. To provide inspectors with a means to verify that dishwashers have integral backflow protection, the 2015 IRC requires compliance to ASME A112.1.3 or A112.1.2. The standards identify methods of providing protection against back-siphonage through means of an air gap and establish physical requirements and methods of testing air gaps. The inspector is now able to identify those standard numbers on either the machine or in the literature for the machine to verify compliance with the code.

The text regarding connection of the dishwasher discharge to either the kitchen sink tailpiece or the food waste disposer has been consolidated into one section, with mostly editorial revisions to clarify the requirements. The provision allowing the sink, dishwasher, and food waste disposer (if one is provided) to drain through a single 1½-inch trap has been retained. The kitchen sink is more precisely defined as a one- or two-compartment sink, as the code has always intended, to ensure there is no misunderstanding that the provision might be limited to a single-compartment sink. The text setting the size of the dishwasher discharge line has been removed because machines are manufactured with different discharge diameters and configurations, and the applicable standards do not provide clear guidance on the discharge connection to the sink tailpiece or food waste disposer.

For some time the code has required the dishwasher discharge hose to loop up high in the under-counter space before connecting to the sink tailpiece or disposer. This configuration prevents the discharge contents, sink backup, and contaminants from flowing back to the dishwashing machine. The language in the 2012 edition of the IRC required a secure connection to the underside of the counter for this discharge loop. Following this requirement to the letter is not always easily accomplished or even feasible. The underside of the counter is oftentimes difficult to reach through small spaces around the kitchen sink, and counters made of granite and similar materials are not suitable for fastening to. The new language simply requires the loop to be held in position and serves the intent of the code without requiring fastening.

P2801

Water Heater Drain Valves and Pans

CHANGE TYPE: Modification

CHANGE SUMMARY: The code now specifically requires drain valves with a threaded outlet for water heaters. The water heater pan requirements have been expanded to accept aluminum and plastic pans of the prescribed thickness. The code clarifies that a pan drain is not required when a water heater is replaced and there is no existing drain.

2015 CODE: P2801.1 Required. ~~Each dwelling shall have an approved automatic water heater or other type of domestic water-heating system sufficient to supply h~~Hot water shall be supplied to plumbing fixtures and plumbing appliances intended for bathing, washing or culinary purposes. ~~Storage tanks shall be constructed of noncorrosive metal or shall be lined with noncorrosive material.~~

P2801.2 Drain Valves. Drain valves for emptying shall be installed at the bottom of each tank-type water heater and hot water storage tank. The drain valve inlet shall be not less than ¾ inch nominal iron pipe size and the outlet shall be provided with a male hose thread.

~~P2801.5~~ P2801.6 Required Pan. Where a storage tank-type water heater or a hot water storage tank is installed in a location where water leakage from the tank will cause damage, the tank shall be installed in a pan constructed of one of the following:~~.~~

1. Galvanized steel or aluminum ~~pan having a material thickness~~ of not less than 0.0236 inch (0.6010 mm) ~~(No. 24 gage)~~ in thickness.
2. Plastic not less than 0.036 inch (0.9 mm) in thickness.
3. Other ~~pans~~ approved materials ~~for such use~~.

~~Listed pans shall comply with CSA LC3.~~ A plastic pan shall not be installed beneath a gas-fired water heater.

~~P2801.5.1~~ P2801.6.1 Pan Size and Drain. The pan shall be not less than 1½ inches (38 mm) deep and shall be of sufficient size and shape to receive all dripping or condensate from the tank or water heater. The pan shall be drained by an indirect waste pipe of not less than ¾ inch (19 mm) diameter. Piping for safety pan drains shall be of those materials listed in Table ~~P2905.5~~ P2906.5. Where a pan drain was not previously installed, a pan drain shall not be required for a replacement water heater installation.

CHANGE SIGNIFICANCE: Previous editions of the code did not specifically require drain valves on water heaters, although manufacturers do provide such drains on storage-tank-type water heaters. The *International Plumbing Code* (IPC) has required a drain valve at the bottom of each tank-type water heater and hot water storage tank, but stated that the drain valve had to comply with a referenced standard. The standard has been discontinued and has been removed from the 2015 IPC. Both codes

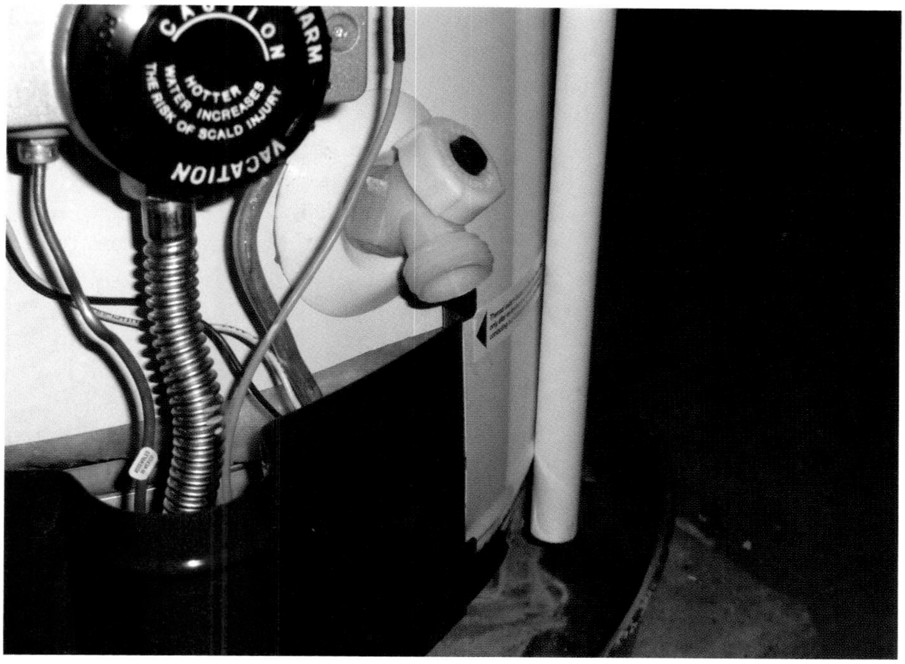

© International Code Council

Water heater drain valve with threaded outlet and drain pan

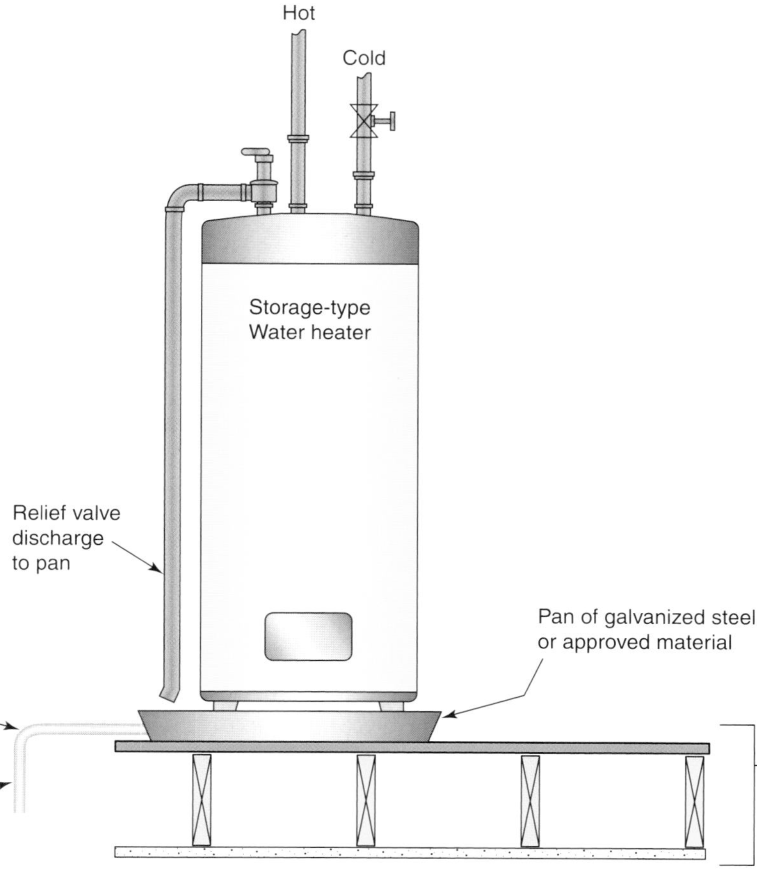

© International Code Council

Water heater pan installation

P2801 continues

P2801 continued

now prescribe drain valves with inlets of not less than ¾-inch nominal iron pipe size and outlets provided with male hose threads to connect a garden hose. Water heater drain valves are necessary to drain water heaters for replacement or maintenance.

Safety pans are required under storage-tank-type water heaters if a leak in the tank will cause damage. Previously, the code prescribed 24-gage galvanized steel pans or other pans approved by the building official. Aluminum and plastic water heater pans are common in the marketplace and are installed all across the United States. The intent of the expanded language is to recognize manufactured products that are widely accepted in the marketplace. In addition, the new text sets appropriate minimum thicknesses for aluminum and plastic pans that are considered sufficiently durable for the use. The provision for listed pans to comply with CSA LC3 has been deleted because there is not any pan produced in the United States that complies with that standard. CSA withdrew the standard in November 2011. A new provision prohibits the use of plastic pans under gas-fired water heaters. Although there was no testimony indicating there had been problems with such installations, there was a concern that radiant heat coming from the bottom of a gas-fired water heater could make a plastic pan more susceptible to puncturing.

A replacement water heater must be installed to the current code, the same as any new installation. If the water heater is in a location where leakage will cause damage, a safety pan is also required even if the original installation did not require or have a safety pan. It is typically not a problem to install a replacement water heater with a pan, but the obstacles to installing a drain for that pan can be significant. Many times, there is no feasible way to provide for a suitable disposal point for the pan drain. The new language provides an exception for replacement water heaters that waives the pan drain requirements. Consensus was that a pan with no drain is better than no pan at all. If the water heater tank begins to leak and there is a pan present, the occupant may notice water in the pan and realize that it is not a normal condition. This is opposed to a situation where there is not a pan and the leaking water flows to an unobservable location and does so for a long time, creating damage and mold issues. A pan without a drain will provide a contained area that might allow more time for the leak to be detected. An optional water alarm sensor also can be installed in the pan without a drain to alert the occupant to an accumulation of leaking water.

P2804.6.1

Water Heater Relief Valve Discharge Piping

CHANGE TYPE: Modification

CHANGE SUMMARY: The temperature and pressure (T&P) relief valve discharge pipe termination must have an air gap suitable to protect the potable water supply distribution system of the building. PEX and PE-RT tubing used for relief valve discharge piping must be one size larger than the T&P valve discharge outlet, and the outlet end of the tubing must be fastened in place.

2015 CODE: ~~P2803.6.1~~ P2804.6.1 Requirements for Discharge Piping. The discharge piping serving a pressure relief valve, temperature relief valve or combination valve shall:

(*Items 1–9 remain unchanged.*)

10. ~~Not~~ ~~t~~Terminate not more than 6 inches (152 mm) and not less than two times the discharge pipe diameter above the floor or waste receptor flood level rim.

(*Items 11–13 remain unchanged.*)

14. Be one nominal size larger than the size of the relief valve outlet, where the relief valve discharge piping is constructed of PEX or PE-RT tubing. The outlet end of such tubing shall be fastened in place.

CHANGE SIGNIFICANCE: The outlet of a temperature and pressure (T&P) relief valve is considered a potable water outlet that must be protected against backflow conditions. Item 2 of Section P2804.6.1 has always required the T&P piping to discharge through an air gap in the same room as the water heater, but the code has not specifically addressed the minimum dimension for the air gap in the water heater section. The revised text in Item 10 in the list of requirements now prescribes a minimum air gap of two times the discharge pipe diameter to provide the appropriate backflow protection.

P2804.6.1 continues

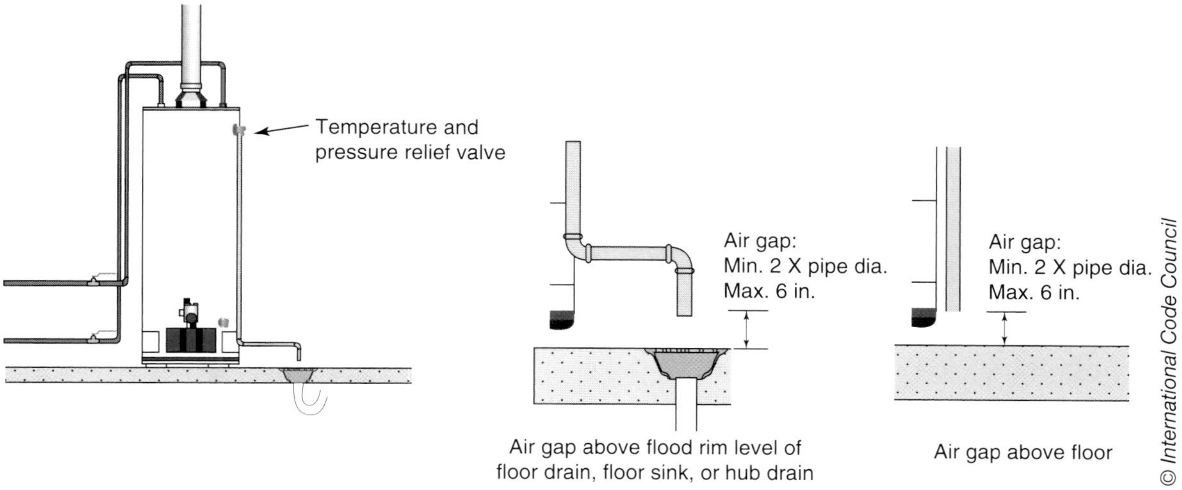

Air gap for temperature and pressure (T&P) relief valve discharge pipe

P2804.6.1 continued

Item 13 of Section P2804.6.1 requires the discharge piping serving the relief valve to be constructed of materials listed in Section P2906.5 and Table P2906.5, Water Distribution Pipe, or materials tested, rated, and approved for such use in accordance with ASME A112.4.1. There are many materials approved for use as water distribution pipe, and any of these can be used for the water heater discharge pipe. Cross-linked polyethylene (PEX) plastic tubing and polyethylene of raised temperature (PE-RT) plastic tubing are examples of approved piping. Some in the industry have been concerned that these two materials use fittings that reduce the inside diameter to less than the nominal pipe size. Item 3 of this section requires that the discharge piping must not be smaller than the diameter of the outlet of the water heater T&P relief valve. In most cases, this outlet is ¾-inch nominal pipe size. PEX and PE-RT tubing use insert fittings for connections. The bore size for a ¾-inch male adapter fitting reduces the internal diameter of the pipe, and there is concern that the discharge from a T&P valve could be restricted and be a safety concern. The new language requires that PEX and PE-RT tubing used for relief valve discharge piping be one size larger so that the insert fitting has a larger bore and does not cause a safety concern. For a typical discharge outlet of ¾-inch diameter, 1-inch PEX or PE-RT tubing would be required.

The other concern regarding PEX and PE-RT tubing is that the material is very flexible and where supplied from a coil, the tubing has a "memory" to return to a coil shape. This flexibility and memory to stay in a coil shape can present installation problems including keeping the discharge end of the tubing in its proper location. Therefore, new language is added to require that the outlet end of the tubing be fastened in place.

P2901, P2910 through P2913

Nonpotable Water Systems

CHANGE TYPE: Modification

CHANGE SUMMARY: Nonpotable water outlets, such as hose connections, that utilize nonpotable water must be identified with a warning and a symbol that nonpotable water is being used. The color purple is established for identifying distribution piping conveying nonpotable water. New Sections P2910 through P2913 are extracted from the *International Green Construction Code* (IgCC) and intend to provide guidance on the collection, storage, and distribution of various types of nonpotable water for residential buildings.

P2901, P2910 through P2913 continues

Nonpotable water is utilized for ________.
CAUTION: NONPOTABLE WATER. DO NOT DRINK
© International Code Council

Nonpotable water outlets such as hose connections require warning signs with a pictograph.

P2901, P2910 through P2913 continued

2015 CODE:

SECTION R202
DEFINITIONS

© International Code Council

Purple piping is required for nonpotable water distribution.

RECLAIMED WATER. Nonpotable water that has been derived from the treatment of wastewater by a facility or system licensed or permitted to produce water meeting the jurisdiction's water requirements for its intended uses. Also known as "Recycled Water."

ON-SITE NONPOTABLE WATER REUSE SYSTEMS. Water systems for the collection, treatment, storage, distribution, and reuse of nonpotable water generated on-site, including but not limited to gray water systems. This definition does not include rainwater harvesting systems.

P2901.1 Potable Water Required. ~~Dwelling units shall be supplied with potable water in the amounts and pressures specified in this chapter. Where a nonpotable water-distribution system is installed, the nonpotable system shall be identified by color marking, metal tags or other appropriate method. Where color is used for marking, purple shall be used to identify municipally reclaimed water, rainwater and graywater distribution systems. Nonpotable outlets that could inadvertently be used for drinking or domestic purposes shall be posted.~~ Potable water shall be supplied to plumbing fixtures and plumbing appliances except where treated rainwater, treated gray water or municipal reclaimed water is supplied to water closets, urinals and trap primers. The requirements of this section shall not be construed to require signage for water closets and urinals.

P2901.2 Identification of Nonpotable Water Systems. Where nonpotable water systems are installed, the piping conveying the nonpotable water shall be identified either by color marking, metal tags or tape in accordance with Sections P2901.2.1 through P2901.2.2.3.

P2901.2.1 Signage Required. Nonpotable water outlets such as hose connections, open-ended pipes and faucets shall be identified with signage that reads as follows: "Non-potable water is utilized for [application name]. CAUTION: NON-POTABLE WATER. DO NOT DRINK." The words shall be legibly and indelibly printed on a tag or sign constructed of corrosion-resistant waterproof material or shall be indelibly printed on the fixture. The letters of the words shall be not less than 0.5 inches (12.7 mm) in height and in colors in contrast to the background on which they are applied. In addition to the required wordage, the pictograph shown in Figure P2901.2.1 shall appear on the required signage.

P2901.2.2 Distribution Pipe Labeling and Marking. Nonpotable distribution piping shall be of purple in color and shall be embossed or integrally stamped or marked with the words: "CAUTION: NONPOTABLE WATER—DO NOT DRINK" or the piping shall be installed with a purple identification tape or wrap. Pipe identification shall include the contents of the piping system and an arrow indicating the direction of flow. Hazardous piping systems shall also contain information addressing the nature of the hazard. Pipe identification shall be repeated at intervals

TABLE P2901.2.2.2 Size of Pipe Identification

Pipe Diameter (inches)	Length Background Color Field (inches)	Size of Letters (inches)
¾ to 1¼	8	0.5
1½ to 2	8	0.75
2½ to 6	12	1.25
8 to 10	24	2.5
over 10	32	3.5

For SI: 1 inch = 25.4 mm.

not exceeding 25 feet (7620 mm) and at each point where the piping passes through a wall, floor or roof. Lettering shall be readily observable within the room or space where the piping is located.

P2901.2.2.1 Color. The color of the pipe identification shall be discernable and consistent throughout the building. The color purple shall be used to identify reclaimed, rain and gray water distribution systems.

P2901.2.2.2 Lettering Size. The size of the background color field and lettering shall comply with Table P2901.2.2.2.

P2901.2.2.3 Identification Tape. Where used, identification tape shall be not less than 3 inches (76.2 mm) wide and have white or black lettering on a purple field stating "CAUTION: NONPOTABLE WATER—DO NOT DRINK." Identification tape shall be installed on top of nonpotable rainwater distribution pipes and fastened not greater than every 10 feet (3048 mm) to each pipe length, and run continuously the entire length of the pipe.

SECTION P2910
NONPOTABLE WATER SYSTEMS

SECTION P2911
ON-SITE NONPOTABLE WATER REUSE SYSTEMS

P2911.6.1 Gray Water Used for Fixture Flushing. Gray water used for flushing water closets and urinals shall be disinfected and treated by an on-site water reuse treatment system complying with NSF 350.

SECTION P2912
NONPOTABLE RAINWATER COLLECTION AND DISTRIBUTION SYSTEMS

SECTION P2913
RECLAIMED WATER SYSTEMS

(The text of new Sections P2910 through P2913 is too extensive to be included in this publication. Please refer to the 2015 IRC for the complete code text.)

P2901, P2910 through P2913 continues

P2901, P2910 through P2913 continued

CHANGE SIGNIFICANCE: New provisions in the IRC for collecting, storing, and using various types of nonpotable water recognize the growing need for water conservation and the increase in the development of water conservation programs in many regions of the United States. The 2012 IRC introduced gray water recycling systems into the body of the code. Gray water recycling systems conserve water by collecting and using the discharge of lavatories, bathtubs, showers, clothes washers, and laundry trays for flushing water closets and for subsurface landscape irrigation. Water conservation practices create a need to identify various alternate sources of water, and the code now is much broader in its scope of nonpotable water systems. The intent of new Sections P2910 through P2913 is to provide guidance on the collection, storage, identification, and distribution of nonpotable water, including rainwater, reclaimed water, and on-site nonpotable water reuse, for designers or builders who choose to utilize such systems. The provisions are extracted from the *International Green Construction Code* (IgCC).

Section P2901.2 more precisely describes identification requirements of nonpotable water systems to prevent cross-contamination with potable water and to adequately caution building occupants that the water is nonpotable and has specific limited uses. The 2012 IRC did require reclaimed, rain and gray water piping to be identified with the color purple. However, because there are other alternate sources of water that need identification to protect the safety of the public, the code now includes all types of nonpotable water. The basis for the new language is text from the IgCC and is written to be in alignment with the IgCC requirements. Nonpotable water distribution piping must be purple in color or be labeled in accordance with the code provisions. Nonpotable water outlets, such as hose connections, that utilize nonpotable water must be identified with a pictograph and a warning to not drink the water because it is nonpotable water. Signage is not required for water closets or urinals that are being supplied with a nonpotable water source for flushing purposes.

Gray water recycling systems previously were found in Section 3009 of the sanitary drainage provisions. The provisions for the collection, storage, and distribution of nonpotable water are located in Chapter 29 of the 2015 IRC because they are related to the water distribution provisions of this chapter. The exception is subsurface landscape irrigation systems connected to nonpotable water from on-site water reuse systems. Provisions that apply to drain, waste, and vent piping for subsurface landscape irrigation systems are still found in Section P3009, although they also have been revised to reflect current practices and acceptable standards.

There are a couple of notable changes to the gray water recycling provisions that appeared in the 2012 IRC. The use of nonpotable water, including gray water, for flushing water closets and urinals no longer requires introduction of a blue or green food-grade dye to identify the gray water. This was considered an outdated and unnecessary practice that often resulted in staining of fixtures and finishes. Identification of the nonpotable water distribution piping is considered sufficient. On the other hand, gray water used for fixture flushing purposes now requires disinfection and treatment by an on-site water reuse treatment system complying with the NSF 350 standard *Onsite Residential and Commercial*

Water Reuse Treatment Systems. In addition to microbiological contaminants that need disinfection, gray water contains organic compounds, suspended solids, and other contaminants that have the potential to accumulate and negatively impact the functioning of water closets and urinals if not treated properly. The 2012 IRC did not require disinfection or treatment of gray water used for flushing purposes.

P2905

Heated Water Distribution Systems

CHANGE TYPE: Addition

CHANGE SUMMARY: Pointers have been added to the IRC plumbing provisions to direct the user to the applicable energy conservation provisions of IRC Chapter 11 related to heated water distribution systems. Section N1103.5 requires automatic controls to maintain hot water temperature for heated water circulation systems and for heat trace temperature maintenance systems when such systems are installed.

2015 CODE:

SECTION P2905
HEATED WATER DISTRIBUTION SYSTEMS

P2905.1 Heated Water Circulation Systems and Heat Trace Systems. Circulation systems and heat trace systems, that are installed to bring heated water in close proximity to one or more fixtures, shall meet the requirements of Section N1103.5.1.

P2905.2 Demand Recirculation Systems. Demand recirculation water systems shall be in accordance with Section N1103.5.2.

CHANGE SIGNIFICANCE: Although the origin of these requirements are in the *International Energy Conservation Code* (IECC) and they are reprinted in IRC Chapter 11, plumbing system designers and contractors are frequently responsible for selecting systems for hot water temperature maintenance using either circulating pumps or heat trace systems. Part of the selection of such systems might include the associated operating controls. For example, some pump systems can be supplied with integral controls. Heated water circulation and temperature maintenance systems are not required, but when installed they must meet the mandatory requirements of Section N1103.5. These systems use circulation pumps or heat trace components to maintain the desired temperature of hot water for the convenience of the user and to conserve water that would otherwise be drawn until hot water reached the fixture outlet. The new provisions in Section N1103.5 do not permit a continuously

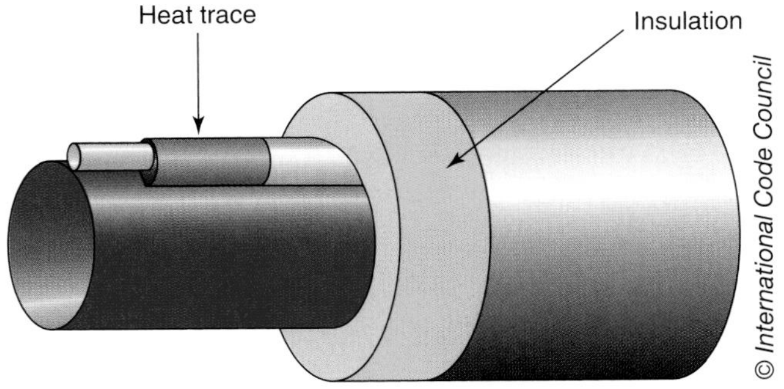

Heat trace temperature maintenance systems must comply with the energy efficiency provisions of Chapter 11 of the IRC.

operating circulating pump. The pump must operate on automatic controls activated when the hot water in the system falls below the desired temperature or when there is a demand for hot water. Pipe insulation is required for hot water circulation systems and the water in the circulation piping can stay hot for an extended time depending on the diameter of the piping. Because the pump only operates intermittently when needed, demand-activated circulation is significantly more energy efficient than a continuously operating heated water circulation system.

A heat trace system is the other energy-efficient means for maintaining the desired temperature in the service hot water system. The energy provisions require heat trace systems to comply with one of the referenced standards and to have automatic controls to conserve energy. As with circulation systems, piping in a heat trace system requires pipe insulation.

P2906.2

Lead Content of Drinking Water Pipe and Fittings

CHANGE TYPE: Modification

CHANGE SUMMARY: The code has a more stringent limitation for lead content in pipe, pipe fittings, joints, valves, faucets, and fixture fittings that convey water used for drinking and cooking.

2015 CODE: **~~P2905.2~~ P2906.2 Lead Content.** The lead content in pipe and fittings used in the water-supply system shall ~~have lead content of~~ be not greater than 8 percent ~~lead~~.

P2906.2.1 Lead Content of Drinking Water Pipe and Fittings. Pipe, pipe fittings, joints, valves, faucets, and fixture fittings utilized to supply water for drinking or cooking purposes shall comply with NSF 372 and shall have a weighted average lead content of 0.25 percent lead or less.

CHANGE SIGNIFICANCE: A U.S. federal law was enacted to go into effect on January 4, 2014, requiring that pipe, pipe fittings, joints, valves, faucets, and fixture fittings that are used to supply water for drinking or

Pipe fittings used to supply water for drinking or cooking purposes must comply with NSF 372 and have not greater than 0.25 percent lead content.

ICC ES PMG™ LOW LEAD

Low lead third-party certification

cooking purposes have not more than 0.25 percent lead content, based on a weighted average of wetted surface areas. The 0.25 percent limitation does not apply to portions of the water distribution system that do not supply water for drinking or cooking. The existing 8 percent limitation is still in effect for those portions of the system. Realistically, in residential construction, entire water distribution systems will likely comply with the new threshold. Contractors are not likely to select products meeting different standards for a residential application. The new limit does not affect materials already in place. Products that are installed as part of a renovation must comply with the code and federal law.

The 2015 IRC reflects the new federal law by requiring that the indicated products comply with the National Sanitation Foundation's standard NSF 372, which matches the lead content limitations set by federal law. It was developed as the basis for third-party certification agencies to verify and certify that products comply with the lead content limitation. Federal law does not require compliance with NSF 372 but sets the maximum lead content. Manufacturers, suppliers, distributers, and installers who do not comply with the federal law could face penalties for violations. Manufacturers are quickly working toward getting products third-party certified to NSF 372, as they are aware of the 2015 code requirements and compliance with NSF 372 will also demonstrate compliance with federal law.

Neither NSF 372 nor the federal law require low-lead-compliant products to be marked or identified in any particular manner, and identification markings are not standardized between manufacturers or third-party certification agencies. However, the 2015 IRC and IPC requirements for third-party certification to NSF 372 will make it easier to verify compliance with the code.

Although the federal law changes the previous requirement of the Safe Drinking Water Act with a threshold of 8 percent lead, to a maximum 0.25 percent lead, the 2015 IRC and IPC do not eliminate the 8 percent lead limitation. Products must still comply with the requirement of not more than 8 percent lead as determined by standard NSF 61. That standard requires evaluation of the product through chemical analysis of prepared test water that has been exposed to the product for a specified length of time. Evaluation of products for compliance with the low-lead federal law (and NSF 372) is by calculation method only.

P3003.9

Solvent Cementing of PVC Joints

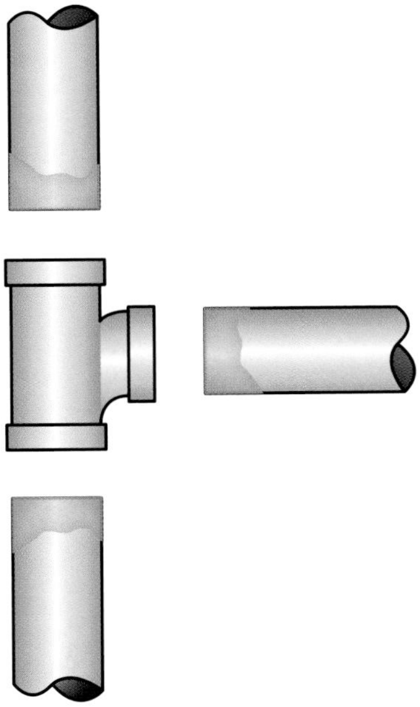

© International Code Council

Purple primer is no longer required for joints of non-pressure PVC DWV piping 4 inches or less in diameter.

CHANGE TYPE: Modification

CHANGE SUMMARY: The application of a primer to drain, waste, and vent PVC pipe and fittings prior to solvent cementing is not required for 4-inch pipe size and smaller, provided that the piping is for a non-pressure application.

2015 CODE: P3003.9 ~~Coextruded Composite P3003.14~~ PVC Plastic. Joints between ~~coextruded composite pipe with a PVC outer layer or~~ PVC plastic pipe or fittings shall comply with Sections P3003.9.1 through P3003.9.3.

P3003.9.1 Mechanical Joints. (*No change to text.*)

P3003.9.2 Solvent Cementing. Joint surfaces shall be clean and free from moisture. A purple primer that conforms to ASTM F 656 shall be applied. Solvent cement not purple in color and conforming to ASTM D 2564, CSA B137.3 or CSA B181.2 shall be applied to all joint surfaces. The joint shall be made while the cement is wet and shall be in accordance with ASTM D 2855. Solvent-cement joints shall be ~~permitted~~ installed above or below ground.

> **Exception:** A primer shall not be required where all of the following conditions apply:
>
> 1. The solvent cement used is third-party certified as conforming to ASTM D 2564.
> 2. The solvent cement is used only for joining PVC drain, waste, and vent pipe and fittings in non-pressure applications in sizes up to and including 4 inches (102 mm) in diameter.

CHANGE SIGNIFICANCE: Recent testing by NSF International has indicated that where solvent cement conforming to ASTM D 2564 is used without primer to join PVC pipe and fittings 4 inches in diameter and smaller, the bonding forces of the connection are in excess of what is required for gravity drainage and waste systems, and vent systems for gravity drainage systems. The strength of joints made without primer often exceeds the pipe and fitting pressure capacity for both solid wall and cellular core types of pipes.

The option to omit purple primer in assembling PVC DWV piping will simplify the installation and result in a more professional-looking finished product. The use of purple primer prior to solvent cementing PVC DWV fittings is often problematic where finished surfaces are in the vicinity. For example, tubs and shower stalls are often set in place during the plumbing rough-in stage. Work on plumbing rough-in piping above these finished products can be challenging when trying to avoid damaging the products. During the fixture set-out stage, pipe and fitting connections are often necessary in the vicinity of cabinetry, flooring, and other finish materials that could be damaged by purple primer. From an aesthetics point of view, piping covered with streaks of purple primer makes a piping job appear less than professional. Purple primer on piping visible to the occupant from within finished areas of the building is not appreciated by the building owner or occupant.

The separate pipe joining provisions for PVC solid wall piping (formerly in Section P3003.14) and PVC cellular core piping have been merged into a single Section P3003.9 titled "PVC Plastic." PVC pipe is manufactured using several different methods. The manufacturing method of a PVC pipe does not affect how the pipe is joined. All forms of PVC pipe are joined by the same methods.

P3005.2

Cleanouts

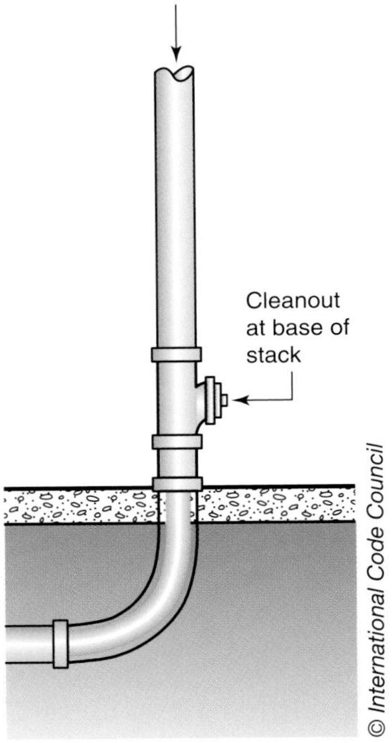

Cleanout on a stack

CHANGE TYPE: Modification

CHANGE SUMMARY: The section on cleanouts has been completely reorganized and reworded for clarity. Brass cleanout plugs are only permitted for metallic piping. Where located at a finished wall, the cleanout must be within 1½ inches of the finished surface. A cleanout is no longer required at the base of each waste or soil stack.

2015 CODE: P3005.2 ~~Drainage Pipe Cleanouts~~ Cleanouts Required. ~~Drainage pipe c~~Cleanouts shall ~~comply~~ be provided for drainage piping in accordance with Sections P3005.2.1 through P3005.2.11.

P3005.2.~~7~~3 Building Drain and Building Sewer Junction. ~~There shall be a cleanout near the junction of the building drain and building sewer. This cleanout shall be either inside or outside the building wall, provided that it is brought up to finish grade or to the lowest floor level. An approved two-way cleanout shall be permitted to serve as the required cleanout for both the building drain and the building sewer. The cleanout at the junction of the building drain and building sewer shall not be required where a cleanout on a 3-inch (76 mm) or larger diameter soil stack is located within a developed length of 10 feet (3048 mm) of the building drain and building sewer junction.~~ The junction of the building drain and the building sewer shall be served by a cleanout that is located at the junction or within 10 feet (3048 mm) developed length of piping upstream of the junction. For the requirements of this section, the removal of a water closet shall not be required to provide cleanout access.

~~**P3005.2.6 Base of Stacks.** A cleanout shall be provided at the base of each waste or soil stack.~~

P3005.2.6 Cleanout Plugs. Cleanout plugs shall be copper alloy, plastic or other approved materials. Cleanout plugs for borosilicate glass piping systems shall be of borosilicate glass. Brass cleanout plugs shall conform to ASTM A 74 and shall be limited for use only on metallic piping systems. Plastic cleanout plugs shall conform to the referenced standards for plastic pipe fittings as indicated in Table P3002.3. Cleanout plugs shall have a raised square head, a countersunk square head or a countersunk slot head. Where a cleanout plug will have a trim cover screw installed into the plug, the plug shall be manufactured with a blind end threaded hole for such purpose.

P3005.2.10 Cleanout Access. Required cleanouts shall not be installed in concealed locations. For the purposes of this section, concealed locations include, but are not limited to, the inside of plenums, within walls, within floor/ceiling assemblies, below grade and in crawl spaces where the height from the crawl space floor to the nearest obstruction along the path from the crawl space opening to the cleanout location is less than 24 inches (610 mm). Cleanouts with openings at a finished wall shall have the face of the opening located within 1½ inches (38 mm) of the finished wall surface. Cleanouts located below grade shall be extended to grade level so that the top of the cleanout plug is at or above grade. A cleanout installed in a floor or walkway that will not have a trim cover installed shall have a counter-sunk plug installed so the top surface of the plug is flush with the finished surface of the floor or walkway.

(Only portions of Section P3005.2 are shown for brevity and clarity.)

CHANGE SIGNIFICANCE: The reorganization and expansion of this section presents the material in a logical format that clarifies the requirements for cleanouts. There are several minor technical changes to the text to reflect current industry-accepted practices and to clarify the application of the cleanout provisions. The code has always allowed removal of a fixture trap or removal of a fixture with an integral trap, such as a water closet, in satisfying the location requirements for cleanouts. The code no longer describes specifically what may be used for cleanouts provided the location and access requirements are met. The industry-accepted practice of removing a water closet for cleanout purposes will continue. However, the code does now specifically prohibit removal of a water closet to serve as the required cleanout for the junction of the building drain to the building sewer.

Brass cleanout plugs are limited to use with metallic piping and fittings. Over-tightening of a brass plug in a threaded plastic cleanout opening can easily crack the fitting. Where a cleanout is located in a wall, the face of the cleanout opening must be within 1½ inches of the face of the wall for ease of access and to prevent damage to the wall finishes during rodding operations. Where a wall face is located farther away from the cleanout opening face, a large cleanout access opening panel could serve the same intent.

Cleanouts in floors do not necessarily require specialized cleanout cover assemblies provided a counter-sunk cleanout plug is installed flush with the floor. However, where vehicular traffic is anticipated, cleanout assemblies in accordance with ASME A112.36.2M must be used.

The requirement to provide a cleanout at the base of each waste or soil stack has been removed. Where multiple stacks discharge to a horizontal drain pipe, there is no need to require a cleanout for the base of every stack. There only needs to be one cleanout access at the most upstream end of the horizontal drain pipe (and every 100 feet from that point). The intent of the previous requirement was to make sure there was an access point for rodding every length of horizontal piping connected to the base of a stack. The requirement had nothing to do with stacks.

P3008.1

Backwater Valves

CHANGE TYPE: Modification

CHANGE SUMMARY: For existing buildings, fixtures that are located above the next upstream manhole cover are allowed to discharge through a backwater valve.

2015 CODE: **P3008.1 Sewage Backflow.** Where the flood level rims of plumbing fixtures are below the elevation of the manhole cover of the next upstream manhole in the public sewer, the fixtures shall be protected by a backwater valve installed in the building drain, branch of the building drain or horizontal branch serving such fixtures. Plumbing fixtures having flood level rims above the elevation of the manhole cover of the next upstream manhole in the public sewer shall not discharge through a backwater valve.

> **Exception:** In existing buildings, fixtures above the elevation of the manhole cover of the next upstream manhole in the public sewer shall not be prohibited from discharging through a backwater valve.

CHANGE SIGNIFICANCE: A new exception specifically addresses a common problem encountered with installation of a backwater valve for an existing building. Existing buildings built before the code began requiring backwater valves for fixtures on floor levels below the elevation of the next upstream manhole cover are at risk for sewage backflows caused by public sewer problems. In some cases, many years will pass without the public sewer creating a fixture overflow in an existing building. As more building sewer connections are made to the public sewer and as stormwater infiltration increases as the public sewer ages, surcharging and clogs in the public sewer can develop. Usually, a building owner will experience only one sewage overflow in the building before consulting with a plumbing contractor to provide a solution to protect against these costly and unsettling events.

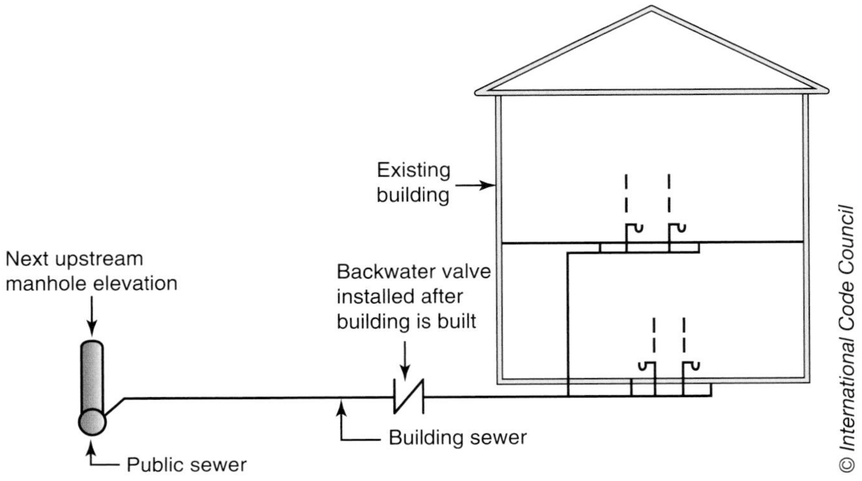

The exception allows this backwater valve arrangement only for existing buildings.

However, installation of a backwater valve after a building is built presents the problem of how to separate the drainage flow from fixtures on floors below the next upstream manhole cover from the fixtures on floors above the next upstream manhole cover. The drainage piping within the building is so integral to the construction of the building that separation of the drainage flows for installation of a backwater valve in accordance with the code is impossible. The new exception allows, for existing buildings only, installation of a backwater valve for all fixtures in a building, even if those fixtures are on a floor above the next upstream manhole cover elevation. A building owner should have the ability to protect his or her property from public sewer surcharging that could cause backflow and damage in the building. Without a backwater valve installed in these situations, multiple overflow events and property damage could continue to occur unabated.

P3103.1, P3103.2

Vent Terminals

CHANGE TYPE: Modification

CHANGE SUMMARY: Where a minimum 3-inch diameter vent terminal is required to prevent frost blockage in cold climates, the 3-inch diameter pipe must extend at least 12 inches inside the building's thermal envelope. The minimum 7-foot height requirement for vent terminations applies only to roofs used for purposes similar to residential decks, patios and balconies.

2015 CODE: P3103.1 Roof Extension. Open vent pipes that extend through a roof shall be terminated not less than 6 inches (152 mm) above the roof or 6 inches (152 mm) above the anticipated snow accumulation, whichever is greater~~, except that~~. ~~w~~<u>W</u>here a roof is to be used for ~~any purpose other than weather protection,~~ <u>assembly, as a promenade, observation deck, sunbathing deck or for similar purposes, open</u> vent ~~extension~~ <u>pipes</u> shall ~~be run~~ <u>terminate</u> not less than 7 feet (2134 mm) above the roof.

P3103.2 Frost Closure. Where the 97.5-percent value for outside design temperature is 0°F (−18°C) or less, ~~every~~ vent extension<u>s</u> through a roof or wall shall be not less than 3 inches (76 mm) in diameter. Any increase in the size of the vent shall be made ~~inside the structure~~ not less than 1 foot ~~below the roof or inside the wall~~ <u>inside the thermal envelope of the building</u>.

CHANGE SIGNIFICANCE: Section P3103.1 requiring a 7-foot vent extension above the roof when the roof was used for any purpose other than weather protection was sometimes interpreted literally to require extension of roof vent terminals where the roof was used for mounting equipment such as HVAC units, solar panels, or antennas. The original intent of the section was to only require vent extension to 7 feet where

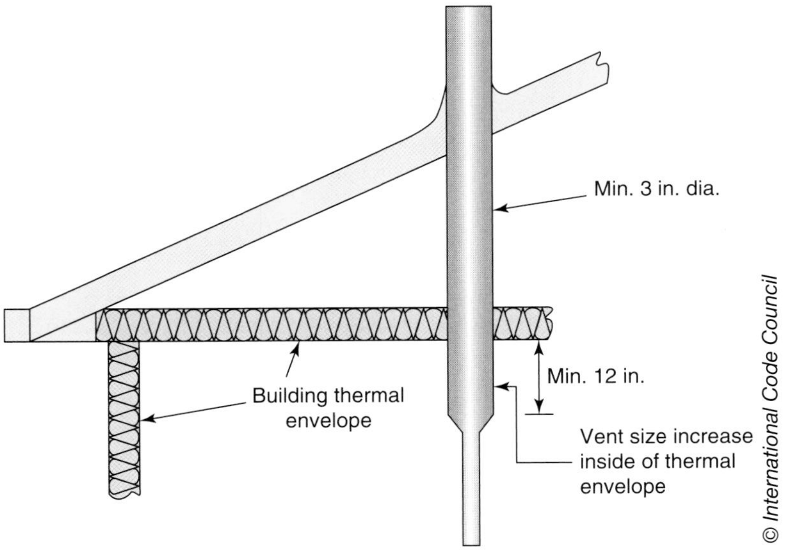

Vent terminal requirements to prevent frost closure in cold climates

the roof was to be used by people, similar to how they use a deck or patio. The purpose was to elevate the vent terminal and the level of discharge of sewer gases above people's heads. This requirement is not necessary for where workers are temporarily installing, repairing or replacing rooftop-mounted equipment or where roofs might be used for people escaping the interior of a building in an emergency. The code is now clear that the intent is to require the vent extension if the roof is used for people to gather or enjoy the outdoors such as occurs with an observation deck or sunbathing deck.

Section P3103.2 has long required vent terminals of not less than 3 inches in diameter where the outside design temperature is 0 degrees Fahrenheit or less. The intent is to prevent frost blockage, and 3-inch-diameter pipe performs well for that purpose. Where the code does not require a 3-inch vent—for example, a 2-inch kitchen or bathroom vent—the vent size must be increased before the vent penetrates through the roof. In this case, the code has always prescribed that the transition to a larger pipe size must occur at least 1 foot below the roof. However, recent reports from building owners in cold areas have indicated vent pipe blockage and damage from freezing condensate when the transition from a smaller pipe size to a 3-inch vent terminal occurs in an unconditioned attic area below the roof. In most attics, the attic temperature is very near the outdoor temperature, and frost closure can occur in smaller vent pipes. In these cold climates, the 2015 IPC requires smaller vent pipes to transition to the 3-inch diameter starting at not less than 1 foot inside the building's thermal envelope. In other words, the vent enlargement must occur at least 1 foot inside the heated zone of the building, typically measured from the insulated ceiling of the topmost story.

P3201.2

Trap Seal Protection Against Evaporation

CHANGE TYPE: Modification

CHANGE SUMMARY: Trap seal protection against evaporation can now be accomplished in a variety of ways, including trap seal primer valves supplied with nonpotable water and barrier-type trap seal protection devices.

2015 CODE: P3201.2 Trap Seals. ~~and Trap Seal Protection.~~ Each fixture trap shall have a liquid seal of not less than 2 inches (51 mm) and not more than 4 inches (102 mm). ~~Traps for floor drains shall be fitted with a trap primer or shall be of the deep seal design. Trap seal primer valves shall connect to the trap at a point above the level of the trap seal.~~

P3201.2.1 Trap Seal Protection. Trap seals of emergency floor drain traps and traps subject to evaporation shall be protected by one of the methods in Sections P3201.2.1.1 through P3201.2.1.4.

P3201.2.1.1 Potable Water-Supplied Trap Seal Primer Valve. A potable water-supplied trap seal primer valve shall supply water to the trap. Water-supplied trap seal primer valves shall conform to ASSE 1018. The discharge pipe from the trap seal primer valve shall connect to the trap above the trap seal on the inlet side of the trap.

P3201.2.1.2 Reclaimed or Gray-Water-Supplied Trap Seal Primer Valve. A reclaimed or gray-water-supplied trap seal primer valve shall supply water to the trap. Water-supplied trap seal primer valves shall conform to ASSE 1018. The quality of reclaimed or gray water supplied to trap seal primer valves shall be in accordance with the requirements

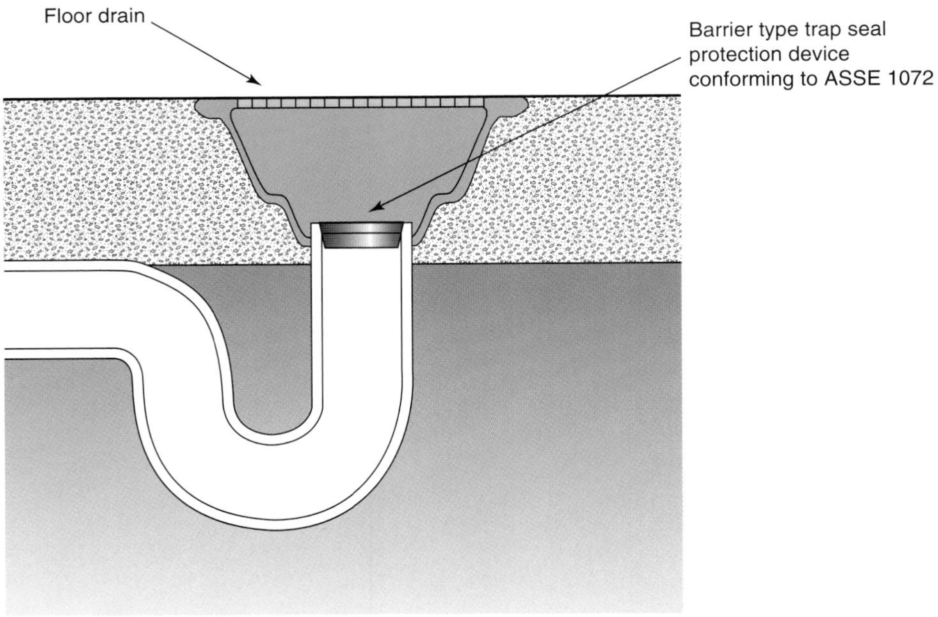

A barrier-type trap seal protection device is one of four methods of protecting the floor drain trap seal from evaporation.

of the manufacturer of the trap seal primer valve. The discharge pipe from the trap seal primer valve shall connect to the trap above the trap seal on the inlet side of the trap.

P3201.2.1.3 Waste-Water-Supplied Trap Primer Device. A waste-water-supplied trap primer device shall supply water to the trap. Waste-water-supplied trap primer devices shall conform to ASSE 1044. The discharge pipe from the trap seal primer device shall connect to the trap above the trap seal on the inlet side of the trap.

P3201.2.1.4 Barrier-Type Trap Seal Protection Device. A barrier-type trap seal protection device shall protect the floor drain trap seal from evaporation. Barrier-type floor drain trap seal protection devices shall conform to ASSE 1072. The devices shall be installed in accordance with the manufacturer's instructions.

CHANGE SIGNIFICANCE: Section P3201.2 was expanded to cover two additional types of trap seal protection devices and to distinguish between the different types of water-supplied trap seal protection devices. Potable-water-supplied trap seal devices have been an industry standard for decades. However, with greater reliance on alternate sources of nonpotable water such as reclaimed water and gray water, the quality of those types of water has a bearing on the performance of water-supplied trap devices. The manufacturer of those devices must be consulted where alternate sources of water are supplied to the devices.

Waste-water-supplied trap primer devices have been used in the plumbing industry for some time, and because they are covered in standard ASSE 1044, they have been code-approved for over a decade. However, these simple and effective trap primer devices were overlooked because the code identified the products only by referring to the standard that they complied with.

The latest trap seal protection device is for floor drains only and utilizes a specially designed and tested insert below the floor drain strainer plate. When water runs into the floor drain, the insert allows the water to pass and then closes to significantly reduce evaporation of the trap seal.

PART 8

Electrical

Chapters 34 through 43

The electrical part of the IRC is extracted, by permission, from NFPA 70 *National Electrical Code* (NEC) published by the National Fire Protection Association (NFPA). The corresponding NEC section number appears in brackets at the end of each IRC section. Appendix Q of the IRC also provides a cross reference for the section numbers of each code. Similar to the mechanical, fuel gas, and plumbing parts of the IRC, Part 8 is divided into several chapters, starting with general requirements applicable to all residential electrical systems and followed by chapters of technical provisions covering design and installation. Chapter 34 covers general requirements such as component identification, equipment location, clearances, protection from damage, and conductor connections. Chapter 35 of the IRC provides definitions specific to electrical installations and supplements (and in some cases supersedes) the general definitions found in Chapter 2. Subsequent chapters cover electrical services, branch circuits, feeders, wiring methods, outlet locations, receptacles, lighting fixtures, and appliance installation for electrical systems of buildings under the scope of the IRC. A separate chapter covers the unique hazards and special requirements related to electrical installations for swimming pools, hot tubs, and whirlpool bathtubs. Limited-voltage circuits are addressed in Chapter 43. ■

E3901.9

Receptacle Outlets for Garages

CHANGE TYPE: Modification

CHANGE SUMMARY: Garage receptacle outlets must be served by a separate branch circuit that does not supply other outlets. At least one receptacle outlet is required for each car space in a garage.

2015 CODE: E3901.9 Basements, Garages and Accessory Buildings. ~~At least~~ Not less than one receptacle outlet, in addition to any provided for specific equipment, shall be installed in each separate unfinished portion of a basement ~~and~~, in each attached garage, and in each detached garage or accessory building that is provided with electrical power. ~~Where a portion of the basement is finished into one or more habitable room(s), each separate unfinished portion shall have a receptacle outlet installed in accordance with this section.~~ The branch circuit supplying the receptacle(s) in a garage shall not supply outlets outside of the garage and not less than one receptacle outlet shall be installed for each motor vehicle space. [210.52(G)(1), (2), and (3)]

CHANGE SIGNIFICANCE: Section E3901 sets requirements for the minimum number and locations of 15- and 20-ampere general-purpose receptacle outlets. Proper outlet distribution is important not only for convenience of the occupants but also for safety by preventing overloading of circuits and eliminating the need for extension cords. Previously, the code required at least one receptacle outlet in each garage. The minimum number of receptacle outlets is now determined by the size of the garage and requires one outlet for every car space. A typical two-stall garage now requires not less than two receptacle outlets. Also new to the 2015 IRC, garage receptacle outlets must be served by a separate branch circuit that does not supply other outlets of the dwelling unit or premises.

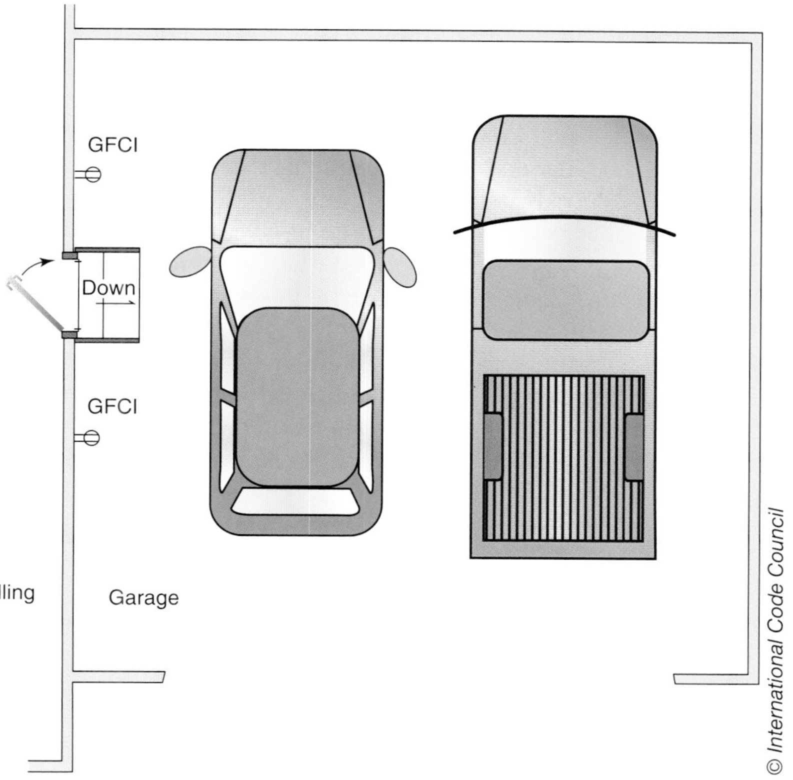

A minimum of two receptacle outlets are required for a two-stall garage.

E3902.8, E3902.9, E3902.10

Ground-Fault Circuit Interrupter Protection

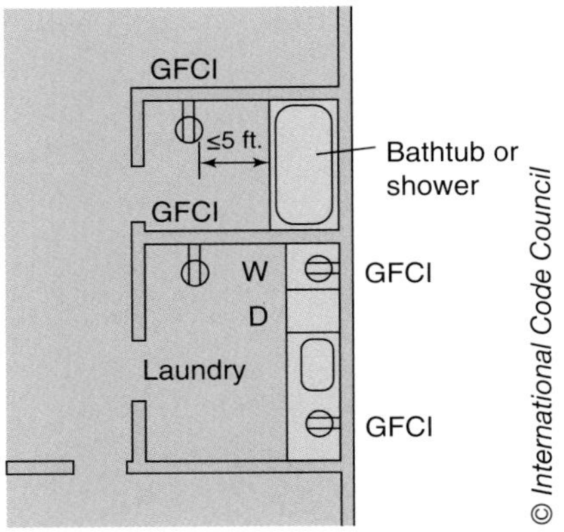

GFCI protection required for 125-volt, 15- and 20-amp receptacle outlets in laundry areas and near showers or bathtubs

CHANGE TYPE: Modification

CHANGE SUMMARY: Laundry areas have been added to the list of locations requiring ground-fault circuit interrupter (GFCI) protection. Receptacles within 6 feet of bathtubs and showers, and receptacles for dishwashers also require GFCI protection.

2015 CODE: **E3902.8 Bathtub or Shower Stall Receptacles.** All 125-volt, single-phase, 15- and 20-ampere receptacles that are located within 6 feet (1829 mm) of the outside edge of a bathtub or shower stall shall have ground-fault circuit interrupter protection for personnel. [210.8(A)(9)]

E3902.9 Laundry Areas. 125-volt, single-phase, 15- and 20-ampere receptacles installed in laundry areas shall have ground-fault circuit interrupter protection for personnel. [210.8(A)(10)]

E3902.10 Kitchen Dishwasher Branch Circuit. Ground-fault circuit interrupter protection shall be provided for outlets that supply dishwashers in dwelling unit locations. [210.8(D)]

CHANGE SIGNIFICANCE: Ground-fault circuit interrupter (GFCI) devices protect people from shock hazards by de-energizing a circuit or receptacle when a fault current to ground is detected. The code has required GFCI protection for 15- and 20-ampere receptacles in locations where occupants are most susceptible to shock hazards, typically wet or damp locations, areas where water is used, or areas with concrete floor surfaces that might provide a path to ground. For example, GFCI protection is required for receptacles in bathrooms, kitchens, garages, unfinished basement areas, and outdoor locations. Three new items appear in the list of locations requiring GFCI protection for the 2015 IRC. Receptacle outlets located within 6 feet of the edge of a bathtub or shower now require GFCI protection. Although the GFCI provisions already apply to receptacles in bathrooms, the additional coverage intends to apply to the occasional bathtub or shower that is not located in a bathroom. The 6-foot distance is consistent with the rule for sinks other than kitchen sinks. The area within this distance of a bathtub, shower, or sink is considered a shock hazard zone because of the presence of water.

The code also now requires GFCI protection for 125-volt, single-phase, 15- and 20-ampere receptacles that are located in laundry rooms. Again, wet and damp conditions in laundry rooms coupled with conductive floor materials can pose a shock hazard to people using appliances in these areas. Section E3902.15 requires GFCI devices to be readily accessible. The 125-volt receptacle serving the clothes washing machine will typically be protected by an upstream device—either a GFCI circuit breaker or a GFCI receptacle outlet. Where the washing machine outlet is a GFCI device, the outlet must be in a readily accessible location, such as in the wall above the level of the top of the machine.

Also new to the 2015 code, ground-fault circuit interrupter protection must be provided for outlets that supply kitchen dishwashers. Previously, the only kitchen outlets requiring GFCI protection were those serving counter tops. Because dishwasher outlets may not be readily accessible, GFCI protection will typically be provided by an upstream GFCI receptacle or by a GFCI circuit breaker.

E4203.4.3

Location of Low-Voltage Luminaires Adjacent to Swimming Pools

CHANGE TYPE: Modification

CHANGE SUMMARY: Listed low-voltage luminaires meeting the prescribed conditions are permitted to be located less than 5 feet from the water's edge of swimming pools, spas, and hot tubs.

2015 CODE: **E4203.4.3 Low-Voltage Luminaires.** Listed low-voltage luminaires not requiring grounding, not exceeding the low-voltage contact limit, and supplied by listed transformers or power supplies that comply with Section E4206.1 shall be permitted to be located less than 5 feet (1524 mm) from the inside walls of the pool. [680.22(B)(6)]

CHANGE SIGNIFICANCE: The new provision adds flexibility to the design and installation of yard and landscape lighting around outdoor swimming pools. The code sets a number of minimum separation requirements for electrical equipment including receptacle outlets and luminaires (complete light fixtures) installed in the vicinity of swimming pools, spas and hot tubs to protect people from shock hazards associated with damp and wet environments. Luminaires must be installed at least 12 feet above the water surface or at least 5 feet horizontally from the water's edge. The new provision recognizes that listed low-voltage lighting can be safely installed less than 5 feet from the pool or hot tub when all of the prescribed conditions are met. Transformers and power supplies for the low-voltage luminaires must meet the same requirements as those used for the supply of underwater luminaires and must be listed for swimming pool and spa use.

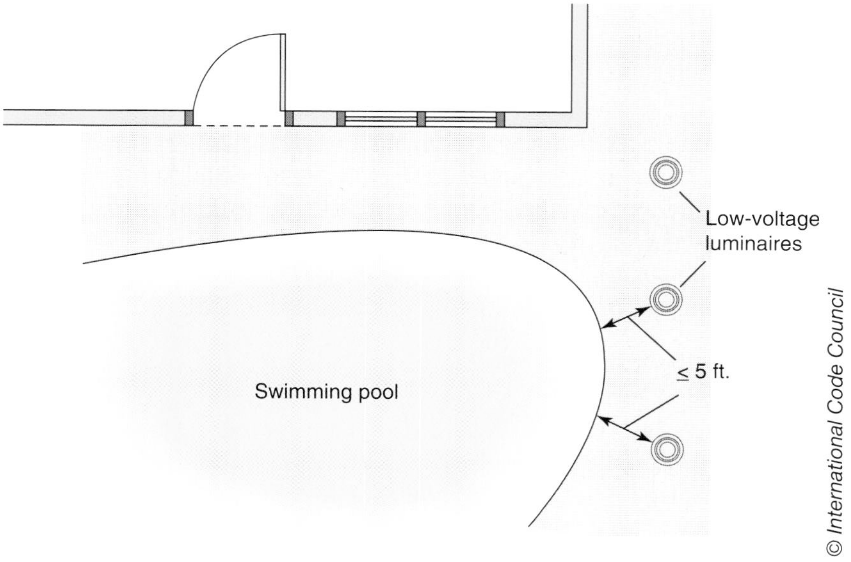

Listed low-voltage luminaires meeting all conditions are allowed within 5 feet of swimming pools.

E4204.2

Bonding of Outdoor Hot Tubs and Spas

CHANGE TYPE: Modification

CHANGE SUMMARY: A new exception to Section E4204.2 amends the perimeter equipotential bonding requirements for listed self-contained hot tubs under the prescribed conditions.

2015 CODE: E4204.2 Bonded Parts. The parts of pools, spas, and hot tubs specified in Items 1 through 7 shall be bonded together using insulated, covered or bare solid copper conductors not smaller than 8 AWG or using rigid metal conduit of brass or other identified corrosion-resistant metal.

(No changes to remainder of paragraph.)

1. Conductive pool shells. *(No significant changes to text.)*
2. Perimeter surfaces. *(No significant changes to text.)*

   Exceptions:

   1. Equipotential bonding of perimeter surfaces shall not be required for spas and hot tubs where all of the following conditions apply:
      - **1.1.** The spa or hot tub is listed as a selfcontained spa for aboveground use.
      - **1.2.** The spa or hot tub is not identified as suitable only for indoor use.
      - **1.3.** The installation is in accordance with the manufacturer's instructions and is located on or above grade.
      - **1.4.** To top rim of the spa or hot tub is not less than 28 in. (711 mm) above all perimeter surfaces that are within 30 in. (762 mm), measured horizontally from the spa or hot tub. The height of nonconductive external steps for entry to or exit from the self-contained spa is not used to reduce or increase this rim height measurement.
   2. The equipotential bonding requirements for perimeter surfaces shall not apply to a listed self-contained spa or hot tub located indoors and installed above a finished floor.

(No significant changes to remainder of Section E4204.2.)

CHANGE SIGNIFICANCE: Provisions for equipotential bonding of metal parts of hot tubs and spas generally match the bonding requirements for swimming pools. The purpose of bonding is to ensure equal electrical potential in the wet environment in and around pools, spas and hot tubs to protect users from shock hazards due to stray voltage. The new provisions

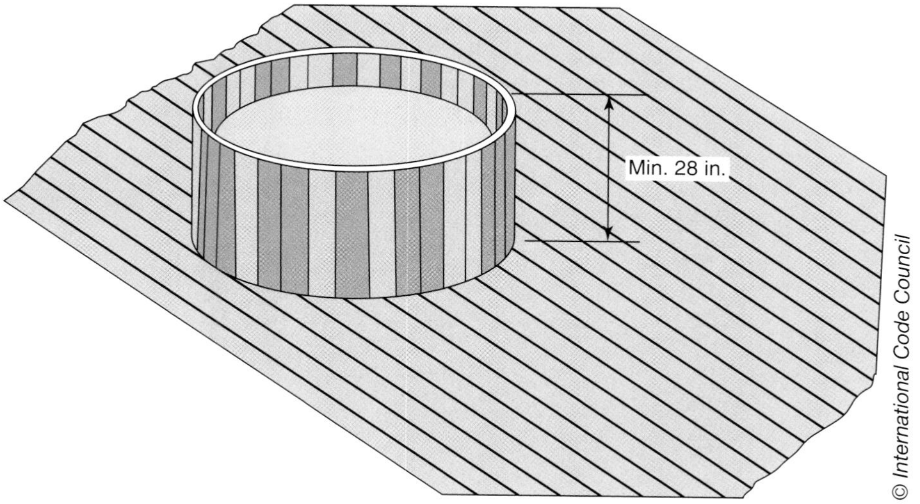

Equipotential bonding is not required for self-contained hot tubs that are installed above ground in accordance with the manufacturer's instructions and the code requirements.

for the 2015 IRC recognize that listed self-contained spas and hot tubs installed outdoors under the prescribed conditions adequately address safety concerns. The exemption from additional on-site bonding applies only to manufactured self-contained units that are installed above ground in accordance with the manufacturer's instructions. In addition, the top of the spa or hot tub must be at least 28 inches higher than the surrounding surface. For example, a hot tub recessed into a deck would not meet the conditions and would require bonding in accordance with Section E4204.2.

PART 9

Appendices

Appendix A through S

- **Appendix A through Q** No changes addressed
- **Appendix R** Light Straw-Clay Construction
- **Appendix S** Strawbale Construction

As stated in Chapter 1 of the IRC, provisions in the appendices do not apply unless specifically referenced in the adopting ordinance. The appendices are developed in much the same manner as the main body of the model code. However, the appendix information is judged to be outside the scope and purpose of the code at the time of code publication. Many times an appendix offers supplemental information, alternative methods, or recommended procedures. The information may also be specialized and applicable or of interest to only a limited number of jurisdictions. Although an appendix may provide some guidelines or examples of recommended practices or assist in the determination of alternative materials or methods, it will have no legal status and cannot be enforced until it is specifically recognized in the adopting legislation. Appendix chapters or portions of such chapters that gain general acceptance over time can move into the main body of the model code through the code-development process. The 2015 IRC introduces two new appendix chapters that reflect the growing awareness and acceptance of green and sustainable construction practices. Appendix R covers light straw-clay construction and Appendix S covers strawbale construction. ■

APPENDIX R

Light Straw-Clay Construction

APPENDIX S

Strawbale Construction

Appendix R
Light Straw-Clay Construction

CHANGE TYPE: Addition

CHANGE SUMMARY: Prescriptive provisions for light straw-clay construction have been added as an appendix to the 2015 IRC. Light straw-clay walls are nonbearing infill around a structural frame.

2015 CODE:

APPENDIX R
LIGHT STRAW-CLAY CONSTRUCTION

The provisions contained in this appendix are not mandatory unless specifically referenced in the adopting ordinance.

SECTION AR101
GENERAL

AR101.1 Scope. This appendix shall govern the use of light straw-clay as a nonbearing building material and wall infill system.

SECTION AR102
DEFINITIONS

AR102.1. General. The following words and terms shall, for the purposes of this appendix, have the meanings shown herein. Refer to Chapter 2 of the *International Residential Code* for general definitions.

CLAY. Inorganic soil with particle sizes of less than 0.00008 inch (0.002 mm) having the characteristics of high to very high dry strength and medium to high plasticity.

Appendix R continues

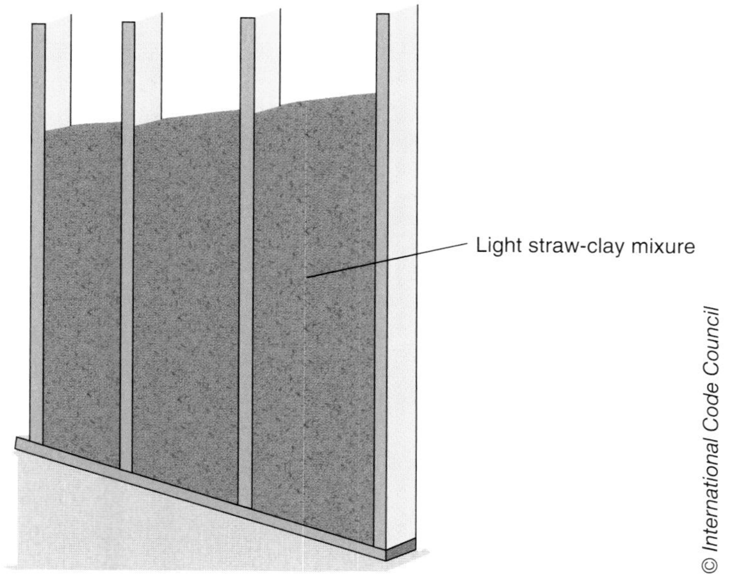

Light straw-clay wall construction

Appendix R continued

CLAY SLIP. A suspension of clay soil in water.

CLAY SOIL. Inorganic soil containing 50 percent or more clay by volume.

INFILL. Light straw-clay that is placed between the structural members of a building.

LIGHT STRAW-CLAY. A mixture of straw and clay compacted to form insulation and plaster substrate between or around structural and non-structural members in a wall.

NONBEARING. Not bearing the weight of the building other than the weight of the light straw-clay itself and its finish.

STRAW. The dry stems of cereal grains after the seed heads have been removed.

VOID. Any space in a light straw-clay wall in which a 2-inch (51-mm) sphere can be inserted.

SECTION AR103
NONBEARING LIGHT STRAW-CLAY CONSTRUCTION

AR103.1 General. Light straw-clay shall be limited to infill between or around structural and nonstructural wall framing members.

AR103.2 Structure. The structure of buildings using light straw-clay shall be in accordance with the *International Residential Code* or shall be in accordance with an approved design by a registered design professional.

AR103.2.1 Number of Stories. Use of light straw-clay infill shall be limited to buildings that are not more than one story above grade plane.

> **Exception:** Buildings using light straw-clay infill that are greater than one story above grade plane shall be in accordance with an approved design by a registered design professional.

AR103.2.2 Bracing. Wind and seismic bracing shall be in accordance with Section R602.10 and shall use Method LIB. The required length of bracing shall comply with Section R602.10.3, with the additional requirements that Table R602.10.3(3) shall be applicable to buildings in Seismic Design Category C, and that the minimum total length of bracing in Table R602.10.3(3) shall be increased by 90 percent. In lieu of these prescriptive requirements, wind and seismic bracing shall be in accordance with an approved design by a registered design professional. Walls with light straw-clay infill shall not be sheathed with solid sheathing.

AR103.2.3 Weight of Light Straw-Clay. Light straw-clay shall be deemed to have a design dead load of 40 pounds per cubic foot (640 kg per cubic meter) unless otherwise demonstrated to the building official.

AR103.2.4 Reinforcement of Light Straw-Clay. Light straw-clay shall be reinforced as follows:

1. Vertical reinforcing shall be not less than nominal 2-inch by 6-inch (51 mm by 152 mm) wood members at not more than 32 inches (813 mm) on center where the vertical reinforcing is non-load-bearing and at 24 inches (610 mm) on center where it is load-bearing. The vertical reinforcing shall not exceed an unrestrained height of 10 feet (3048 mm) and shall be attached at top and bottom in accordance with Chapter 6 of the *International Residential Code*. In lieu of these requirements, vertical reinforcing shall be in accordance with an approved design by a registered design professional.
2. Horizontal reinforcing shall be installed in the center of the wall at not more than 24 inches (610 mm) on center and shall be secured to vertical members. Horizontal reinforcing shall be of any of the following: ¾ inch (19 mm) bamboo, ½ inch (13 mm) fiberglass rod, 1-inch (25 mm) wood dowel or nominal 1-inch by 2-inch (25 mm by 51 mm) wood.

AR103.3 Materials. The materials used in light straw-clay construction shall be in accordance with Sections AR103.3.1 through AR103.3.4.

AR103.3.1 Straw. Straw shall be wheat, rye, oats, rice or barley, and shall be free of visible decay and insects.

(Portions of Appendix R not shown for brevity and clarity. See 2015 IRC for full text.)

CHANGE SIGNIFICANCE: Light straw-clay is added to the 2015 IRC as a non-load-bearing building material and wall infill system in a nonmandatory appendix.

Light straw-clay construction has been used in a number of states in the western United States. To date, most permitting of light straw-clay construction has been approved by building officials on a case-by-case basis. Two exceptions are the State of New Mexico and the State of Oregon. Since 1998 the State of New Mexico has approved straw-clay construction using its standard "Clay Straw Guidelines." Building officials in surrounding states have also approved straw-clay construction in their jurisdictions based on these guidelines. In October of 2011 the Oregon Reach Code (ORC) was amended to include light straw-clay construction.

Inclusion in the IRC makes straw-clay provisions accessible to designers and builders interested in using this material and to building officials evaluating and enforcing its proper use. Currently, residential and non-residential projects using straw-clay have been completed in 17 states, most with full permits and inspections.

The adopted mixture of clay and straw as a monolithic non-load-bearing building enclosure has been successfully used in Europe since 1950 and in the United States since 1990. Prior to this, a heavier form of clay, straw and woven wood construction known as wattle and daub was in common use throughout Europe, Africa, Asia, and North and South America.

Appendix R continues

Appendix R continued

The European predecessors and light straw-clay buildings built to date in North America have all been constructed without the use of a moisture barrier. The proposed light straw-clay materials are vapor permeable and do not require a moisture barrier. Code precedents for vapor-permeable construction exist for adobe construction, log construction and half-timber construction. In these systems, as in light straw-clay construction, there is sufficient water-retaining capacity to hold and re-release moisture without damage to structural members or degradation of the wall due to weather-related moisture fluctuations.

In 2004 the Canada Mortgage and Housing Corporation (CMHC) funded a study to explore the material characteristics of Straw Light Clay (SLC) construction. This study is used as a basis for this appendix. The CMHC study includes issues of thermal performance, fire resistance, moisture, and vapor permeability.

Appendix S

Strawbale Construction

© Guillermo del Olmo/Shutterstock.com.

Straw

CHANGE TYPE: Addition

CHANGE SUMMARY: Prescriptive provisions for strawbale construction have been added as an appendix to the 2015 IRC. Strawbale walls may be bearing walls or nonbearing infill around a structural frame depending upon the method of construction and detailing. Appendix S contains requirements for both construction methods.

2015 CODE:

APPENDIX S
STRAWBALE CONSTRUCTION

The provisions contained in this appendix are not mandatory unless specifically referenced in the adopting ordinance.

SECTION AS101
GENERAL

AS101.1 Scope. This appendix provisions prescriptive and performance-based requirements for the use of baled straw as a building material. Other methods of strawbale construction shall be subject to approval in accordance with Section R104.11 of this code. Buildings using strawbale walls shall comply with this code, except as otherwise stated in this appendix.

Appendix S continues

© Ljupco Smokovski/Shutterstock.com.

Straw bales

Appendix S continued

Section AS102 Definitions

Section AS103 Bales

Section AS104 Finishes

Section AS105 Strawbale Walls – General

Section AS106 Strawbale Walls – Structural

Section AS107 Fire Resistance

Section AS108 Thermal Insulation

Section AS109 Referenced Standards

(Portions of Appendix S not shown for brevity and clarity. See 2015 IRC for full text.)

CHANGE SIGNIFICANCE: Strawbale construction is added to the 2015 IRC appendices as a nonmandatory option. Strawbale may be used as a load-bearing structural material or as non-load-bearing infill.

First used in Nebraska in the late 1800s, strawbale construction increased in the 1980s in southwestern states. Since then it has been further developed and explored, including testing and research of structural performance (under vertical and lateral loads) and moisture, fire, thermal, and acoustic properties.

New Mexico and Oregon have adopted statewide strawbale building provisions. California has legislated strawbale construction guidelines for voluntary adoption by local jurisdictions. In addition, nine U.S. cities or counties have adopted strawbale building requirements.

Most of the strawbale building provisions that do exist are derived from the first guideline, created for and adopted by Tucson/Pima County, Arizona, in 1996. The 2015 IRC provisions are an expanded version that considers lateral and gravity loads, fire resistance, moisture resistance and energy efficiency.

Strawbale buildings are now found in 49 of the 50 states, and strawbale construction is practiced in over 45 countries throughout the world and in every climate. There are an estimated 600 strawbale buildings in California alone. The strawbale buildings include residences, public and private schools, libraries, office and retail buildings, wineries, multi-story buildings, load-bearing strawbale structures and structures in areas of high seismic risk (plastered strawbale walls are particularly resistant to earthquakes because they are energy-absorbing structures).

For nonstructural, infill walls, the provisions limit unrestrained wall size, require mesh and pins, and limit the wind speed and Seismic Design Category (SDC) in which different methods of construction may be used. Wind speeds listed are V_{asd} and not the new V_{ult} values. Strawbale construction may be used in regions with V_{asd} wind speeds up to 120 mph and with SDCs A through D_2.

For structural walls, the provisions allow for testing of plaster, require a minimum compressive strength, and set limits for steel mesh, support of plaster skins and transfer of loads through the plaster skin. Buildings using the strawbales structurally are limited to a single story. The walls may be used as braced wall panels; minimum bracing length is included within the provisions.

Index

A

B

T

U

V

W

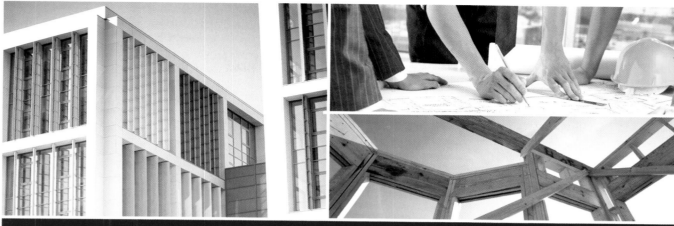

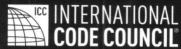

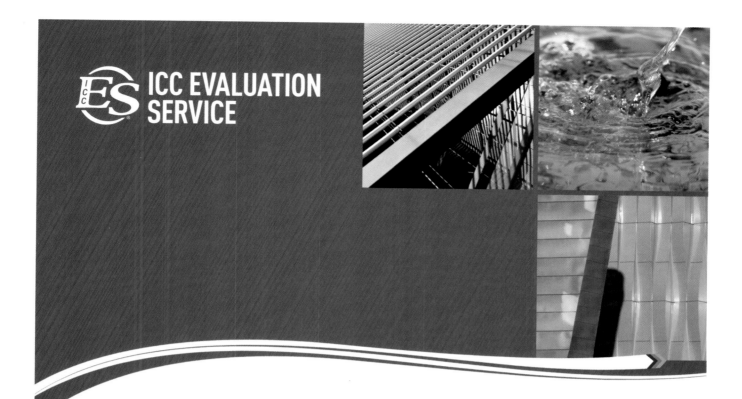

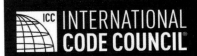